T0137133

Studies in Applied Philosophy, Epistemology and Rational Ethics

Volume 44

Studies in Applied Philosophy, Epistemology and Rational Ethics (SAPERE) publishes new developments and advances in all the fields of philosophy, epistemology, and ethics, bringing them together with a cluster of scientific disciplines and technological outcomes: ranging from computer science to life sciences, from economics, law, and education to engineering, logic, and mathematics, from medicine to physics, human sciences, and politics. The series aims at covering all the challenging philosophical and ethical themes of contemporary society, making them appropriately applicable to contemporary theoretical and practical problems, impasses, controversies, and conflicts. Our scientific and technological era has offered "new" topics to all areas of philosophy and ethics – for instance concerning scientific rationality, creativity, human and artificial intelligence, social and folk epistemology, ordinary reasoning, cognitive niches and cultural evolution, ecological crisis, ecologically situated rationality, consciousness, freedom and responsibility, human identity and uniqueness, cooperation, altruism, intersubjectivity and empathy, spirituality, violence. The impact of such topics has been mainly undermined by contemporary cultural settings, whereas they should increase the demand of interdisciplinary applied knowledge and fresh and original understanding. In turn, traditional philosophical and ethical themes have been profoundly affected and transformed as well: they should be further examined as embedded and applied within their scientific and technological environments so to update their received and often old-fashioned disciplinary treatment and appeal. Applying philosophy individuates therefore a new research commitment for the 21st century, focused on the main problems of recent methodological, logical, epistemological, and cognitive aspects of modeling activities employed both in intellectual and scientific discovery, and in technological innovation, including the computational tools intertwined with such practices, to understand them in a wide and integrated perspective.

Advisory Board

More information about this series at http://www.springer.com/series/10087

Vincent C. Müller
Editor

Philosophy and Theory of Artificial Intelligence 2017

Editor
Vincent C. Müller
Interdisciplinary Ethics Applied (IDEA) Centre
University of Leeds
Leeds, West Yorkshire
UK

ISSN 2192-6255 ISSN 2192-6263 (electronic)
Studies in Applied Philosophy, Epistemology and Rational Ethics
ISBN 978-3-030-07194-3 ISBN 978-3-319-96448-5 (eBook)
https://doi.org/10.1007/978-3-319-96448-5

Softcover re-print of the Hardcover 1st edition 2018

This Springer imprint is published by the registered company Springer Nature Switzerland AG
The registered company address is: Gewerbestrasse 11, 6330 Cham, Switzerland

Editorial Note

The papers in this volume result from the 3rd conference on the "Philosophy and Theory of Artificial Intelligence" (PT-AI) 4–5 November 2017 which I organised in Leeds where I am a university fellow—for details on the conference, see http://www.pt-ai.org/.

For this conference, we had 77 extended abstract submissions by the deadline, which were reviewed double-blind by two to four referees. A total of 28 submissions, i.e. 36%, were accepted for presentation. We also accepted 18 posters to be presented. The invited speakers were Thomas Metzinger, Mark Sprevak, José Hernández-Orallo, Yi Zeng, Susan Schneider, David C. Hogg and Peter Millican. All papers and posters were submitted in January at full length (two pages for posters) and reviewed another time by at least two referees among the authors. In the end, we have 32 papers here that represent the current state of the art in the philosophy of AI. We grouped the papers broadly into three categories: "Cognition–Reasoning–Consciousness", "Computation–Intelligence–Machine Learning" and "Ethics–Law". This year, we see a significant increase in ethics, more work on machine learning, perhaps less on embodiment or computation—and a stronger "feel" that our area of work has entered the mainstream. This is also evident from more papers in "standard" journals, more book publications with mainstream philosophy presses and more philosophers from neighbouring fields joining us, such as philosophy of mind or philosophy of science. These are encouraging developments, and we are looking forward to PT-AI 2019!

We gratefully acknowledge support from the journal *Artificial Intelligence*, and the *IDEA Centre* at the University of Leeds.

May 2018 Vincent C. Müller

Contents

Cognition - Reasoning - Consciousness

Artificial Consciousness: From Impossibility to Multiplicity 3
Chuanfei Chin

Cognition as Embodied Morphological Computation 19
Gordana Dodig-Crnkovic

"The Action of the Brain": Machine Models and Adaptive Functions in Turing and Ashby 24
Hajo Greif

An Epistemological Approach to the Symbol Grounding Problem 36
Jodi Guazzini

An Enactive Theory of Need Satisfaction 40
Soheil Human, Golnaz Bidabadi, Markus F. Peschl, and Vadim Savenkov

Agency, Qualia and Life: Connecting Mind and Body Biologically 43
David Longinotti

***Dynamic Concept Spaces* in Computational Creativity for Music** 57
René Mogensen

Creative AI: Music Composition Programs as an Extension of the Composer's Mind 69
Caterina Moruzzi

How Are Robots' Reasons for Action Grounded? 73
Bryony Pierce

Artificial Brains and Hybrid Minds 81
Paul Schweizer

Huge, but Unnoticed, Gaps Between Current AI and Natural Intelligence 92
Aaron Sloman

Social Cognition and Artificial Agents 106
Anna Strasser

Computation - Intelligence - Machine Learning

Mapping Intelligence: Requirements and Possibilities 117
Sankalp Bhatnagar, Anna Alexandrova, Shahar Avin, Stephen Cave, Lucy Cheke, Matthew Crosby, Jan Feyereisl, Marta Halina, Bao Sheng Loe, Seán Ó hÉigeartaigh, Fernando Martínez-Plumed, Huw Price, Henry Shevlin, Adrian Weller, Alan Winfield, and José Hernández-Orallo

Do Machine-Learning Machines Learn? 136
Selmer Bringsjord, Naveen Sundar Govindarajulu, Shreya Banerjee, and John Hummel

Where Intelligence Lies: Externalist and Sociolinguistic Perspectives on the Turing Test and AI 158
Shlomo Danziger

Modelling Machine Learning Models 175
Raül Fabra-Boluda, Cèsar Ferri, José Hernández-Orallo, Fernando Martínez-Plumed, and M. José Ramírez-Quintana

Is Programming Done by Projection and Introspection? 187
Sam Freed

Supporting Pluralism by Artificial Intelligence: Conceptualizing Epistemic Disagreements as Digital Artifacts 190
Soheil Human, Golnaz Bidabadi, and Vadim Savenkov

The Frame Problem, Gödelian Incompleteness, and the Lucas-Penrose Argument: A Structural Analysis of Arguments About Limits of AI, and Its Physical and Metaphysical Consequences 194
Yoshihiro Maruyama

Quantum Pancomputationalism and Statistical Data Science: From Symbolic to Statistical AI, and to Quantum AI 207
Yoshihiro Maruyama

Getting Clarity by Defining Artificial Intelligence—A Survey 212
Dagmar Monett and Colin W. P. Lewis

Epistemic Computation and Artificial Intelligence 215
Jiří Wiedermann and Jan van Leeuwen

Will Machine Learning Yield Machine Intelligence? 225
Carlos Zednik

Ethics - Law

In Critique of RoboLaw: The Model of SmartLaw 231
Paulius Astromskis

AAAI: An Argument Against Artificial Intelligence 235
Sander Beckers

Institutional Facts and AMAs in Society 248
Arzu Gokmen

A Systematic Account of Machine Moral Agency 252
Mahi Hardalupas

A Framework for Exploring Intelligent Artificial Personhood 255
Thomas B. Kane

Against Leben's Rawlsian Collision Algorithm for Autonomous Vehicles 259
Geoff Keeling

Moral Status of Digital Agents: Acting Under Uncertainty 273
Abhishek Mishra

Friendly Superintelligent AI: All You Need Is Love 288
Michael Prinzing

Autonomous Weapon Systems - An Alleged Responsibility Gap 302
Torben Swoboda

Author Index 315

Cognition - Reasoning - Consciousness

Artificial Consciousness: From Impossibility to Multiplicity

Chuanfei Chin(✉)

Department of Philosophy, National University of Singapore,
Singapore 117570, Singapore
phiccf@nus.edu.sg

Abstract. How has multiplicity superseded impossibility in philosophical challenges to artificial consciousness? I assess a trajectory in recent debates on artificial consciousness, in which metaphysical and explanatory challenges to the possibility of building conscious machines lead to epistemological concerns about the multiplicity underlying 'what it is like' to be a conscious creature or be in a conscious state. First, I analyse earlier challenges which claim that phenomenal consciousness cannot arise, or cannot be built, in machines. These are based on Block's Chinese Nation and Chalmers' Hard Problem. To defuse such challenges, theorists of artificial consciousness can appeal to empirical methods and models of explanation. Second, I explain why this naturalistic approach produces an epistemological puzzle on the role of biological properties in phenomenal consciousness. Neither behavioural tests nor theoretical inferences seem to settle whether our machines are conscious. Third, I evaluate whether the new challenge can be managed through a more fine-grained taxonomy of conscious states. This strategy is supported by the development of similar taxonomies for biological species and animal consciousness. Although it makes sense of some current models of artificial consciousness, it raises questions about their subjective and moral significance.

Keywords: Artificial consciousness · Machine consciousness
Phenomenal consciousness · Scientific taxonomy · Subjectivity

1 Introduction

I want to trace a trajectory in recent philosophical debates on artificial consciousness. In this trajectory, metaphysical and explanatory challenges to the possibility of building conscious machines are supplanted by epistemological concerns about the multiplicity underlying 'what it is like' to be a conscious creature or be in a conscious state. Here *artificial consciousness* refers, primarily, to phenomenal consciousness in machines built from non-organic materials. Like most of the philosophers and scientists whom I discuss, I will follow Block (1995) in using the concept of phenomenal consciousness to refer subjective experience. By Block's definition, the sum of a state's phenomenal properties is what it is like to be in that conscious state, and the sum of a creature's phenomenal states is what it is like to be that conscious creature. The paradigms of such conscious states include having sensations, feelings, and perceptions.

V. C. Müller (Ed.): PT-AI 2017, SAPERE 44, pp. 3–18, 2018.
https://doi.org/10.1007/978-3-319-96448-5_1

Many surveys on artificial consciousness stress that this sub-field in artificial intelligence research has multiple interests (Gamez 2008; Holland and Gamez 2009; Reggia 2013; Scheutz 2014). Its research programmes aim to build machines which mimic behaviour associated with consciousness, machines with the cognitive structure of consciousness, or machines with conscious states. Often a distinction is drawn between *strong* artificial consciousness, which aims for conscious machines, and *weak* artificial consciousness, which builds machines that simulate some significant correlates of consciousness. Of course, a research programme may nurture interests in both strong and weak artificial consciousness; and the same model may be used to investigate both strong and weak artificial consciousness.

I shall focus on philosophical challenges to strong artificial consciousness. First, in the next section, I will analyse two earlier challenges which claim that phenomenal consciousness cannot arise, or cannot be built, in machines. These are based on Block's Chinese Nation and Chalmers' Hard Problem. To defuse such challenges, we can appeal to empirical methods and models of explanation. Second, I will explain why this naturalistic approach leads to an epistemological puzzle on the role of biological properties in phenomenal consciousness. Neither behavioural tests nor theoretical inferences seem to settle whether our machines are conscious. Third, I will evaluate whether the new challenge can be handled by a more fine-grained taxonomy of conscious states. This strategy is supported by the development of more fine-grained taxonomies for biological species and animal consciousness. Although it makes sense of some current models of artificial consciousness, it raises questions about their subjective meaning and moral status.

2 The Impossibility of Artificial Consciousness

The literature on artificial consciousness contains several philosophical challenges to the possibility of building conscious machines (Bishop 2009; Gamez 2008; McDermott 2007; Prinz 2003; Reggia 2013; Scheutz 2014). Such challenges draw on philosophical arguments about the nature of consciousness and our access to it. One set of challenges is against the *metaphysical possibility* of artificial consciousness. These are based on the provocative thought experiments in Block (1978), Searle (1980), Maudlin (1989), which suggest that machines, however sophisticated in functional or computational terms, cannot be conscious. Another set of challenges is directed at the *practical possibility* of building conscious machines. They are based on philosophical claims, made by McGinn (1991), Levine (1983), Chalmers (1995), about our ignorance of how conscious states arise from physical states. According to these challenges, we can hardly expect to produce consciousness in machines if we cannot explain it in human brains.

Most theorists of artificial consciousness are not troubled by such challenges. In his survey, Scheutz (2014) describes two attitudes that support this stance. Here is how I understand them. First, some theorists hold a *pragmatic attitude* towards the concept of consciousness. They define this concept in an operational way, in terms of the processes and principles which psychologists take to underlie consciousness. Their aim is to use these processes and principles to improve performance in machines. They do not want to replicate consciousness, so they need not worry if consciousness can arise, or be produced, in machines. This attitude particularly suits those whose research lies in weak artificial consciousness. Second, other theorists hold a *revisionary attitude*. They want to refine or replace the concept of consciousness through their empirical investigation of the underlying processes and principles identified by psychologists. In doing so, they wish to contribute to both psychology and philosophy. For instance, their models of the relevant processes and principles may enable new psychological experiments and produce new theories of consciousness. These may, in turn, influence philosophical intuitions and views about consciousness.

I take this last point to mean that empirical research into strong artificial consciousness need not be halted by the intuitions and views current in philosophy. To demonstrate this, I will show that theorists of artificial consciousness can appeal to empirical methods and models of explanation to defuse some philosophical challenges. In particular, I will look at how we can respond to two challenges to the possibility of building conscious machines – one based on Block's Chinese Nation thought experiment, the other on Chalmers' Hard Problem of consciousness.[1] Even those theorists who are less inclined to take philosophical challenges seriously can clarify their methodological commitments by considering these responses. Moreover, in the next section, I will show why the commitments underlying these responses lead to an epistemological puzzle which should interest all theorists of artificial consciousness.

(a) The first challenge centres on the nature of consciousness. It suggests that conscious machines cannot be built since machines cannot be conscious. More precisely, it suggests that the functional properties realisable by machines are not sufficient for consciousness. In Block's thought experiment, a billion people in China are instructed to duplicate the functional organisation of mental states in a human mind. Through radio connections and satellite displays, they control an artificial body just as neurons control a human body. They respond to various sensory inputs into the body with appropriate behavioural outputs. But, according to Block (1978), we are loath to attribute consciousness to this system: 'there is prima facie doubt whether it has any mental states at all – especially whether it has what philosophers have variously called "qualitative states," "raw feels," or "immediate phenomenological qualities"' (73). If our intuition about the Chinese Nation is sound, then consciousness requires more than the functional properties discovered in psychology. If so, the machines that realise only these functional properties cannot be conscious.

I do not think that we need to defer to this intuition about the Chinese Nation. Rather we should use empirical methods to uncover more about the nature of consciousness. Our best research – in psychology, neuroscience, and artificial consciousness – may determine that functional properties at a coarse-grained psychological level are sufficient for consciousness. Or it may determine that functional properties at a more fine-grained neurological level are necessary too. Whether the relevant properties are realisable in our machines is a further question, also to be determined by empirical

[1] I learnt especially from the responses offered in Prinz (2003) and Gamez (2008). I have put aside challenges based on Searle's Chinese Room thought experiment: they are analysed exhaustively in the literature on artificial consciousness, with what looks to be diminishing returns. One response to these challenges can be modelled after my response in (a).

investigation. None of this research should be pre-empted *a priori* by what our intuition says in a thought experiment and what that supposedly implies about the possibility of conscious machines.

Even Block would agree on this methodological point. He notes that, intuitively, the human brain also does not seem to be the right kind of system to have what he calls 'qualia', the subjective aspect of experience. So our intuition, on its own, cannot be relied on to judge which system does or does not have qualia. According to Block, we can overrule intuition if we have independent reason to believe that a system has qualia, and if we can explain away the apparent absurdity of believing this. Here his qualm about a system like the Chinese Nation rests mainly on our lack of a theoretical ground to believe that it has qualia. No psychological theory that he considers seems to explain qualia. That is why he insists of the system: 'any doubt that it has qualia is a doubt that qualia are in the domain of psychology' (84). To assuage this qualm, we need to build an empirical theory of consciousness which explains qualia and evaluates whether Chinese Nations, machines, and other systems have them.

(b) The second challenge directly addresses our explanation of consciousness. It suggests that we cannot build machines to be conscious even if machines can be conscious. According to Chalmers (1995), the Hard Problem we face is to explain how conscious experiences arise from physical processes and mechanisms in the brain. He distinguishes this from easy problems which require us to explain various psychological functions and behaviours in terms of computational or neural mechanisms. We have yet to solve the Hard Problem because we do not know how consciousness is produced in the human brain. But, until we do so, we cannot produce consciousness in a machine except by accident. Here is how Gamez (2008) sums up this line of reasoning based on our ignorance: "if we don't understand how human consciousness is produced, then it makes little sense to attempt to make a robot phenomenally conscious" (892).

I find two related reasons to reject this challenge. First, the production of consciousness may not require its explanation. Through empirical investigation, we may be able to produce consciousness without explaining it in terms of physical processes and mechanisms in the brain. If so, it suffices for us to create in machines the conditions which give rise to consciousness in humans; we need not understand, in philosophically satisfying terms, how the conditions do this. Our research to produce consciousness in machines may then help our research to explain consciousness in humans. This cross-fertilisation between research programmes would be in keeping with the revisionary attitude that Scheutz highlights.

Second, even if we need some kind of explanation to enable production, the explanation of consciousness in empirical terms may not require a solution to the Hard Problem. Through their empirical theories, scientists do not aim to explain, in some metaphysically intelligible way, how the properties of a phenomenon 'arise from' other properties at lower levels. Instead, they aim to establish a theoretical identity for the phenomenon in terms of its underlying properties (Block and Stalnaker 1999; McLaughlin 2003; Prinz 2003; Shea and Bayne 2010). (I say more about how this applies to consciousness science in the next section.) To build their theories, scientists draw correlations between levels, tying together some higher-level and lower-level properties. In the biological and psychological sciences, what requires this kind of explanation between levels depends on context: it is often determined by which properties, at higher or lower levels, appear anomalous (Wimsatt 1976; Craver 2009, Chap. 6; Prinz 2012, 287–8). These practices suggest that an empirically successful theory of consciousness need not fill in the gap between phenomenal and physical properties – at least, not in the terms defined by Chalmers' Hard Problem.

3 The Multiplicity in Phenomenal Consciousness

I have shown how empirical methods and models of explanation can defuse philosophical challenges to the possibility of artificial consciousness. They allow us to counter intuitions drawn from thought experiments on the nature of consciousness, and to undercut arguments derived from our ignorance of how conscious states arise from physical states. By appealing to these empirical methods and models, we adopt a naturalistic approach to the study of artificial consciousness. We use empirical methods, as far as possible, to answer questions about the nature of consciousness and our access to it. We thereby allow empirical discoveries about phenomenal consciousness to inform our conceptual understanding of artificial consciousness. But that naturalistic approach produces a different philosophical challenge, arising from what we discover to be the multiplicity underlying consciousness. This new challenge to artificial consciousness is epistemological: it suggests that, even if we can build conscious machines, we cannot tell that the machines are conscious.

The challenge rests on our difficulty in determining the role of biological properties in phenomenal consciousness. Unless we determine their role, we cannot discover whether our machines, lacking at least some of these properties, are conscious. Several philosophers analyse this difficulty (Block 2002; Papineau 2002, Chap. 7; Prinz 2003, 2005; Tye 2016, Chap. 10). Yet their arguments are largely ignored by theorists of artificial consciousness. I will focus on Prinz's arguments – since they arise naturally from his work on an empirical theory of consciousness and are addressed directly to theorists of artificial consciousness.

Prinz begins by analysing, at the psychological level, the *contents* of our conscious states and the *conditions* under which they become conscious. Following Nagel, he considers having a perspective to be fundamental to consciousness: 'We cannot have a conscious experience of a view from nowhere' (2003, 118). In his analysis, humans experience the world, through our senses, 'from a particular vantage point'. So the contents of our consciousness are both perceptual and perspectival. These contents become conscious when we are paying attention. When these contents become available for our deliberation and deliberate control of action, they enable our flexible responses to the world. Putting together these hypotheses, Prinz proposes that consciousness arises in humans when we attend to phenomena such that our perspectival perceptual states become available for deliberation and deliberate control of action.

Next, by drawing on empirical studies, Prinz maps these contents and conditions of conscious states onto the computational and neural levels. In information processing, the contents of consciousness seem to lie at the intermediate level. Our intermediate-level representations are 'vantage-point specific and coherent' (2003, 119). They are distinct from higher-level representations which are too abstract to preserve perspective, and lower-level representations which are too local to be coherent. In computational models of cognition, attention is a process that filters representations onto the next stage, while deliberate control is handled by working memory, a short-term storage capacity with executive abilities. In the human brain, these computational processes are implemented by a neural circuit between perceptual centres in the temporal cortex, attentional centres in the parietal cortex, and working memory centres in the frontal cortex (2003, 119; 2005, 388). Prinz (2012) cites several lines of evidence indicating that gamma vectorwaves play the crucial role in these brain regions. So, according to his latest theory, consciousness arises in us 'when and only when vectorwaves that realize intermediate-level representations fire in the gamma range, and thereby become available to working memory' (293). That is, in empirical terms, a good candidate for the neurofunctional basis of consciousness in humans.

Despite this progress, Prinz (2003, 2005) highlights an epistemological limitation, which is independent of whatever empirical theory of consciousness we settle on. He argues that we cannot determine if our biological properties are constitutive of consciousness. So we cannot discover if our machines, which will lack at least some of these properties, are conscious. This is the basis of his pessimism about research in strong artificial consciousness: 'It simply isn't the case that scientific investigations into the nature of consciousness will make questions of machine consciousness disappear. Even if scientific theories of consciousness succeed by their own standards, we must remain agnostic about artificial experience' (117).

Like others who share his pessimism, Prinz cites the in-principle failure of behavioural tests to settle these questions (Prinz 2003, IV; Block 2002; Papineau 2002, Chap. 7, 2003). How do we find constitutive properties of consciousness? The standard method is to test for what Prinz calls 'difference-makers' (121). It involves changing processes at a tested level while keeping constant processes at other levels. If this change makes a difference to conscious behaviour in humans, then some properties at this tested level are constitutive of consciousness. Suppose that it is technically possible to substitute silicon chips for neurons in the human brain. And suppose that it is nomologically possible to do so while keeping constant the relevant processes at the psychological and computational levels.[2] This surgically altered person will become a 'functional duplicate of a normal person with a normal brain' (123). By design, the functional duplicate will behave exactly as conscious humans do – reporting pain, showing signs of anger, apparently 'seeing sunsets and smelling roses'. Yet our current tests for consciousness centre on behaviour. So we do not have a genuine test for consciousness in the duplicate. We cannot, by these tests, tell if our properties at the biological level are constitutive of consciousness.

I agree with Prinz (2003) that this thought experiment highlights a 'serious epistemological problem' (130). Indeed, I believe that he and others understate the depth of

[2] This is a common idealisation in the thought experiment. In reality, we will find more than one psychological level and more than one computational one (Prinz 2003, 120–1). During chip replacement, we are more likely to keep constant processes at less fine-grained psychological and computational levels. The epistemological difficulty with testing remains, though it is made more complicated. Elsewhere, in Chin (2016), I analyse more complicated versions of the multiple-kinds problem in consciousness science; see also Irvine (2013), Chap. 6.

the problem. They focus on the failure of behavioural tests to discover if biological properties make a difference to conscious states. Prinz claims that this 'method of difference-makers seems to be the only way to find out what levels matter' (130). Yet, like other philosophers, he also recommends that we use inference to the best explanation to establish a theoretical identity for consciousness (116).[3] He does not explain why this theoretical inference cannot clarify the role of biological properties in consciousness and, thereby, improve the current tests for consciousness.

Let me make these connections explicit through the multiple-kinds problem shown in Fig. 1. As the thought experiment suggests, we will discover at least two functional structures responsible for conscious behaviour in humans. One is a neurofunctional structure, such as that identified in Prinz's theory. Another is a functional structure that abstracts away from some biological mechanisms in the neurofunctional structure. Therefore, the kind defined by the neurofunctional structure ($kind_2$) is nested within the kind defined by the more abstract functional structure ($kind_1$). $Kind_1$ includes conscious humans and our functional duplicates, while $kind_2$ excludes the functional duplicates. So which is *the* structure of consciousness? Which structure defines a kind formed by all and only conscious beings?

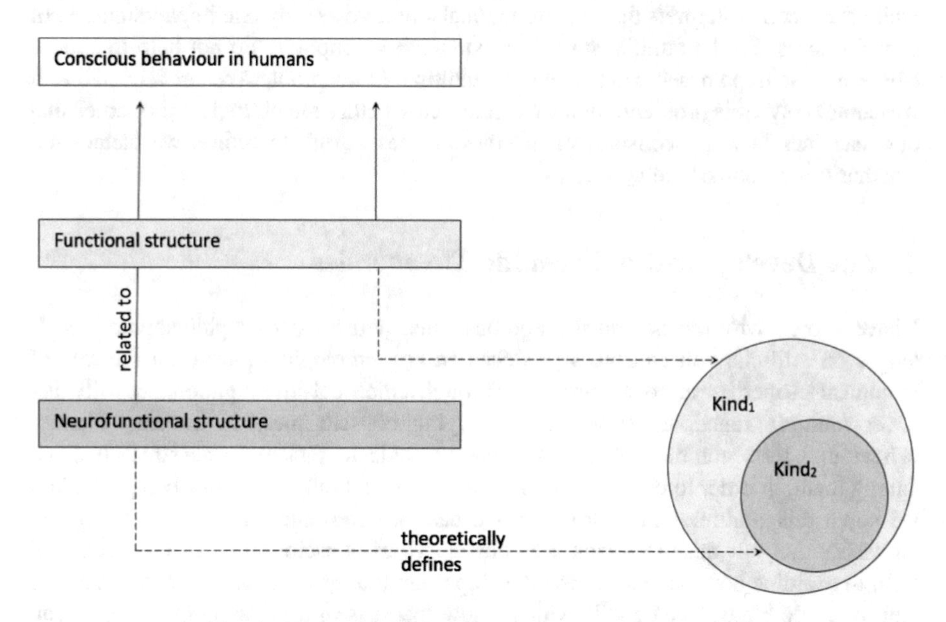

Fig. 1. The multiple kinds in phenomenal consciousness

Prinz's argument shows that current tests, based on behaviour, cannot solve this multiple-kinds problem. I want to extend this argument, to show why inference to the best explanation does not help. Both the neurofunctional structure and the more abstract structure are correlated with consciousness in humans. Both are also

[3] Other philosophers include Block and Stalnaker (1999), McLaughlin (2003), Shea and Bayne (2010), Allen and Trestman (2016), Sect. 4.3.

systematically related to conscious behaviour in humans. By focusing on the systematic relations between the neurofunctional structure, consciousness in humans, and their conscious behaviour, we can support an identity between consciousness and the neurofunctional structure. But this move is *ad hoc*, classifying our functional duplicates by fiat as not conscious. On the other hand, by focusing on the equally systematic relations between the more abstract functional structure, consciousness in humans, and their conscious behaviour, we can support an identity between consciousness and that structure. Yet this is equally *ad hoc*, re-classifying the duplicates by fiat as conscious.

Neither hypothesis offers a simpler explanation. Whether we identify consciousness with the neurofunctional structure or the more abstract structure, we must invoke both structures to account for the total explananda. If we identify consciousness with the neurofunctional structure, then we must use the more abstract structure to explain why the duplicates share the same behaviour as humans even though the duplicates do not have human brains. If we identify consciousness with the more abstract structure, then we must use the neurofunctional structure to explain how the more abstract structure is implemented differently in conscious humans and their duplicates. The first hypothesis interprets consciousness as only one implementation of the more abstract structure, while the second interprets the neurofunctional structure as only one implementation of consciousness. So the familiar norms of explanatory simplicity do not help to choose between these hypotheses. That is why the multiple-kinds problem seems intractable. If we cannot solve this problem, then we cannot tell whether the biological properties that our machines lack are constitutive of consciousness. And, therefore, we cannot tell whether our machines are conscious.

4 The Development of Scientific Taxonomies

I have shown why the naturalistic approach that defuses earlier philosophical challenges on artificial consciousness produces an epistemological puzzle on the role of biological properties in consciousness. Through empirical investigation, we will discover multiple functional structures underlying consciousness in humans. Neither behavioural tests nor theoretical inferences are able to pick out one structure from among them, in order to define a kind formed by all and only conscious beings. Unless we solve this multiple-kinds problem, we cannot determine whether the biological properties that our machines lack are constitutive of consciousness. In this section, I want to examine how other scientists develop more fine-grained taxonomies to manage their multiple kinds. Then I will evaluate how theorists of artificial consciousness can use this taxonomic strategy.

How does the multiple-kinds problem arise elsewhere? One prominent instance is what biologists call the 'species problem'.[4] When biologists try to classify organisms

[4] This problem is analysed by both biologists and philosophers: see the surveys in Coyne and Orr (2004); Cracraft (2000); Ereshefsky (2010, 2017); Richards (2010). I also learnt from the analysis in LaPorte (2004), though we come to different conclusions. Richards (2010) argues that the problem goes back to pre-Darwinian times: Darwin himself was confronted by 'a multiplicity of species concepts' (75).

into species, they discover multiple structures underlying biodiversity. These structures centre on interbreeding, genetic or phenotypic similarity, ecological niche, evolutionary tendency, or phylogeny. They lead to conflicting definitions of what a species is. Different structures define overlapping kinds, consisting of different populations of organisms. According to the biologists Coyne and Orr (2004), at least nine species definitions remain 'serious competitors'. Three of them are often mentioned in the philosophical literature: the Biological Species Concept (BSC), the Phylogenetic Species Concept (PSC), and the Ecological Species Concept (ESC).[5] They focus, respectively, on three primary processes involved in evolution: sexual reproduction, descent from common ancestry, and environmental selection pressures. Of the three, which defines the nature of species?

Proponents of the BSC, the PSC, and the ESC sometimes claim that their definition of species is the 'best'.[6] But, in practice, biologists choose between these definitions according to their empirical interests. As de Queiroz (1999) explains, 'they differ with regard to the properties of lineage segments that they consider most important, which is reflected in their preferences concerning species criteria' (65). Their choice of the BSC, the PSC, or the ESC allows them to investigate the wider explanatory structures associated, respectively, with sexual reproduction, descent from common ancestry, or ecological niche. For instance, those who are interested in the history of life prefer the PSC over the BSC because they believe that reproductive isolation is 'largely irrelevant to reconstructing history' (Coyne and Orr 2004, 281). Those who are interested in the explanation of biodiversity reject the PSC because they see phylogeny as 'largely irrelevant to understanding the discreteness of nature'. Instead they use the BSC to study populations that sexually reproduce or use the ESC to study adaptive zones in ecology.

The result is a more fine-grained taxonomy of species, which can be used to manage the multiple kinds found within biodiversity. Biologists now distinguish between species which arise from interbreeding, species which arise from phylogenetic connection, and species which arise from environmental selection (Ereshefsky 2010). As Fig. 2 shows, the BSC and the PSC tend to define overlapping kinds of populations. When genealogically distinct populations can reproduce with each other, the populations of a phylogenetic species are nested within the populations of an interbreeding species. Through their taxonomy, biologists can clarify the relations between these kinds and demarcate the explanatory structures involving these kinds.

---

[5] The BSC defines species as 'groups of interbreeding natural populations that are reproductively isolated from other such groups' (Mayr 1969). The PSC defines them as the 'smallest diagnosable cluster of individual organisms within which there is a parental pattern of ancestry and descent' (Cracraft 1983). The ESC defines them as 'a lineage (or a closely related set of lineages) which occupies an adaptive zone minimally different from that of any other lineage in its range and which evolves separate from all lineages outside its range' (Van Valen 1976).

[6] As Cracraft (2000) warns, 'the notion of "best" is always relative' (10). He urges us to 'look hard at the context of what *best* might mean', including how general in application a definition is meant to be, and whether a more general definition is always more useful.

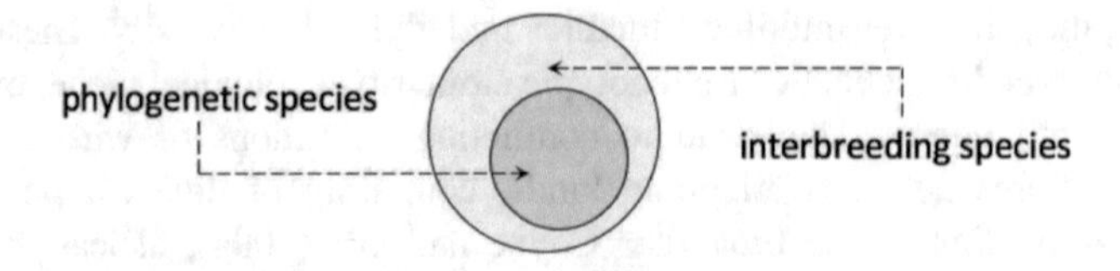

Fig. 2. Two overlapping kinds of biological species

With the more fine-grained taxonomy in place, what matters to biological explanation is not whether the BSC or the PSC offers the 'best' definition of species. Rather biologists have to ensure that those who are interested in interbreeding species not confuse classifications with those who are interested in phylogenetic species. In a context with shared interests, such confusion is unlikely to arise. For instance, most biologists interested in sexual reproduction and its effects focus on interbreeding species. Their interests already pick out these relevant kinds from the overlapping ones associated with sexual reproduction, descent from ancestry, and environmental selection pressures. In a context with competing interests, biologists can avoid misunderstanding by making explicit reference to either interbreeding species, phylogenetic species, or ecological species. However, in some general contexts, biologists need not specify the kinds to which they refer. They may be keen to make generalisations across different branches of biology (Brigandt 2003). So their claims apply uniformly to interbreeding species, phylogenetic species, and ecological species.

The multiple-kinds problem also afflicts debates on animal consciousness. Here it lies closer to our epistemological puzzle on artificial consciousness. For animal consciousness, the problem arises because we discover at least two cognitive structures underlying consciousness in humans. Both structures are responsible, in different ways, for conscious behaviour in humans. I will follow how Godfrey-Smith (2016a, b) distinguishes these structures. The first involves simple modes of information processing associated with pain and other primitive bodily feelings, such as thirst and feeling short of breath. This structure enables us to respond to actual and potential injury with flexible non-reflexive behaviour. The second structure involves more sophisticated modes of information processing which integrate information from different senses and bodily feelings, through the use of memory, attention, and executive control. According to some theories of cognition, this structure allows us to model the world before responding to it.

Figure 3 shows that these two cognitive structures define two overlapping kinds of animals. The kind of animals with cognitive integration is nested within the kind with primitive bodily feelings, because cognitive integration requires more machinery, such as memory, attention, and executive control. So which is *the* cognitive structure of consciousness? Which structure defines a kind formed by all and only conscious animals? If cognitive integration is necessary for consciousness, then only animals with memory, attention, and executive control count as conscious. But if primitive bodily feelings are sufficient for consciousness, many more animals count as conscious, so long as they have the sensorimotor capacities associated with primitive bodily feelings.

Faced with this multiple-kinds problem, Godfrey-Smith (2016a, b) proposes a more fine-grained taxonomy of subjective experiences in animals. There are at least two

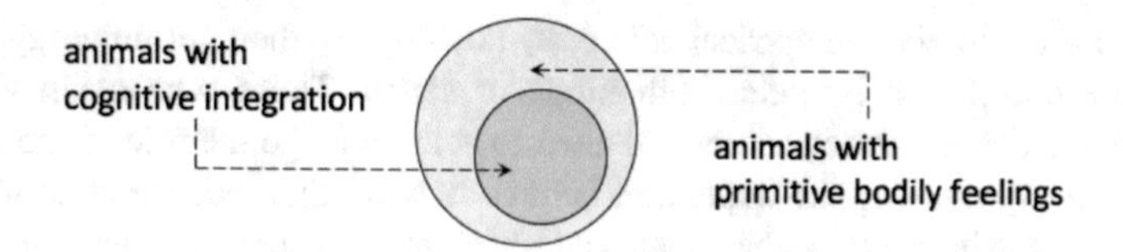

Fig. 3. Two overlapping kinds of animals

kinds of subjective experiences. The basic kind, which evolved first, consists of experiences of pain and other primitive bodily feelings; the complex kind, which evolved later, consists of experiences which integrate information from different senses and bodily feelings. Both kinds of subjective experiences are found in conscious humans: 'Much human experience does involve the integration of different senses, integration of the senses with memory, and so on, but there is also an ongoing role for what seems to be old forms of experience that appear as intrusions into more organized kinds of processing' (2016b, 500). Through his taxonomy, we can clarify the relations between both kinds of experiences and demarcate the explanatory structures involving both kinds.

With the more fine-grained taxonomy in place, we can see that what matters in the explanation of animal behaviour is not whether the basic or complex kind of subjective experiences counts as conscious. Rather theorists of animal consciousness can focus on either kind of experiences according to their empirical interests, so long as their terminology does not obscure the differences between both kinds. For instance, Godfrey-Smith classifies only experiences with cognitive integration as conscious: '"Consciousness" is something beyond mere subjective experience, something richer or more sophisticated' (2016a, 53). Animals which experience pain and other primitive bodily feelings have qualia; it feels like something to be them. But, without cognitive integration, they do not count for him as conscious: 'I wonder whether squid feel pain, whether damage feels like anything to them, but I do not see this as wondering whether squid are conscious' (2016b, 484). As he acknowledges, other theorists with different interests tend to equate qualia with phenomenal consciousness: 'If there is something it *feels like to be* a system, then the system is said to have a kind of consciousness' (483–4). In turn, these theorists have to distinguish phenomenal consciousness from other, more sophisticated, kinds of consciousness that require cognitive integration.

How might this taxonomic strategy address the epistemological puzzle on artificial consciousness? We can develop a more fine-grained taxonomy of conscious states, in order to manage the multiplicity that troubles theorists of artificial consciousness. If Prinz is right, then we need to distinguish at least two kinds of states. The first consists of neurofunctional states, such as those specified in his theory of consciousness. Our functional duplicates do not have this kind of states. The second consists of functional states that abstract away from some biological mechanisms in the neurofunctional states; both humans and the duplicates share this kind of states. With this taxonomy, we can clarify the relations between the neurofunctional and functional states, then demarcate the explanatory structures involving both kinds of states. What matters in explaining humans and duplicates is not whether the neurofunctional or functional states count as conscious. Rather theorists of consciousness can focus on either kind of states according to their empirical interests, so long as their terminology does not obscure the differences between both kinds of states. Those who classify only the neurofunctional states as conscious still need to acknowledge the role of the functional states, which explain why the duplicates behave in ways that indicate consciousness in humans. Those who classify the functional states as conscious still need to acknowledge the role of the neurofunctional states; they explain how the functional states are realised in humans.

This analysis brings out an epistemological difference between the case of biological species and that of artificial consciousness. Biologists are now confident that interbreeding species, phylogenetic species, and ecological species play significant explanatory roles. They know that the kinds associated with the BSC, the PSC, and the ESC are involved in different explanatory structures associated with sexual reproduction, ancestral descent, and ecological niche. In contrast, we do not yet know, in any precise terms, the states that will play significant explanatory roles in research on artificial consciousness. However, this difference does not invalidate our use of the taxonomic strategy. We need only begin with a provisional taxonomy of conscious states to explore the different explanatory structures that interest us. As we discover more about these explanatory structures, we can refine the taxonomy so that it reflects, in more precise terms, the computational and biological processes cited in our explanations. That is similar to how biologists developed their taxonomy for species.

Indeed, this taxonomic strategy can already make sense of some current models of artificial consciousness. Some theorists suggest that building the right computational processes into machines is sufficient to make them conscious. For instance, Dehaene et al. (2017) propose that machines are conscious if they can select information for global broadcasting, making it flexibly available for computations, and if they can self-monitor those computations. To support their proposal, they claim that a machine with both computational processes will behave ‘as though it were conscious’ (492). They also cite evidence suggesting that subjective experience in humans ‘appears to cohere with’ global broadcasting and self-monitoring (492). Other theorists believe that building the right biological processes into machines is necessary to make them conscious. Haladjian and Montemayor (2016) connect consciousness to biological processes in humans that endow them with emotion and empathy. So, in their view, machines designed purely to compute with artificial intelligence will not have subjective experiences. According to Godfrey-Smith (2016b), machines can have subjective experiences only if they have some functional properties associated with ‘living activity’ (505). For him, these properties include the robustness and adaptability typical of complex biological systems in humans.

From our perspective, these models of artificial consciousness need not come into conflict. Rather we can see them as jointly clarifying the more fine-grained taxonomy of conscious states needed in research on artificial consciousness. On one hand, Dehaene et al. (2017) are investigating the kind of states which are defined purely in computational terms without reference to biological mechanisms; in particular they are interested in the explanatory structures associated with global broadcasting and self-monitoring. On the other hand, Haladjian and Montemayor (2016), Godfrey-Smith (2016b) are interested in another kind of states, defined partly in biological terms; they raise different difficulties for realising such states in machines.

5 Conclusion

In this paper, I assessed a trajectory in which multiplicity superseded impossibility in philosophical challenges to artificial consciousness. First, I tackled two earlier challenges which claim that phenomenal consciousness cannot arise, or cannot be built, in machines. The first challenge, from the nature of consciousness, is based on Block's Chinese Nation thought experiment. The second challenge, from the explanation of consciousness, is based on Chalmers' Hard Problem. I showed how a naturalistic approach, appealing to empirical methods and models of explanation, can defuse these challenges. To discover if machines can be conscious, we should rely on theories of consciousness developed through empirical methods, rather than the intuitions about consciousness provoked by thought experiments. To explain consciousness in empirical terms, we need not supply a philosophically satisfying account of how phenomenal properties arise from physical ones.

Second, I explained why this naturalistic approach leads to an epistemological puzzle on the role of biological properties in phenomenal consciousness. Through empirical investigation, we will discover multiple functional structures underlying consciousness in humans. As several philosophers argued, behavioural tests cannot pick out one structure from among them, in order to define a kind formed by all and only conscious beings. I argued that inference to the best explanation cannot help too. If we cannot solve this multiple-kinds problem, then we cannot determine whether the biological properties that our machines lack are constitutive of consciousness. We also cannot determine whether these machines are conscious.

Third, I evaluated whether a taxonomic strategy used in other sciences can address this new challenge. To manage the overlapping kinds which they cite in explanations, theorists of biological species and animal consciousness develop more fine-grained taxonomies. I argued that, similarly, theorists of artificial consciousness can develop a more fine-grained taxonomy of conscious states, which distinguishes between the neurofunctional states specified in an empirical theory of consciousness and the functional states that abstract away from some biological mechanisms in the neurofunctional states. Such a taxonomy enables us to clarify the relations between both kinds of states and demarcate the explanatory structures involving both kinds. In addition, I argued that this taxonomic strategy helps to make sense of current models of artificial consciousness, including those which require only computational states and those which require partly biological states. We can interpret them as models for investigating different kinds of conscious states.

This strategy presents us with three related challenges, on the explanatory, subjective, and moral significance of the kinds in any new taxonomy. First, we need to establish that these kinds of states play significant explanatory roles in research on artificial consciousness. This is primarily an empirical challenge, depending on theorists of artificial consciousness to explore different explanatory structures that interest us. Second, we need to examine the subjective significance of these kinds of states. Thus far, we have construed a conscious state's phenomenal properties as capturing 'what it is like to be' in that state. But this construal does not help to discriminate what the multiple kinds mean in subjective terms. We may do so by investigating the capacities and interactions made possible by the underlying structures that define these kinds. For instance, some basic structures may support what it is like to be an artificial patient, while others may support what it is like to be an artificial agent. Third, we need to explore the moral significance of these kinds of states. In what ways do the artificial patients count as moral patients whose suffering we must ameliorate? In what ways do the artificial agents count as moral agents whose lives we must attend to?

Acknowledgements. I thank Arzu Gokmen, Michael Prinzing, and Kaine Yeo for their suggestions. Abhishek Mishra, Susan Schneider, Paul Schweitzer, and Alexandra Serrenti commented on the talk. This research was supported by an NUS Early Career Award.

References

Allen, C., Trestman, M.: Animal Consciousness. In: Zalta, E.N. (ed.) The Stanford Encyclopedia of Philosophy, Winter 2016 (2016). https://plato.stanford.edu/archives/win2016/entries/consciousness-animal/. Accessed 10 Jan 2018

Bishop, J.M.: Why computers can't feel pain. Minds Mach. **19**(4), 507–516 (2009)

Block, N.: Troubles with functionalism. In: Block, N. (ed.) Consciousness, Function, and Representation: Collected Papers, Volume 1 (2007), pp. 63–101. MIT Press, Cambridge (1978)

Block, N.: On a confusion about the function of consciousness. In: Block, N. (ed.) Consciousness, Function, and Representation: Collected Papers, Volume 1 (2007), pp. 159–213. MIT Press, Cambridge (1995)

Block, N.: The harder problem of consciousness. In: Block, N. (ed.) Consciousness, Function, and Representation: Collected Papers, Volume 1 (2007), pp. 397–433. MIT Press, Cambridge (2002)

Block, N., Stalnaker, R.: Conceptual analysis, dualism, and the explanatory gap. Philos. Rev. **108** (1), 1–46 (1999)

Brigandt, I.: Species pluralism does not imply species eliminativism. Philos. Sci. **70**(5), 1305–1316 (2003)

Chalmers, D.J.: Facing up to the problem of consciousness. J. Conscious. Stud. **2**(3), 200–219 (1995)

Chin, C.: Borderline Consciousness, Phenomenal Consciousness, and Artificial Consciousness: A Unified Approach, DPhil thesis. University of Oxford (2016)

Coyne, J.A., Orr, H.A.: Speciation: a catalogue and critique of species concepts. In: Rosenberg, A., Arp, R. (eds.) Philosophy of Biology: An Anthology, pp. 272–292. Wiley-Blackwell, Oxford (2004)

Cracraft, J.: Species concepts and speciation analysis. In: Johnston, R.F. (ed.) Current Ornithology, pp. 159–187. Springer, New York (1983)

Cracraft, J.: Species concepts in theoretical and applied biology: a systematic debate with consequences. In: Wheeler, Q.D., Meier, R. (eds.) Species Concepts and Phylogenetic Theory: A Debate, pp. 3–14. Columbia University Press, New York (2000)

Craver, C.F.: Explaining the Brain. Oxford University Press, Oxford (2009)

Dehaene, S., Lau, H., Kouider, S.: What is consciousness, and could machines have it? Science **358**(6362), 486–492 (2017)

Ereshefsky, M.: Species, taxonomy, and systematics. In: Rosenberg, A., Arp, R. (eds.) Philosophy of Biology: An Anthology, pp. 255–271. Wiley-Blackwell, Oxford (2010)

Ereshefsky, M.: Species. In: Zalta, E.N. (ed.) The Stanford Encyclopedia of Philosophy, Fall 2017 (2017). https://plato.stanford.edu/archives/fall2017/entries/species/. Accessed 10 Jan 2018
Gamez, D.: Progress in machine consciousness. Conscious. Cogn. **17**(3), 887–910 (2008)
Godfrey-Smith, P.: Animal evolution and the origins of experience. In: Smith, D.L. (ed.) How Biology Shapes Philosophy: New Foundations for Naturalism, pp. 51–71. Cambridge University Press, Cambridge (2016a)
Godfrey-Smith, P.: Mind, matter, and metabolism. J. Philos. **113**(10), 481–506 (2016b)
Haladjian, H.H., Montemayor, C.: Artificial consciousness and the consciousness-attention dissociation. Conscious. Cogn. **45**, 210–225 (2016)
Holland, O., Gamez, D.: Artificial intelligence and consciousness. In: Banks, W.P. (ed.) Encyclopedia of Consciousness, pp. 37–45. Academic Press, Oxford (2009)
Irvine, E.: Consciousness as a Scientific Concept: A Philosophy of Science Perspective. Springer, Dordrecht (2013)
LaPorte, J.: Natural Kinds and Conceptual Change. Cambridge University Press, Cambridge (2004)
Levine, J.: Materialism and qualia: the explanatory gap. Pac. Philos. Q. **64**(October), 354–361 (1983)
Maudlin, T.: Computation and consciousness. J. Philos. **86**(8), 407–432 (1989)
McDermott, D.: Artificial intelligence and consciousness. In: Zelazo, P.D., Moscovitch, M., Thompson, E. (eds.) The Cambridge Handbook of Consciousness, pp. 117–150. Cambridge University Press, Cambridge (2007)
McGinn, C.: The Problem of Consciousness: Essays Towards a Resolution. Blackwell, Oxford (1991)
McLaughlin, B.P.: A naturalist-phenomenal realist response to block's harder problem. Philos. Issues **13**(1), 163–204 (2003)
Papineau, D.: Thinking about Consciousness. Clarendon Press, Oxford (2002)
Prinz, J.: Level-headed mysterianism and artificial experience. In: Holland, O. (ed.) Machine Consciousness, pp. 111–132. Imprint Academic, Exeter (2003)
Prinz, J.: A neurofunctional theory of consciousness. In: Brook, A., Akins, K. (eds.) Cognition and the Brain: The Philosophy and Neuroscience Movement, pp. 381–396. Cambridge University Press, Cambridge (2005)
Prinz, J.: The Conscious Brain: How Attention Engenders Experience. Oxford University Press, Oxford (2012)
de Queiroz, K.: The general lineage concept of species and the defining properties of the species category. In: Wilson, R.A. (ed.) Species: New Interdisciplinary Essays, pp. 49–89. MIT Press, Cambridge (1999)
Reggia, J.A.: The rise of machine consciousness: studying consciousness with computational models. Neural Netw. **44**, 112–131 (2013)
Richards, R.A.: The Species Problem: A Philosophical Analysis. Cambridge University Press, Cambridge (2010)
Scheutz, M.: Artificial Emotions and Machine Consciousness. In: Frankish, K., Ramsey, W.M. (eds.) The Cambridge Handbook of Artificial Intelligence, pp. 247–266. Cambridge University Press, Cambridge (2014)
Searle, J.R.: Minds, brains and programs. Behav. Brain Sci. **3**(3), 417–457 (1980)
Shea, N., Bayne, T.: The vegetative state and the science of consciousness. Br. J. Philos. Sci. **61** (3), 459–484 (2010)

Tye, M.: Tense Bees and Shell-Shocked Crabs: Are Animals Conscious?. Oxford University Press, New York (2016)
Valen, L.V.: Ecological species, multispecies, and oaks. Taxon **25**(2/3), 233–239 (1976)
Wimsatt, W.C.: Reductionism, levels of organization, and the mind-body problem. In: Globus, G.G., Maxwell, G., Savodnik, I. (eds.) Consciousness and the Brain, pp. 205–267. Springer, Dordrecht (1976)

Cognition as Embodied Morphological Computation

Gordana Dodig-Crnkovic[1,2](✉)

[1] Chalmers University of Technology, Gothenburg, Sweden
dodig@chalmers.se
[2] Gothenburg University, Gothenburg, Sweden

Abstract. Cognitive science is considered to be the study of mind (consciousness and thought) and intelligence in humans. Under such definition variety of unsolved/unsolvable problems appear. This article argues for a broad understanding of cognition based on empirical results from i.a. natural sciences, self-organization, artificial intelligence and artificial life, network science and neuroscience, that apart from the high level mental activities in humans, includes sub-symbolic and sub-conscious processes, such as emotions, recognizes cognition in other living beings as well as extended and distributed/social cognition. The new idea of cognition as complex multiscale phenomenon evolved in living organisms based on bodily structures that process information, linking cognitivists and EEEE (embodied, embedded, enactive, extended) cognition approaches with the idea of morphological computation (info-computational self-organisation) in cognizing agents, emerging in evolution through interactions of a (living/cognizing) agent with the environment.

1 Understanding Cognition

Cognitive science is currently defined as a study of processes of *knowledge generation* through perception, thinking (reasoning), memory, learning, problem solving, and similar. Thagard (2013) makes an extension of the idea of "thinking" to include emotional experience. This move bridges some of the distance between cognition as thinking and its (sub-)processes, but the fundamental problem of *generative mechanisms* that can dynamically overarch the chasm between matter and mind remains. The definition of cognitive science does not mention biology, chemistry, (quantum- nano-, etc.) physics or chaos theory, self-organisation, and artificial life, artificial intelligence or data science, extended mind, or distributed cognition as studied with help of network science, sociology or ecology.

On the current view, cognition is about *high-level processes* remote from physical-chemical-biological substrate. It is modeled either by classical sequential computation, understood as symbol manipulation, or by neural networks. On the other hand, historically, behaviorism offered an alternative view of cognition with the focus on the *observable behavior* of a subject. This divide is mirrored in the present day schism between cognitivism/computationalism on one side and EEEE (embodied, embedded, enactive, extended) cognition on the other. There have been numerous attempts to bridge this gap

V. C. Müller (Ed.): PT-AI 2017, SAPERE 44, pp. 19–23, 2018.
https://doi.org/10.1007/978-3-319-96448-5_2

(Clark 2013), (Scheutz 2002), (Pfeifer and Iida 2005) and others, offering connection between lower level sub-symbolic signal processing and higher-level processes of (classical, mental) cognition.

The most frequent view of cognition is still human-centric and not evolutionary, generative model. Thagard (2014) lists open philosophical problems of this approach to cognition. Majority of those problems can only be solved on the basis of empirical data, experiments and adequate generative models and simulations.

The idea of morphological computing has been proposed by (Paul 2004) (Pfeifer and Iida 2005), (Hauser et al. 2014) and (Müller and Hoffmann 2017) defining computation in a more general way than the traditional symbol manipulation, or connectionist models. It is taking into account *physical embodiment of computational mechanisms*, thus presenting suitable tool for modeling of a broader range of cognitive phenomena. In a related approach, (Dodig-Crnkovic 2014) takes cognition in a cognitive agent to be morphological computation, defined as information processing performed by morphology on several levels of organization. Cognition in this framework is capacity possessed by all living organisms, as (Maturana and Varela 1980) and (Stewart 1996) argued. Every single cell, while alive, constantly cognizes. It registers inputs from the world and its own body, ensures continuous existence through morphological processes run on metabolic production of energy. It is avoiding dangers that could cause disintegration or damage, adapting its morphology to the environmental constraints. Physico-chemical-biological processes present morphological computation on different levels of organization. They depend on the morphology of the organism: its material, form and structure.

Morphological computation is modeled as a dynamics of a structure of nodes (agents) that exchange (communicate) information. Single living cell presents such a structure. Groups of unicellular organisms (such as bacteria) communicate and build swarms or films through morphological computation that presents social/distributed cognition. Groups of cells through morphological computation cluster into multicellular assemblies with specific control mechanisms, forming the tissues, organs, organisms and groups of organisms. This layered organization of networks within networks provides information processing speed-up.

A new quality in morphological computing in living organisms emerges with the development of nervous system. With it, multicellular organisms as cognizing agents acquire ability of self-representation, which enables distinction between “me” and the “other” and presents basic functionality that supports locomotion. Animals that possess nervous systems with centralized control connected to sensors and actuators, are capable of locomotion which increases probability of survival. Brains in animals consist of large number of mutually communicating cells. A single neuron is a relatively simple information processor, while the whole brain possesses advanced information processing/computational capacities. We see the similar mechanism as in bacteria swarms with distributed cognition implemented as morphological computation.

Besides the ability to model cognition as embodied, embedded, enactive, and extended through interactions with the environment, morphological computing provides means of understanding how this capacity evolved and how it develops during the life of an organism.

2 Problems Solutions with a Broader View of Cognition

Revisiting the list of unsolved/unsolvable problems of cognitive science under the current idea of cognition (Thagard 2014) we can see their natural solution under a more general concept of cognition as morphological computation:

The Emotion Challenge: Morphological computing of embodied cognition has layered computational architecture. Sub-symbolic electro-chemical processes present the basic layer in the information processing related to emotion (von Haugwitz and Dodig-Crnkovic 2015).

The Consciousness Challenge: Consciousness is proposed as information integration that has central role in the control of behavior (Tononi 2004) (Freeman 2009).

The World Challenge: Distributed morphological computation processes representing hierarchies of computation solves this problem (Abramsky and Coecke 2007) (Sloman 2011) (Piccinini and Shagrir 2014) (Dodig-Crnkovic 2016, 2017).

The Body Challenge: Explicit modeling of a body is a consequence of the inclusion of morphological computational processes in the substrate as an integral part of cognition (Matsushita et al. 2005) (Pfeifer and Bongard 2006) (MacLennan 2010).

The Dynamical Systems Challenge: Dynamical systems are a very important class of computational systems, as argued in (van Leeuwen and Wiedermann 2017) (Burgin and Dodig-Crnkovic 2015).

The Social Challenge: Adopting cognition that is not only individual but also distributed/social, solves this problem (Epstein 2007) (Barabasi 2010).

The Mathematics Challenge (brain cannot be conventional computer): Morphological computing in living beings (unconventional computing) starts at quantum level and propagates to higher levels of organisation as different kinds of physical, chemical, biological, cognitive and social computing. (Cooper 2012) (Zenil 2012).

This short account presents an outline of an argument for the adoption of a broader view of cognition then the one that presents the current received view. For the future work, it remains to study the exact mechanisms of morphological computation at variety of levels of organisation of living organisms in terms of computation as information self-structuring (Dodig-Crnkovic 2016 and 2017). At the same time, cognitive computational models are being tested in artifactual cognitive systems with artificial intelligence and cognitive computing.

References

Abramsky, S., Coecke, B.: Physics from computer science. Int. J. Unconv. Comput. **3**(3), 179–197 (2007)

Barabasi, A.-L.: Bursts: The Hidden Pattern Behind Everything We Do. Dutton, London (2010)

Burgin, M., Dodig-Crnkovic, G.: A taxonomy of computation and information architecture. In: ECSA 2015 ASDS Workshop, Proceedings of the 2015 European Conference on Software Architecture Workshops (ECSAW 2015). ACM, New York (2015)

Clark, A.: Whatever next? Predictive brains, situated agents, and the future of cognitive science. Behav. Brain Sci. **36**(3), 181–204 (2013)

Cooper, S.B.: The mathematician's bias and the return to embodied computation. In: Zenil, H. (ed.) A Computable Universe: Understanding and Exploring Nature as Computation. World Scientific Pub. Co Inc. (2012)

Epstein, J.M.: Generative Social Science: Studies in Agent-Based Computational Modeling. Princeton University, Princeton (2007)

Dodig-Crnkovic, G.: Modeling life as cognitive info-computation. In: Beckmann, A., Csuhaj-Varjú, E., Meer, K. (eds.) Computability in Europe 2014, Proceedings of the 10th Computability in Europe 2014, Language, Life, Limits. LNCS, Budapest, Hungary, 23–27 June 2014. Springer (2014)

Dodig-Crnkovic, G.: Information, computation, cognition. Agency-based hierarchies of levels. (author's draft). In: Müller, V.C. (ed.) Fundamental Issues of Artificial Intelligence. Synthese Library, vol. 377, pp. 139–159. Springer International Publishing, Cham (2016). https://doi.org/10.1007/978-3-319-26485-1_10, http://arxiv.org/abs/1311.0413

Dodig-Crnkovic, G.: Nature as a network of morphological infocomputational processes for cognitive agents. Eur. Phys. J. Spec. Top. **226**, 181–195 (2017). https://doi.org/10.1140/epjst/e2016-60362-9

Freeman, W.J.: The neurobiological infrastructure of natural computing: intentionality. New Math. Nat. Comput. **5**, 19–29 (2009)

Hauser, H., Füchslin, R.M., Pfeifer, R. (eds.): Opinions and Outlooks on Morphological Computation (2014). ISBN (Electronic) 978-3-033-04515-6. http://www.merlin.uzh.ch/contributionDocument/download/7499. Accessed 28 Jan 2018

MacLennan, B.J.: Morphogenesis as a model for nano communication. Nano Commun. Netw. **1**, 199–208 (2010)

Matsushita, K., Lungarella, M., Paul, C., Yokoi, H.: Locomoting with less computation but more morphology. In: Proceedings of the 2005 IEEE International Conference on Robotics and Automation, pp. 2008–2013 (2005)

Maturana, H.R., Varela, F.J.: Autopoiesis and cognition - the realization of the living. In: Cohen, R.S., Wartofsky, M.W. (eds.) Boston Studies in the Philosophy of Science, vol. 42. D. Reidel Publishing, Dordrecht (1980)

Müller, V.C., Hoffmann, M.: What is morphological computation? On how the body contributes to cognition and control. Artif. Life **23**, 1–24 (2017)

Paul, C.: Morphology and computation. In: Proceedings of the International Conference on the Simulation of Adaptive Behaviour Los Angeles, CA, USA, pp. 33–38 (2004)

Pfeifer, R., Bongard, J.: How the Body Shapes the Way We Think. A New View of Intelligence. MIT Press, Cambridge (2006)

Pfeifer, R., Iida, F.: Morphological computation: connecting body, brain and environment. Jpn. Sci. Mon. **58**(2), 48–54 (2005)

Piccinini, G., Shagrir, O.: Foundations of computational neuroscience. Curr. Opin. Neurobiol. **25**, 25–30 (2014)

Scheutz, M. (ed.): Computationalism: New Directions. Cambridge University Press, Cambridge (2002)

Sloman, A.: What's information, for an organism or intelligent machine? How can a machine or organism mean? (Book chapter). In: Dodig-Crnkovic, G., Burgin, M. (eds.) Information and Computation. Series in Information Studies. World Scientific Publishing Co (2011)

Stewart, J.: Cognition = Life: implications for higher-level cognition. Behav. Proc. **35**, 311–326 (1996)

Thagard, P.: Cognitive science. In: Encyclopedia Britannica (2013). https://www.britannica.com/science/cognitive-science. Accessed 28 Jan 2018

Thagard, P.: Cognitive science. In: Zalta, E.N. (ed.) The Stanford Encyclopedia of Philosophy (Fall 2014 Edition) (2014). https://plato.stanford.edu/archives/fall2014/entries/cognitive-science/. Accessed 28 Jan 2018

Tononi, G.: An information integration theory of consciousness. BMC Neurosci. **5**, 42 (2004)

van Leeuwen, J., Wiedermann, J.: Knowledge, representation and the dynamics of computation. In: Dodig-Crnkovic, F., Giovagnoli, R. (eds.) Representation and Reality in Humans, Other Living Organisms and Intelligent Machines, p. 69. Springer, Cham (2017)

von Haugwitz, R., Dodig-Crnkovic, G.: Probabilistic computation and emotion as self-regulation. In: ECSA 2015 ASDS Workshop, Proceedings of the 2015 European Conference on Software Architecture Workshops (ECSAW 2015). ACM, New York (2015)

Zenil, H. (ed.): A Computable Universe. Understanding Computation & Exploring Nature As Computation. World Scientific Publishing Company/Imperial College Press, Singapore (2012)

"The Action of the Brain"
Machine Models and Adaptive Functions in Turing and Ashby

Hajo Greif[1,2(✉)]

[1] Munich Center for Technology in Society, Technical University of Munich, Munich, Germany
hajo.greif@tum.de

[2] Department of Philosophy, University of Klagenfurt, Klagenfurt, Austria

Abstract. Given the personal acquaintance between Alan M. Turing and W. Ross Ashby and the partial proximity of their research fields, a comparative view of Turing's and Ashby's works on modelling "the action of the brain" (in a 1946 letter from Turing to Ashby) will help to shed light on the seemingly strict symbolic/embodied dichotomy: while it is a straightforward matter to demonstrate Turing's and Ashby's respective commitments to formal, computational and material, analogue methods of modelling, there is no unambiguous mapping of these approaches onto symbol-based AI and embodiment-centered views respectively. Instead, it will be argued that both approaches, starting from a formal core, were at least partly concerned with biological and embodied phenomena, albeit in revealingly distinct ways.

Keywords: Artificial Intelligence · Cybernetics · Models
Embodiment · Darwinian evolution · Morphogenesis · Functionalism

1 Introduction

Not very much has been written to date on the relation between Alan M. Turing and W. Ross Ashby, both of whom were members of the "Ratio Club" (1949–1958).[1] Not much of the communication between the two seems to have been preserved or discovered either, the major exception being a letter from Turing to Ashby that includes the following statement:

> In working on the ACE [an early digital computer] I am more interested in the possibility of producing models of the action of the brain than in the practical applications of computing. [...]
> It would be quite possible for the machine to try out variations of behaviour and accept or reject them in the manner you describe and I have been

[1] The best historical accounts of the Ratio Club and Turing's and Ashby's roles therein are Husbands and Holland (2008), Holland and Husbands (2011).

V. C. Müller (Ed.): PT-AI 2017, SAPERE 44, pp. 24–35, 2018.
https://doi.org/10.1007/978-3-319-96448-5_3

hoping to make the machine do this. This is possible because, without altering the design of the machine itself, it can, in theory at any rate, be used as a model of any other machine, by making it remember a suitable set of instructions. (Turing 1946, 1 f)

A comparative view of Turing's and Ashby's work on modelling "the action of the brain" (Turing 1946) will help to elucidate the modelling properties of machines with respect to human thinking and behaviour. It will be safe to say that Turing was committed to formal "symbolic simulations" and Ashby to material "working models", with corresponding modes of reference to their target systems. This part of the analysis will be largely in line with Peter Asaro's account of "Computers as Models of the Mind" (2011). However, in terms of Turing's and Ashby's fundamental views of the nature of what is modelled, the picture gets more complex: despite the respective foci of their models on the functions of machines and biological systems, both approaches were in some important respects concerned with biological and embodied phenomena. Both relied on theories of these phenomena, but they relied on competing theories in distinct ways. I will go through Turing and Ashby twice in order to make these points clear, first outlining their takes on modelling (Sect. 2), then their biological credentials (Sect. 3) and finally their implications (Sect. 4).

2 Formal and Material Models

There are various key motives shared between Turing's and Ashby's work that would figure in either AI or cybernetics. Both Turing and Ashby believed that "the action of the human brain" can be subject to a method of modelling that casts it in a strict mathematical description and breaks it down into elementary routines in such a way that the model could be implemented in some kind of machine, in principle at least.

The shared motive of devising machine-implementable models rests on the premise that the behaviour of some system can be described or imitated by a system of altogether different physical make-up. The notion that an identical set of logical operations can be realised in physically variant systems has its paradigm in Turing's universal computing machines, which were initially *theoretical* machines (1936). This proposition has come to be known as "machine state" or "Turing Machine" functionalism (this terminology being introduced by Putnam 1975).

However, first, while Ashby certainly embraced multiple realisability for his machine models, the functionalism he employed and the analogies it implied ultimately were different in kind from Turing's machine state functionalism. Instead, Ashby's functionalism was a biological, essentially Darwinian one, as shall be demonstrated in Sect. 3. Second, there is a number of ways in which Turing and Ashby differed with respect to the manner in which machines shall serve as models, and what the paradigm of machines to do that modelling is. These differences will be briefly outlined in the present section.

Turing: Turing based his models on his mathematical theory of computation. His original quest was for solving the "Entscheidungsproblem" (decision problem) in Gödel (1931): Turing's idea was that the operations required for evaluating whether a given logical proposition can be proven true or false within the calculus to which that proposition belongs could be implemented in an appropriate kind of theoretical machinery, christened the "Logical Computing Machine", by means of interchangeable sets of formal instructions, that is, programs (Turing 1936). This method could be applied to any field of inquiry that allows for a translation of complex logical propositions into a set of elementary, machine-executable operations. The physical characteristics of any real-world machine to implement these functions are underspecified by this requirement, as long as the requisite formal characteristics are in place. As a matter of historical fact, however, the machines to accomplish this task turned out to be digital, electronic, programmable computers. While Turing's own work made major conceptual and practical contributions to the development of these machines, and while computers are the paradigmatic logical computing machines, this does not imply that digital computers are the only conceivable machines of this sort.

The theoretical import of Turing's models lies fully within the realm of mathematics, while their empirical import lies in demonstrating the scope and force of his theory of computability in (thought-) experimental fashion in a variety of fields. His self-ascribed primary empirical interest was in the action of the brain, but his most substantial contribution to any field outside computer science was his mathematical theory of morphogenesis, that is, the patterns of organic growth (Turing 1952),[2] to which I will pay detailed attention in Sect. 3. With respect to cognitive phenomena, Turing placed his inquiry on two separate levels: in "Computing Machinery and Intelligence", he engaged in "drawing a fairly sharp line between the physical and the intellectual capacities of a man" (1950, p. 434) in the design of his "imitation game". In order to grant fair play to machines in this game of simulating human conversational behaviour, he suggested to disconnect conversational abilities from any underlying organic traits. However, when Turing moved on to a consideration of possible mechanisms responsible for these conversational abilities, he introduced his proto-connectionist "B-type unorganised machines" that exemplify structures and processes in the brain on an abstract level (Turing 1948; see also Copeland and Proudfoot 1996).

Either way, Turing considered the phenomena in question chiefly in their *form*, and thereby to the extent they are accessible to the computational method.

---

[2] In the introduction to a posthumous collection of Turing's writings on morphogenesis, Peter T. Saunders claims that Turing (1952) "is still very frequently cited (more than the rest of Turing's works taken together [...])" (Saunders 1992, p. xvi). If *Google Scholar* and citation counts are resources to go by, the parenthetical part of this statement is an exaggeration, but Turing (1952) still ranks approximately 10 and 20% higher in number of citations respectively than the other two of his most-referenced works, Turing (1950) and (1936): https://scholar.google.de/citations?user=VWCHlwkAAAAJ&hl=en&oi=ao (accessed March 28th, 2018). The *Thomson Reuters* and *Scopus* databases have an incomplete record of the original editions, hence cannot be used for comparison.

More precisely, his quest was for descriptions of the behaviour of a target system that can either directly serve as, or be transformed into, input variables for a set of equations which can then be solved by applying computational routines, so that the output either directly describes or predicts a further behaviour of that target system, or can be transformed into such a description or prediction. Any system whose behaviour can be formalised so as to be amenable to this kind of procedure could be subject to the computational method.

Moreover, Turing engaged in formally "describing the mode of behaviour" (Turing 1946, p. 2) of a learning system on an *individual* level in the first place, treating it as a self-contained entity connected to an environment through a number of in- and output channels. The environment of that entity remains underspecified, and is mostly conceived of as input from the experimenter. When Turing considered the action of the brain, he purposefully limited his focus on that organ proper (for this observation, see Hodges 2008, p. 85). Such focus on self-contained, individual entities was arguably guided by a methodological presupposition: as the original topic of Turing's inquiry were elementary recursive operations within a calculus, any empirical test for the force and scope of the computational method would, if not necessarily then naturally, commence with relations of this kind in the target system.[3] The notion of arithmetical routines repeatedly using their own output as input for their next round of application (hence "calling on themselves") may count as the paradigm of Turing's computational method. The method's focus is on what happens to an initial, expressly restricted, input over the course of repeated computational steps, unlike, for example, the equations describing the time evolution of a dynamical system, which take an open-ended sequence of states of the environment as their input values. This basic methodological presupposition, rather than the higher-order question of embodiment, might be the first indicator of the schism that would later develop between cognitivist AI and cybernetics.

Ashby: Ashby's models and their target systems differ from Turing's, first, in his quest being for the ORIGINS of adaptive behaviour of organisms and other systems with respect to their environments (Ashby 1947; 1960). He built his homeostat as a system that was supposed to actually learn, and to share a set of core features of functional organisation with any other, natural or artificial, learning system that has to cope with changing environmental variables. The overarching systematic goal of his research was to explain "whence come the patterning properties of the nervous system" (Ashby 1928–1972, p. 6117, entry of June 13th 1959). In shaping those patterning properties, interactions with the environment, including the behaviour of other organisms, played a crucial role, and were thus fully and expressly incorporated in Ashby's machine model. Hence, second, the best available evidence of the validity of the model lies in its ability to function in real, variable environments, and is best exemplified by

[3] For anti-individualistic views of Turing's approach, see the reading of Turing (1948) proposed by Herold (2003) and the claim that Turing machines are situated systems by virtue of their tapes being part of their local environments (Fabry 2018).

a physical machine, the homeostat. There was both a didactic and a systematic purpose in having a physical implementation of the model. Third, Ashby's methodological choice was to describe the time evolution of a system, as defined by the observer. That description consisted in tracking a succession of states as "lines of behaviour" in a phase space or "field" of the values of the selected variables, given certain initial states and a sequence of environmental inputs. This approach has fairly little in common with the recursive operations applied to delimited and largely self-contained systems that was favoured by Turing.

These marks of distinction from Turing's approach find their common roots in the Darwinian paradigm to which Ashby committed himself, and hence in a theory of the evolution of biological systems. Ashby assumed this framework to generalise to all sorts of systems capable of adaptive behaviours or, in Ashby's terminology, attaining equilibrial states. His assumption was that adaptivity and goal-directed organisation emerge from processes that are not goal-directed themselves but include random variation and deterministic selection, and he tried to single out the basic natural mechanisms by which they are accomplished. Ashby's machines modelled organism-environment relations as relations of negative feedback, in which changes in environmental variables provoke countereffects in the machine, and vice versa. If the change in environmental variables pushes some of the variables within the machine beyond a threshold of stability, the machine, by means of "step-mechanisms", randomly produces new states (as the counterpart of variation in Darwinian evolution) and matches them against what the environment provides (as the analogue of natural selection) until it re-enters a domain of stable values (resulting in an analogue of fitness).

Ashby's approach to modelling was formal and mathematical inasmuch as the theory of feedback mechanisms, equilibria, and stability can be articulated in rigorous mathematical fashion. However, Ashby employed a formal, mathematical apparatus primarily in instrumental fashion, making it subserve the broader purpose of a general science of organisms and other systems. Hence, the theoretical import of Ashby's modelling did not lie within the realm of mathematics. Nor were his models *computational* under any interpretation that would approximate Turing's notion. Moreover, Ashby's approach to modelling was also genuinely material, not merely in terms of the model's physical implementation. There are isomorphisms supposed to hold between the functional status of a machine and the functional status of a target system in such a way that transformations in the target system are matched by transformations in the machine model that are analogous in terms of the functional states involved. If there are an irritation and a negative feedback in a target system upon a certain input, there should be an irritation and a negative feedback in the machine model as well, with measurable correspondences. Arguably, the materiality and non-computationality of Ashby's models in conjunction made them less universal and less adaptable than Turing's computer models, at least once they were computer-implementable in fact (which, as Asaro 2011 observes, makes the choice of modelling approaches a matter of available resources in part).

On the background of this discussion, the differences between Turing's and Ashby's approaches can be located on two general levels: *First*, the primary though not the exclusive focus was on formal versus material modelling respectively, with diverging roles assigned to the mathematical methods involved. *Second*, the functionalism implied by Ashby's argument was of a different sort than Turing's: it primarily considered adaptive, biological or biologically based, functions of brains and other systems to be modelled rather than the logical properties of machine states. Ashby expressly focused on, while Turing intentionally skipped over, the question of where and how the goal-directed organisation of organism or machine in their concrete environments originate.

3 Adaptive Functions vs Laws of Form

I will now argue that Turing's and Ashby's views on biology provide a key to understanding their differences in approach on either of the aforementioned levels. Both the preferred type of modelling and the kind of functionalism chosen are deeply informed by their views of the relevance of Darwinian evolution.

The arguably most articulate and most influential computational model devised by Turing concerned the biological processes of morphogenesis (see Turing 1952), which built on Sir D'Arcy Thompson's, at its time, influential work *On Growth and Form* (1942). Turing sought to apply and test Thompson's account of the generation of organic patterns, from an animal's growth to the grown animal's anatomy, from the dappledness or stripedness of furs to the arrangement of florets of a sunflower and the phyllotaxis, that is, the ordering of leaves on a plant's twigs. Turing's question, like Thompson's, was how such intricate and differentiated patterns develop from genetically homogenous cells. Was there a general mechanism of pattern formation that could be formally described? The formalism of linear and non-linear differential equations used by Turing was expressly impartial to the actual biochemical realisation of pattern formation. It would only provide some clues as to what concrete reactants, termed "morphogens" by Turing, one should look out for. Answering questions of de facto realisation would be a central task for the many biologists, chemists and others who followed Turing's lead. Still, his account of morphogenesis was as close to a direct modelling relation to a natural target system as it would get in Turing's entire work – closer in embodied detail than his proto-connectionist endeavours, and certainly much closer than his imitation game.

In his morphogenetic inquiries, Turing did *not* inquire into any adaptive function, in Darwinian terms, of the patterns so produced. These patterns may or may not serve an adaptive function. Computationally modelling their formative processes does not contribute to explaining that form's function. Whether the florets of a sunflower are patterned on a Fibonacci series, as they, in fact, are, or whether they are laid out in grid-like fashion, as they *cannot* possibly be according to the mathematical laws of form, is unlikely to make a difference in terms of selective advantage. In turn, however, natural selection may not offer a path to a grid-like pattern in the first place, while allowing for, and perhaps enabling, but arguably not determining the Fibonacci pattern.

This seeming indifference towards adaptive functions might be referred to Asaro's observation that the computational mode of modelling is distinctly indirect, and cannot establish the modelling relation by itself (2011). It would have to incorporate a reasonably elaborated theory of the target system that assigns a suitable calculus and suitable inputs in order to attain any degree of sophistication. A reasonably elaborated theory of this kind was amenable to Turing's computational method in D'Arcy Thompson's laws of form but apparently not in Darwinian variation and natural selection.

The reasonable methodological choices that entitled Turing to a limited interest in the evolutionary origins of organic patterns have a far-reaching implication in terms of alignment with competing research paradigms: D'Arcy Thompson's laws of form were articulated with an explicit scepticism towards the relevance of adaptation by natural selection in biology, and claimed an autonomy of formative processes in organisms from Darwinian mechanisms (Thompson 1942). More precisely, D'Arcy Thompson argued that an organism's development is subject to constraints on form that have to be explained, and can be sufficiently explained, by reference to mathematical and physical regularities. Hence, pattern formation cannot be subsumed under a Darwinian account of random variation and natural selection. Natural selection does act on biological forms with respect to their environmental fitness but it cannot generate them, nor is it the only or even the primary constraint on the realisation of possible forms. Instead, D'Arcy Thompson referred to Goethe's archetype theory in this context, picking up on laws of symmetry and the expression of identical forms in otherwise variant organisms, and sharing the observation that some biological forms are more probable to develop than others, whereas still others are genuinely impossible. He puts these Goethean claims and observations on a strictly mathematical and physical footing: organic forms and their transformations are both enabled and constrained by physical laws that can be expressed in mathematical terms. Turing picked up on the form of these mathematical expressions and developed them into a dynamic model of pattern formation.

Even if Turing did not actively endorse D'Arcy Thompon's sceptical view of Darwinism, he at least implicitly went along with it, and chose to work under a paradigm that built on it. Notably, as Boden (2006, pp. 1264–1267) observes, Turing's theory of morphogenesis was enthusiastically received by the embryologists of his day, who were more likely to be attached to theories of archetypes and ontogenetic recapitulation than to notions of Darwinian selection. Moreover, Boden highlights that Turing's mathematical theory of morphogenesis was ultimately proven right by more recent and powerful computer simulation methods, and that it informed those branches of Artificial Intelligence that have come to be known as Artificial Life. However, a similar status can be argued to accrue to principles of Darwinian evolution.

Unlike Turing, Ashby repeatedly referred to the Darwinian notions of adaptation and natural selection (for example, Ashby 1960, p. 29). In doing so, he exclusively focused on the adaptive aspects of evolution – which is not a matter of course, as we saw, and which was not at all the dominant view in biology at the time of his writing. Still, Ashby believed that random variation of existing traits of a reproducible (or, in Ashby's models, modifiable) system and selection by differential reproduction (or continued functioning versus dysfunction) of that system under the influence of a given set of environmental conditions will not only be the primary but the exclusive path for a system to attain an adaptive or, in Ashby's terminology, equilibrial state. Given that Darwinian evolution is by and large gradual, variation is limited, first, to minor alterations and, second, to acting only on parts of an existing structure, thereby excluding large-scale changes across most or all parts of the system. Third, variation would typically not affect continuously functioning adaptive traits: "organisms are usually able to add new adaptations without destroying the old" (Ashby 1960, p. 142). With these limiting conditions in place, no other mechanisms than random variation and natural selection were deemed required to explain the goal-oriented structure and behaviour of an organism or other organised system.

Remarkably, however, in his very emphasis on adaptation, Ashby restricted his focus to the origins of adaptive behaviour by learning, not inquiring deeper into "genic" adaptation, so that the organic basis for the production of such behaviour was largely left aside.[4] In fact, Ashby referred to biological evolution and what he called "Darwinian Machinery" only in the more philosophical and speculative of his writings, for example, Ashby (1952, pp. 50–52) and Ashby (1967), and, of course, in many places in his *Journal*. The Darwinian basis appears to be taken for granted in its own, genuinely biological, adaptive organisation by Ashby – while both behavioural and genic adaptation were assumed to be subject to the same general laws of variation and selection.

The first obstacle to an incorporation of the organic level of adaptation into Ashby's model lies in a perennial problem of evolutionary theory: apart from some fast-breeding model populations, observed under laboratory conditions and the partial evidence they provide, one cannot observe natural variation and selection in real-time. Hence, one is not enabled, first, to precisely and unequivocally map organic traits onto environmental conditions so as to define them as adaptations (rather than contingent effects) to precisely these (rather than other) conditions. Second, even if such mapping can be accomplished, there will be no evidence whatsoever on the history and the dynamics of the concrete *process* of adaptation – that is to say, unless one has a very good grasp of population genetics and ecology along with paleontological evidence. These problems were compounded by an interpretation of Darwinian evolution under which "the species is fundamentally aimless (it *finds* its goals as it goes along)" (Ashby no year, no. 5). In a discussion of the DAMS, a post-homeostat machine model, he actually embraces the view that "variables in the brain should be driven actively by the environment" (Ashby 1928–1972, p. 3831, entry of May 19th, 1952). Contemporary biology has come to accept different, more differentiated views of the organism-environment relation. Some but not all present-day views appreciate D'Arcy Thompson and Turing's morphogenetic laws, while many if

[4] For Ashby's discussion of "Darwinian Machinery", see also Asaro (2008, pp. 166–168).

not most of them ascribe a more active role to organisms with respect to their environments.

4 Constraints, Implications, and Potentials

From the preceding observations, one might infer that there is one remarkable lacuna in both Turing's and Ashby's accounts of the action of the brain: despite their references to variant traditions on biology, they appear to pay little systematic attention to the question of the origins of the biological mechanisms responsible for that action. This seeming lacuna is not a mere omission, but, first and foremost a constraint imposed by the state of the biological sciences of their time. It may also serve as a diagnostic of Turing's and Ashby's modes of systematic theorising and their implications, which in turn may offer a potential resolution to their seeming opposition.

With respect to the constraints involved, leaving the mechanisms of biological origins out of the picture may, in Turing's reliance on D'Arcy Thompson's view of biological form as well as in Ashby's focus on behavioural adaptation, partly owe to the fact that, at the time of their writing in the late 1940s and early 1950s, the authors were not in a position to rely on what would become known as the "modern synthesis" in evolutionary biology. That synthesis, developed between 1936 and 1947, paradigmatically stated by Huxley (1942) and later summarised by Mayr (1991, especially Chap. 9),[5] was just about to become the dominant biological paradigm, and took several years more to become part of common knowledge. The modern synthesis amounted to crossing the Darwinian mechanisms of adaptation by random variation and natural selection – which had become somewhat rarefied since Darwin's time – with the statistical laws of mathematical population genetics, so as to produce, for the first time, a comprehensive and strongly empirically grounded paradigm of Darwinian biology. Thereby, a considerable degree of consensus was established in evolutionary biology which, in conjunction with the rise of molecular biology, sidelined D'Arcy Thompson's laws of form along with a variety of epigenetic and vitalist theories.

While Turing did not live to see the full establishment of the modern synthesis, an appreciation of Ernst Mayr's work can be found in Ashby's later writings, (see Ashby 1928–1972, p. 6637, entry of December 27th, 1966). Accordingly, an argument for isomorphisms between machines and human cognitive traits that can rely both on mechanisms of genetic replication and variation in population and on the Darwinian concept of functional analogy between phylogenetically distinct traits was not available to Turing, given his preference for a different, competing and at that time still competitive tradition in biology. To Ashby, Darwinian functional analogy was a desirable and straightforward route but, given the state of biology at his time, one he could not consider entirely safe. The perspectives towards a synthesis between their respective approaches might have much improved with an adoption of some of the key insights of the modern synthesis, and of what followed afterwards.

[5] A lucid secondary source on the modern synthesis is Depew and Weber (1995, Pt. II).

With respect to the systematic implications, if the importance of the organic level of adaptation is generally acknowledged as a precondition of behavioural adaptation in principle, and if the importance of embodiment and environment in adaptive processes is accepted, Ashby's material approach to modelling the action of the brain provides an outline for a biologically informed cognitive science – even though neither his homeostat model nor the state of biological theory available to him were sufficiently equipped to develop it adequately.

If however, as in Turing, adaptive functions are deemed of secondary theoretical importance, while trial-and-error learning and neuronal patterns became topics of his computational models, the possible relevance vs. irrelevance of adaptive functions, for matters of consistency, should be allocated to different levels of his inquiry. The distinction between the formal nature of the computational method and the materiality and embodiment of its target systems will be of methodological importance when it comes to computationally modelling embodied phenomena and their adaptive functions. After all, Turing was right in claiming a degree of universality for his computational method and its applicability in science that could not be attained by other approaches to modelling, including Ashby's. Nothing in his approach rules out the possibility of computationally modelling adaptive processes once a sufficiently elaborated theory is in place. At the same instance, nothing in Turing's arguments requires that the operations of his theoretical machines directly correspond to, let alone are identical with, the action of the brain or the development of organisms. If one prefers to argue that cognition and biological pattern formation *are* computational processes *sensu strictu*, one will have to look somewhere else than Turing.

With these constraints and implications stated, it will be possible venture beyond what Turing and Ashby could de facto accomplish: from opposite angles, they charted routes towards solutions to present-day issues in biology and cognitive science alike. In conjunction, mathematical principles of biological pattern formation and the mechanisms of genetic replication, expression and control discovered since will provide information on the bounds on genetic variation and phenotypical variance. By the same token, they will help to identify mechanisms of transmission and use of both genetic and developmental information in structuring organic patterns and organism-environment relations. Conversely, the effects of the structures thus produced might be subject to natural selection or analogous mechanisms of retention of reproducible properties. Hence, apart from natural selection proper, organisms become able to use information actively and directly, or they become able to modify their environments in such a way as to adapt them to their needs.

It is not too surprising then that, on the one hand, some of the more recent heterodox accounts of evolution, such as Goodwin (1994), expressly supplement Darwinian mechanisms with D'Arcy Thompson and Turing's morphogenetic laws and similarly minded strands of complexity theory. It is perhaps more surprising that even as like-minded an approach as developmental systems theory (Oyama et al. 2001) fails to do so. On the other hand, selection-based self-organisation of Ashby's variety encountered a renaissance in some subfields of Artificial

Intelligence (for example, Beer and Williams 2015; Harvey et al. 1994), or has been integrated with Turing's approach in inquiries into the evolution of information processing, from molecules to human beings, under the heading of "metamorphogenesis" (Sloman 2013; 2018). Hence, what fist may seem like a tension between two diverging approaches may actually converge on various levels of inquiry, and may always have been meant to do so by Ashby and Turing themselves.

Acknowledgements. This paper has been developed from a research proposal currently under review with the Austrian Science Fund (FWF), project ref. P31136. Particular thanks go out to the reviewers of this paper, Aaron Sloman and Torben Swoboda.

References

Asaro, P.M.: From mechanisms of adaptation to intelligence amplifiers: the philosophy of W. Ross Ashby. In: Husbands, P., Holland, O., Wheeler, M. (eds.) The Mechanical Mind in History, pp. 149–184. MIT Press, Cambridge/London (2008)
Asaro, P.M.: Computers as models of the mind: on simulations, brains, and the design of computers. In: Franchi, S., Bianchini, F. (eds.) The Search for a Theory of Cognition. Early Mechanisms and New Ideas, pp. 89–114. Rodopi, Amsterdam/New York (2011)
Ashby, W.R.: Journal. The W. Ross Ashby Digital Archive. In 25 volumes, (1928–1972). http://www.rossashby.info/journal/index.html. Accessed 28 March 2018
Ashby, W.R.: The nervous system as physical machine: with special reference to the origin of adaptive behaviour. Mind **56**(221), 44–59 (1947)
Ashby, W.R.: Can a mechanical chess-player outplay its designer? Br. J. Philos. Sci. **3**, 44–57 (1952)
Ashby, W.R.: Design for a Brain. The Origin of Adaptive Behaviour, 2nd edn. Wiley, New York/London (1960)
Ashby, W.R.: The place of the brain in the natural world. Currents Mod. Biol. **1**, 95–104 (1967)
Ashby, W.R.: Aphorisms index. The W. Ross Ashby Digital Archive (no year). http://www.rossashby.info/aphorisms.html. Accessed 28 March 2018
Beer, R.D., Williams, P.L.: Information processing and dynamics in minimally cognitive agents. Cogn. Sci. **39**, 1–38 (2015)
Boden, M.A.: Mind as Machine: A History of Cognitive Science. Oxford University Press, Oxford (2006)
Copeland, B.J., Proudfoot, D.: On Alan Turing's anticipation of connectionism. Synthese **108**(3), 361–377 (1996)
Depew, D.J., Weber, B.H.: Darwinism Evolving. Systems Dynamics and the Genealogy of Natural Selection. MIT Press, Cambridge (1995)
Fabry, R.E.: Turing redux: enculturation and computation (2018). Under review
Gödel, K.: Über formal unentscheidbare Sätze der Principia Mathematica und verwandter Systeme I. Monatshefte für Mathematik **38**, 173–198 (1931)
Goodwin, B.: How the Leopard Changed its Spots. The Evolution of Complexity. Princeton Science Library Edition. Princeton University Press 2001, Princeton. Scribners, New York (1994)
Harvey, I., Husbands, P., Cliff, D.: Seeing the light: artificial evolution, real vision. In: Cliff, D., Husbands, P. et al. (eds.) From Animals to Animats 3: Proceedings of the Third International Conference on Simulation of Adaptive Behavior, pp. 392–401. MIT Press, Cambridge (1994)

Herold, K.: An information continuum conjecture. Minds Mach. **13**, 553–566 (2003)
Hodges, A.: What did Alan Turing mean by “machine”? In: Husbands, P., Holland, O., Wheeler, M. (eds.) The Mechanical Mind in History, pp. 75–90. MIT Press, Cambridge/London (2008)
Holland, O., Husbands, P.: The origins of British cybernetics: the Ratio Club. Kybernetes **40**, 110–123 (2011)
Husbands, P., Holland, O.: The Ratio Club: a hub of British cybernetics. In: Husbands, P., Holland, O., Wheeler, M. (eds.) The Mechanical Mind in History, pp. 91–148. MIT Press, Cambridge/London (2008)
Huxley, J.: Evolution. The Modern Synthesis. Allen & Unwin, London (1942)
Mayr, E.: One Long Argument. Charles Darwin and the Genesis of Modern Evolutionary Thought. Harvard University Press, Cambridge (1991)
Oyama, S., Griffiths, P.E., Gray, R.D. (eds.): Cycles of Contingency. Developmental Systems and Evolution. MIT Press, Cambridge/London (2001)
Putnam, H.: Mind, Language and Reality. Philosophical Papers, vol. 2. Cambridge University Press, Cambridge (1975)
Saunders, P.T.: Introduction. In: Saunders, P.T. (ed.) Collected Works of A.M. Turing: Morphogenesis, vol. 3, pp. XI–XXIV. North-Holland, Amsterdam (1992)
Sloman, A.: Virtual machinery and evolution of mind, part 3: metamorphogenesis: evolution of information-processing machinery. In: Cooper, S.B., van Leeuwen, J. (eds.) Alan Turing - His Work and Impact, pp. 849–856. Elsevier, Amsterdam (2013)
Sloman, A.: Huge but unnoticed gaps between current AI and natural intelligence. In: Müller, V.C. (ed.) Philosophy and Theory of Artificial Intelligence III (in this volume). SAPERE. Springer, Berlin (2018)
Thompson, D.W.: On Growth and Form, 2nd edn. Cambridge University Press, Cambridge (1942)
Turing, A.M.: On computable numbers, with an application to the Entscheidungsproblem. Proc. London Math. Soc. **s2–42**(1), 230–265 (1936)
Turing, A.M.: Letter to W. Ross Ashby of 19 November 1946 (approx.) The W. Ross Ashby Digital Archive (1946). http://www.rossashby.info/letters/turing.html. Accessed 4 Aug 2017
Turing, A.M.: Intelligent machinery: a report by A.M. Turing. National Physical Laboratory, London (1948)
Turing, A.M.: Computing machinery and intelligence. Mind **59**, 433–460 (1950)
Turing, A.M.: The chemical basis of morphogenesis. Philos. Trans. R. Soc. B **237**, 37–72 (1952)

An Epistemological Approach to the Symbol Grounding Problem

The Difference Between Perception and Demonstrative Knowledge and Two Ways of Being Meaningful

Jodi Guazzini(✉)

University of Trento, Trento, TN, Italy
jodi.guazzini@unitn.it

Abstract. I propose a formal approach towards solving Harnad's "Symbol Grounding Problem" (SGP) through epistemological analogy. (Sect. 1) The SGP and Taddeo and Floridi's "Zero Semantical Commitment Condition" (z-condition) for its solution are both revisited using Frege's philosophy of language, in such a way that the SGP is converted into two circumscribed tasks. (Sect. 2) The ground for studying these tasks within human cognition is that both the human mind and AI are conceivable, as in Newell's "physical symbol systems" (PSSs), and that they share the core of the SGP: the problem of constructing an objective reference. (Sect. 3) After two forms of reference have been identified in the human mind, I then show why the latter may constitute a model for facing the SGP.

1 Harnad's Problem Within Frege's Framework

According to Harnad's definition, the SGP is the problem of grounding symbols processed by computational devices on something other than more strings of symbols equally requiring interpretation (Harnad 1990, 339–340). Moreover, the SGP has to be solved within a purely formal and syntactical use of symbols (Harnad 1990, 336. See points 2–4), and interpretability as a systematic property of the adopted symbolic language (Harnad 1990, 336. See points 7–8) must be independently assured by computational devices.

In Frege's terms, the SGP is, generally speaking, the problem of finding a referent in such a way that what the object represented is (that is, the referent. See Frege 1892, 57), and how it is comprehended (that is, sense. See Frege 1892, 57), are information explicitly available to the computational devices. Further clarification is possible.

(P_1) According to Harnad, what is missing is one or both of the instruments applied by «cryptologists of ancient languages and secret codes» (Harnad 1990, 339. They represents those involved in the solvable version of the SGP): some known starting language or real world experience. We cannot seek a solution within the former here; it would imply breaking the first part of the so-called Taddeo and Floridi's "z-condition": not postulating any form of native semantic competence or resource (Taddeo and Floridi 2005, 421), and as such would be a first known (interpreted) language. Thus, if Harnad's

V. C. Müller (Ed.): PT-AI 2017, SAPERE 44, pp. 36–39, 2018.
https://doi.org/10.1007/978-3-319-96448-5_4

list of tools is exhaustive, the focus must necessarily be on real world experience, hence some kind of reference to the real world has to be found.

(P_2) Merely providing some sensorial device is not enough for this purpose, since if it were, then systematically interpreting symbol strings (both computational or linguistic) would mean simply having an intuition of the corresponding object with all the relevant details. As Harnad said, in this case «we'd hardly need the definienda» (Harnad 1990, 340). It follows that it is within the "linguistic" structure that an indication about what is the proper referent and how reference has to be organized must be found.

(P_3) The z-condition (A) prohibits both nativism of semantic resources and (B) the uploading of semantic elements, and (C) allows a machine to have non-semantic resources (Taddeo and Floridi 2005, 421). In my terms, concerning sense, the z-condition (a) requires both that the interpretation of symbols which express sense be construed and (b) that the referent not be given through an already constructed semantic framework, and (c) authorizes any non-semantic resource to form sense and individuate referents.

To sum up, (P_1) since referents must be found within the external world; (P_2) since constructing reference must stem from sense; (P_3) since symbols expressing sense must acquire and not presuppose a systematic semantic interpretability, it therefore results that (∴) the task at issue consists of constructing a sense in such a way that it will make possible (I) the pointing out of how to find a referent and (II) to actually trace that referent according to these indications, so that successively, the referent may be explicitly designated.

2 The Root of the SGP

Newell defined his concept of a PSS as «a broad class of systems that is capable of having and manipulating symbols, yet it is also realizable within a physical universe», where "symbolic" means that an element has a reference to something else (Newell 1980, 136). Reference capacity is the matter at issue both in natural language and in symbolic strings for cognitive purposes. It follows that both the human mind and AI can be thought of as PSSs, and that the problem of reference is common to both. Concerning the general situation of reference, PSSs are in the situation represented in Fig. 1.

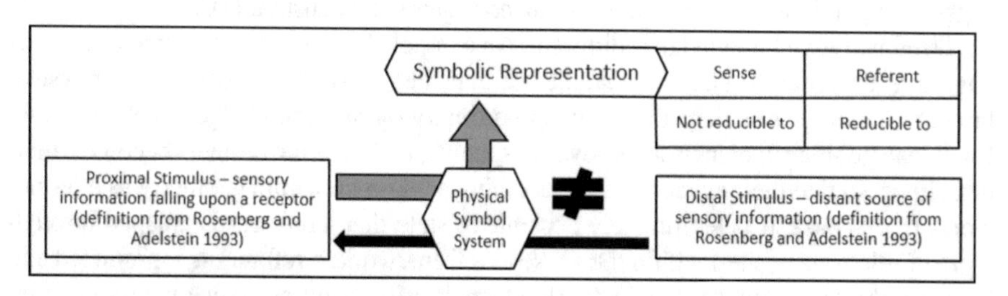

Fig. 1. Schema of PSSs relative to reference

Why does reference generate the SGP? Consider the following argument. (P_1) It is possible to compose comprehensible sentences without a referent, but (P_2) this could not be possible if either objects were the only source of informative content, or if sense alone were not able to convey content by itself. Thus, ($\therefore_1$) what we call "sense" is a semantic content by itself. (P_3) When acquisition of a real object occurs, the information conveyed by sense does not increase, although referent is not analytically included within sense (Cf. Kant 1998, 567): there is a synthetic conjunction, and (P_4) every synthetic conjunction requires a ground for the connection (Kant 1998, 281). ($\therefore_2$) The problem of constructing sentences with both sense and an actual referent, is one of connecting representations conveyed by sense with an appropriate referent, through a middle term suitable as a ground.

The point is identical for both AI and the human mind: PSSs do not produce a sense with an actual referent analytically contained within, hence they must construct the reference; if PSSs could, no SGP would arise, since the production of a symbolic language would immediately provide a reference to actual objects.

3 Proposed Solution

I consider reference as the extent to which it must result from the relation between sense and referent, hence I take into account specifically "perception" and "demonstrative knowledge". The former is that cognitive structure in which reference is grounded on sensible data through the connection of such data with the internal activity of representing (awareness of the result is "intuition"); the latter is that cognitive structure in which reference is grounded on the act of proving consequences, which define the referent. Since the SGP is a problem that derives from the relation between sense and referent, uncategorized sensation and metacognitive capacities are not at issue here.

Perception can provide no solution to the SGP, since this kind of model would lead to breaking constraint A/a of the z-condition. Things are represented as having many general properties (e.g. "white", "cubic" and so forth), which form sense within perception. Their referents are features of facts, but these properties do not exist in this general form. Thus, a further ground is required: namely, intuition, within which the properties are interpreted through the correlation with their particular occurrences. This interpretation is phenomenal, and is accessed as an already-made fact through intuition. Thus, this scenario can suggest no purely logical ground, as is indeed required by constraint A/a.

Demonstrative knowledge is different. An example here is helpful to understand this cognitive structure. In Newton's studies of planetary movement, referents of physical laws are not objects as they are grasped in ordinary terms. Knowledge develops within Euclidean geometry: planets are known as spheres, and their movements become Euclidean lines, so that their existence and becoming are known within geometry as a framework. In this case, it is legitimately possible to state that a referent is grasped through *a priori* relations posited within the sense as a framework: a referent is reproduced and known only within the sense which articulates it. Moreover, reference is grounded on the possibility of individuating true consequences posited by the sense, and it is such a possibility that confirms what are the referent and its presentation. This is particularly evident when parts of sense are formulated in advance of its referent: for example, when Newton predicted the shape of the Earth.

A general conclusion may be reached from this example, and here lies the proposed solution. (P_A) When the referent is subsumed under a system of elements within which (A) *a priori* demonstrable relations are given and (B) these relations convey a sense by themselves from which a referent can be individuated, then the relationship between sense and referent is one of replacement. (P_B) The case reported in (P_A) can be achieved in principle within a purely symbolic language, so that tasks (I) and (II) (see the first passage) would be fulfilled within it. ($\therefore_A$) Under the conditions indicated in (P_A), a purely symbolic language may be intrinsically meaningful, but explicit rules of subsuming proximal stimuli into sense are required.

To sum up, the proposed solution consists of imitating the model of demonstrative knowledge. Stimuli have to be subsumed under a given system of elements with features A-B (for example, mathematics), so that their definition as referents, and their being grasped, may be traced back and explicitly stated starting from the sense; i.e. upon receiving a stimulus, the PSS seeks an adequate referent as if it had been *a priori* defined, instead of immediately subsuming the referent under its representations. The rules for manipulating sense and the relations it posits are not necessarily semantic (mathematics can be implemented in pure symbols), hence the strategy outlined above may, in principle, be developed in purely symbolic terms.

4 Conclusion

In such a scenario, all that would be required would be for a PSS to prove the validity of its symbolic description of referents, and designate them explicitly according to that description. This may constitute a formal solution to the SGP, since conditions I-II (see first part) would be fulfilled.

Refereces

Frege, G.: On sense and reference. In: Geach, P., Black, M. (eds. and trans.) Translations from the Philosophical Writings of Gottlob Frege, pp. 56–78. Blackwell, Oxford (1892)

Harnad, S.: The symbol grounding problem. Phys. D Nonlinear Phenom. (1990). http://doi.org/10.1016/0167-2789(90)90087-6

Kant, I.: Critique of Pure Reason (trans. and ed. by Guyer, P. and Wood, A.). Cambridge University Press, Cambridge (1998)

Müller, V.: Symbol Grounding in Computational Systems: A Paradox of Intentions (2009). https://doi.org/10.1007/s11023-009-9175-1

Newell, A.: Physical symbol systems. Cogn. Sci. (1980). https://doi.org/10.1207/s15516709cog0402_2

Rosenberg, L., Adelstein, B.: Perceptual decomposition of virtual haptic surfaces. In: Proceedings of 1993 IEEE Research Properties in Virtual Reality Symposium (1993). https://doi.org/10.1109/VRAIS.1993.378264

Taddeo, M., Floridi, L.: Solving the symbol grounding problem: a critical review of fifteen years of research. J. Exp. Theor. Artif. Intell. (2005). https://doi.org/10.1080/09528130500284053

An Enactive Theory of Need Satisfaction

Soheil Human[1,2(✉)], Golnaz Bidabadi[3], Markus F. Peschl[1], and Vadim Savenkov[2]

[1] Department of Philosophy and Cognitive Science Research Platform, University of Vienna, Universitätsring 1, 1010 Vienna, Austria
{soheil.human, franz-markus.peschl}@univie.ac.at
[2] Department of Information Systems and Operations, Vienna University of Economics and Business, Welthandelsplatz 1, 1020 Vienna, Austria
{soheil.human, vadim.savenkov}@wu.ac.at
[3] Cisco Systems, Inc., 170 West Tasman Drive, San Jose, CA 95134, USA
golnaz@cisco.com

Abstract. In this paper, based on the predictive processing approach to cognition, an enactive theory of need satisfaction is discussed. The theory can be seen as a first step towards a computational cognitive model of need satisfaction.

1 Introduction

Life can be seen as the constant process of satisfaction of needs, and thus numerous need theories have been proposed in humanities and social sciences, such as psychology, economics, philosophy, sociology, anthropology and social policy over the last century (see Human et al. 2017, for some examples). While no consistency can be found in the usage of the term "need" within or across different disciplines (Gasper 2007), it can be said that most of the conducted research on human needs have been dedicated to development of different categories or lists of needs. Maslow's (1970) hierarchy of needs can be considered as the most famous example of such categorizations of human needs. For sure, such categorizations have had conceptual application in their respective disciplines, however the recent advancements in cognitive science are not reflected in most of them.

In this paper, we reflect on the concepts of *need* and *need satisfaction* from an enactive perspective. Specifically, we take a first step towards development of a theory of need satisfaction in *Predictive Processing* (PP) agents. We are aware that one can draw an intimate connection between need satisfaction, and the classical problem of *planning* that has been tackled throughout the history of AI. However, we hope that our reflection based on PP goes beyond the classical approaches, and will contribute to constructing future novel approaches for development of computational cognitive models of need satisfaction or need-based artificial agents.

V. C. Müller (Ed.): PT-AI 2017, SAPERE 44, pp. 40–42, 2018.
https://doi.org/10.1007/978-3-319-96448-5_5

2 An Enactive PP-based Theory of Need Satisfaction

Over the past decade, there has been a great increase in research based on Bayesian approaches to brain function. According to this approach, which is called predictive processing by Clark (2015), the brain is a probabilistic inference device: a sophisticated hypothesis-testing mechanism that uses hierarchical generative models and seeks to minimise its prediction errors about sensory inputs (Hohwy 2013). In other words, based on PP, the brain continually and at multiple spatiotemporal scales tries to minimise the error between its predictions of sensory input and the actual incoming input. A wide range of anatomical and physiological aspects of the brain and various cognitive processes has been explained and modelled using the predictive processing approach (Clark 2013).

How can the concept of *need* be understood from a PP perspective? We can have two approaches to answer this question:

(1) From a systemic PP-perspective, any living self-organizing system embodies a predictive generative model in order to ensure that free energy is minimised through action (Calvo and Friston 2017). Therefore, one can consider, the minimization of free energy (or minimization of surprise) over time, as the basic *need* of any PP-agent. While this radical standpoint could be very inspiring for a general understanding of notions of *life* and *need*, it seems that grounding a computational cognitive model of *need satisfaction* on this general systemic view would be a very difficult task.

(2) From a top-down perspective, we can consider needs as general priors (hyperpriors). It is important to emphasise that this view does not preclude other general priors which cannot be considered as needs (such as the general regularities in the physics of the world) (see Hohwy 2013, p. 116). While this can be considered as a more conservative view, it seems that it provides an appropriate framework for going beyond a purely conceptual understanding of need. Considering needs as general priors enables us to tackle the fundamental question of *how needs are satisfied in a PP-agent?* In other words, by applying this perspective, elements of the PP formal framework (Hohwy 2012) can be used to model the process of need satisfaction:

  (I) *Hierarchy*: The PP mechanism is a general kind of statistical building block that is repeated throughout different cortical levels. The input of each level is conceived as prediction error and what cannot be predicted at one of the levels is passed on the next level. Lower levels of the hierarchy predict basic attributes and causal regularities at very fast time-scales. More complex regularities, at increasingly slower time scales, are dealt with at higher levels. This can potentially provide a formal solution for dealing with different levels of needs, desires, satisfiers, etc (see Human et al. 2017, for a discussion on these notions).

  (II) *Contextual probabilities*: Predictions at any level of the hierarchy are subject to contextual modulation. This would provide the appropriate key for dealing with the contextual differences in needs satisfaction.

(III) *Empirical Bayes*: In empirical Bayes, priors are extracted from hierarchical statistical learning. This empowers us to not only model the prior beliefs about needs/satisfiers on a moment to moment basis but also through long-term exposure to individual experience. Furthermore, more hard-wired and instantiated needs, e.g. over evolutionary time-scales, can also be modelled based on the empirical Bayes.

(IV) *Active Inference*: Based on the depth of the represented causal hierarchy, the active inference can be a useful tool for modelling short-term and long-term planning for needs satisfaction.

(V) *Top-down and Bottom-up*: Seeing the bottom-up information as prediction-errors and top-down information as causal models of the world, we can develop a model of needs satisfaction that deeply considers the statistical regularities of the world.

We shall consider all these elements as predictive processing is applied to the problem of need satisfaction. If this is done appropriately, it would be possible to model need satisfaction in a way which (a) is consistent with state-of-the-art in cognitive science such as enactivism, and (b) captures different aspects of need satisfaction such as context-dependency and individual heterogeneities.

Acknowledgement. This research was partially supported by the WU Anniversary Fund of the City of Vienna and by the Austrian Federal Ministry for Transport, Innovation and Technology (Grant 855407).

References

Calvo, P., Friston, K.: Predicting green: really radical (plant) predictive processing. J. R. Soc. Interface **14**(131), 20170,096 (2017)

Clark, A.: Whatever next? predictive brains, situated agents, and the future of cognitive science. Behav. Brain Sci. **36**(3), 181–204 (2013)

Clark, A.: Surfing Uncertainty: Prediction, Action, and the Embodied Mind. Oxford University Press, Oxford (2015)

Gasper, D.: Conceptualising human needs and wellbeing. In: McGregor, J.A. (ed.) Gough I, New Approaches and Research Strategies, pp. 47–70. Cambridge University Press, Wellbeing in Developing Countries (2007)

Hohwy, J.: Attention and conscious perception in the hypothesis testing brain. Front. Psychol. **3**, 96 (2012)

Hohwy, J.: The Predictive Mind. Oxford University Press, Oxford (2013)

Human, S., Fahrenbach, F., Kragulj, F., Savenkov, V.: Ontology for representing human needs. In: Różewski, P., Lange, C. (eds.) Knowledge Engineering and Semantic Web, pp. 195–210. Springer, Cham (2017)

Maslow, A.: Motivation and Personality. Harper & Row, New York (1970)

Agency, Qualia and Life: Connecting Mind and Body Biologically

David Longinotti(✉)

Columbia, MD, USA
longinotti@hotmail.com

Abstract. Many believe that a suitably programmed computer could act for its own goals and experience feelings. I challenge this view and argue that agency, mental causation and qualia are all founded in the unique, homeostatic nature of living matter. The theory was formulated for coherence with the concept of an agent, neuroscientific data and laws of physics. By this method, I infer that a successful action is homeostatic for its agent and can be caused by a feeling - which does not motivate as a force, but as a control signal. From brain research and the locality principle of physics, I surmise that qualia are a fundamental, biological form of energy generated in specialized neurons. Subjectivity is explained as thermodynamically necessary on the supposition that, by converting action potentials to feelings, the neural cells avert damage from the electrochemical pulses. In exchange for this entropic benefit, phenomenal energy is spent as and where it is produced - which precludes the objective observation of qualia.

1 Introduction

The thesis of strong artificial intelligence is that the mind is essentially a computer, such that a suitably designed and programmed machine could pursue its own goals and have phenomenal experiences (Johnson-Laird 1988). In this paper, I contend that these claims are analytically and scientifically untenable, and describe a biological solution to the mind body problem. My approach is naturalistic and scientific; I assume that agency and qualia supervene on other phenomena that we take to be natural, and that qualia have regular, discoverable effects on the world. The theory I offer is based on the evaluation of hypotheses for their coherence with the concept of an agent, empirical data and laws of physics. Scientific explanation often requires the postulation of mechanisms, like the events by which an axon conducts an electro-chemical pulse (Machamer et al. 2000). Accordingly, the consideration of mechanisms is central to my method, which leads me to infer that actions and feelings have a common origin in the homeostatic nature of living matter.

The three main sections of the paper concern life, agency and qualia, respectively. I first review the relevant properties of a living system as an entity that is self-organized,

D. Longinotti—Independent.

V. C. Müller (Ed.): PT-AI 2017, SAPERE 44, pp. 43–56, 2018.
https://doi.org/10.1007/978-3-319-96448-5_6

and that maintains itself against thermodynamic decay. The next section concerns the nature and source of agency. From the concept of an agent, I deduce that it is a living substance. Because behavior motivated by a feeling has the homeostatic form of an action, I infer that life is the source of qualia, and that mental causation is based in the regulatory function of affective experiences. In the third main section I address the nature of qualia and the mechanism of their production. Laws of physics are adduced for the hypothesis that a feeling depends on the matter and energy at its location, rather than a causal pattern. From empirical evidence, I surmise that qualia are a distinct form of energy, a property generated in specialized neurons. The subjectivity of qualia is explained as required by thermodynamics if, by producing them, the source of the qualia avoids an increase in its entropy. I conclude with some remarks on the merits of the theory.

I avoid the term "consciousness" in the paper due to its many meanings, one of which involves cognitive attention. Here, my theorizing on consciousness is limited to 'qualia', what Block (1995) describes as 'phenomenal consciousness.' I use "subjective" to mean that a quale is not objectively observable and, in that sense, is private to its subject.

2 Life

2.1 Life Is Self-organizing

The scientific view of life is that it is a natural phenomenon. A living cell is commonly characterized as self-organized, that is, the structure of the cell results from the materials that comprise it, not from an externally imposed design plan. Living matter is similar in this way to other substances that depend on chemical forces for their composition (e.g., crystals, acids, proteins). No outside influence is needed for the internal organization of such substances. As Pross (2003) explains, "living systems are no more than a manifestation of a set of complex chemical reactions and, as such, are governed by the rules of kinetics and thermodynamics." The relevant implication with regard to agency is that the behavior of a living organism in a particular environment is self-determined; its movements result from the way its constituent materials organized themselves.

2.2 Life Is Self-maintaining

Jonas (2001) writes that "in living things, nature springs an ontological surprise in which the world-accident of terrestrial conditions brings to light an entirely new possibility of being: systems of matter that are unities of a manifold … in virtue of themselves, for the sake of themselves, and continually sustained by themselves." Like all systems, a living organism obeys the second law of thermodynamics, which states that the entropy (i.e., disorder) of an isolated system increases with time. That is, every system tends to decay to its equilibrium state of maximum disorganization; for a living thing, this deterioration results in its death. Preventing or slowing this breakdown requires the expenditure of energy from outside the system. In this regard, a living cell functions somewhat like a refrigerator; it consumes energy from external sources to prevent thermal decomposition. However, a refrigerator only slows the decay of things inside it, while a cell sustains its own substance. Schrödinger (1944) views this capability as unique to living matter, and explains that "the device by which an organism maintains itself stationary at a fairly high level of orderliness (=fairly low level of entropy) really consists in continually sucking orderliness from its environment."

The life-supporting order that is obtained from the environment is 'free energy' in various forms, energy at a sufficiently low level of entropy such that it can be metabolized by the organism. For life on earth, the ultimate source of free energy is sunlight, which is used by plants to construct organic complexes that contain chemical energy. Much of the energy and material consumed by a living cell is used in re-synthesizing the numerous proteins required to maintain the cell, as the proteins continually degrade (Pross 2012). Systems that consume energy to maintain themselves in a far-from-equilibrium state are described as 'dissipative' by Prigogine (1978) in that they reduce the amount of free energy in the environment, the energy that can be used for work. Schneider and Kay (1994) hold that "life should be viewed as the most sophisticated (until now) end in the continuum of development of natural dissipative structures, from physical to chemical to autocatalytic to living systems."

Maturana and Varela (1980) characterize a living system as a mechanism that is homeostatic with regard to its own composition. They use the term "autopoietic" (i.e., self-constructing) for such a system: "an autopoietic machine continuously generates and specifies its own organization through its operation as a system of production of its own components … it has its own organization (defining network of relations) as the fundamental variable which it maintains constant."

So, a living cell is self-organized, and its movements are self-determined relative to its environment. Those movements involve the consumption of materials and energy to repair the structure of the cell against the effects of heat and other threats to its biological integrity. A living cell is a homeostatic (i.e., self-maintaining) substance.

3 Agency and Mental Causation

Conceptually, an agent is something that moves itself to realize a goal; such behavior is termed an action. The lack of the goal is the motivation for an action, and the movement for the objective is initiated and controlled by the agent itself. A successful action concludes with the attainment of the goal, which ends the motivation for the behavior.

3.1 An Agent Is a Type of Substance

An agent 'moves itself' in the sense that it determines the way it behaves in response to some stimulus. An agent is 'active'; its movement is powered by energy it contains. In the words of Barandarian et al. (2009), "an agent is a source of activity, not merely a passive sufferer of the effects of external forces."

In general, the two determinants of a system's movement are the characteristics (material and form) of its components, and their organization. Computers and the operations they perform are multiply realizable: the same sequence of computational operations (i.e., the algorithm or software program) can be implemented using a wide variety of materials, and the same material can be used to realize a limitless variety of computational algorithms.

Because a computer is multiply realizable, the specific sequence of operations it performs depends only on its organizational structure. But this structure is not determined by the material of the computer. If it were, the same type of material could not be used to run many different programs. So, the material composition and functional organization that determines how a computer moves is not intrinsic to the computer. It is not an agent, but a tool of its designer.

Accordingly, a necessary property of an agent is that it is self-organized, which makes it the source of its own behavior. This entails that the structure of an agent depends on forces that are intrinsic to its components. Hence, an agent is organized by chemical bonding, forces that inhere in the very nature of the joined materials. But when entities combine chemically, the resulting substance differs in kind from its constituents taken individually. For example, the characteristics of hydrogen and oxygen are lost when they bond to form water.

So, the concept of an agent entails that it is a chemically composed substance, one which consumes energy to move for a goal.

3.2 Agency Depends on Living Matter

What kind of chemical substance is an agent? An action commences with some sort of change *within* its agent, a change that disturbs the agent from its quiescent state. This change 'motivates' (i.e., is the proximate cause of) the action. But because an agent moves for a goal, it must also be the 'want' of the goal that triggers its movement. So, the want of the goal is the motivating change in the agent. Accordingly, the goal of an action is to undo the change in the agent that motivated the movement, thereby returning the agent to its prior 'resting' state. Hence, a successful action has a 'circular' form; it begins and ends in the same entity within the agent. In contrast, a reflex is a 'linear', programmed movement that, once initiated, is carried out irrespective of its effect (if any) on that which triggered it. Unlike a reflex, an action has a homeostatic nature; an agent moves to keep itself in a certain state. And, as argued above, it is a substance that determines its own movement. Hence, an agent is a material having a homeostatic nature.

The concept of an agent accords with the unique character of living matter. A deviation from its self-maintaining activity causes a living cell to expend energy such that, if its movement is effective, the cell returns itself to a more sustainable, dynamic state. I believe that Aristotle recognizes the homeostatic basis of agency where, in Apostle's (1981) translation of *de Anima,* he asserts that "the principle of moving and stopping … is a power of such a nature as to preserve that which has it and to preserve it qua such." Aristotle coins a word for this power: *entelecheia*. In his literal translation, Sachs (2001) takes this term to mean "being at work staying itself". This description of an agent is fully consistent with the scientific characterization of life as reviewed above, wherein a cell is depicted as consuming energy in a manner that maintains its material composition and structure - thereby enabling it to continue this very activity. A living cell is its own goal.

But living *organisms* do not necessarily behave in this way; a moth that flies into a flame apparently moves reflexively, rather than for self-preservation. How, then, do some living organisms move as agents?

3.3 Qualia Originate in Living Matter

A movement of an organism that is motivated by an affective feeling has the homeostatic form of an action; successful behavior ends the painful or pleasurable feeling. The usual response to thirst is an example; the feeling that motivates the movement is extinguished in the organism when it restores itself to its hydrated condition. Similarly, behavior for pleasure ends when satiation is reached. Damasio (2012) remarks that "in brains capable of representing internal states … the parameters associated with a homeostatic range correspond, at conscious levels of processing, to the experiences of pain and pleasure." On the assumption that a feeling is caused by some change in its subject, hedonically motivated movement is an action; attainment of the goal occurs when the part of the organism that produced the feeling is returned to its prior, resting state. In this regard, Spencer (1855) notes that feeling-related movements begin when reflexive motion ends: "…as the psychical changes become too complicated to be perfectly automatic, they become incipiently sensational. Memory, Reason, and Feeling take their rise at the same time." With my earlier inference that an action is a movement of living matter, the observation that hedonic feelings can motivate actions enables a straightforward deduction regarding the origin of at least some types of qualia:

Every action is caused by a change in a living substance.
Some actions are caused by affective feelings.
An affective feeling is caused by a change in a living substance.

This deduction is specific to hedonic feelings. But all qualia are subjective, and I will argue in Sect. 4.4 that subjectivity results from the living nature of the source of qualia. Assuming that is correct, it entails that all qualia - not just the affective types - depend on life. The syllogism above also presumes that qualia can influence behavior in some way. The question of how that occurs is the problem of mental causation.

3.4 Qualia Affect Behavior as Control Signals

For some, the claim that feelings can influence physical movement is equivalent to Cartesian interactionist dualism. This is the view that mind and body are fundamentally different, but that there are causal connections between them. Dualism is not entailed by interactionism, however. In Newton's time, many held that his theory of gravity required the existence of a supernatural phenomenon, because it was widely believed that all forces operated by contact (Gibbon 2002). The current, 'physicalist' view of the world reflects a stance similar to that of Newton's critics; physicalists typically claim that the 'physical' (i.e., non-mental) world is causally closed. But this is contrary to experience. If a phenomenon had no causal relationships with the rest of the world, we would be totally oblivious of it - but we are not oblivious to qualia. In Russell's (1959) view, they are the only sort of thing we know by direct 'acquaintance', rather than through inference.

On the theory of qualia offered here, they can have effects at two levels. At the micro-level of their production, qualia benefit the biological integrity of their living source, as I will posit in accounting for their subjectivity. At a higher level of organization, a phenomenal experience can prompt an organism to act in some way, as assumed in the previous section with the example of thirst.

A possible objection to the view that qualia can cause actions is that, if they were to influence an organism's movement, they would have to do so by exerting a telekinetic force on neural activity - and there is no evidence of such a force. But telekinesis is not necessary for feelings to affect behavior; they can do so as control signals. Analogously, a ship can be steered automatically using light from stars, even though the starlight exerts no relevant force on the ship. All that is required is that the ship be able to *detect* the stars, measure their positions relative to its heading, and adjust its course accordingly. All the force needed to change the direction of the ship is supplied by the ship itself, not by the stars. Similarly, no force on neural activity is needed for a feeling to affect the behaviour of an organism; the organism need only detect the feeling and respond to it in some way – generally, by selecting a type of movement that will influence the feeling (e.g., by eliminating the organism's thirst).

One source of the perceived difficulty in understanding mental causation is a line of reasoning that Kim (2005) calls the "supervenience argument". Let M be the experiential property of a mental state like pain, where M supervenes on its physical base P. M is thought to cause P*, some neural event that results in pain-reducing behavior. But P also appears to be the cause of P*, in which case P* is causally over-determined. Such dual causation is very unlikely so either M is reducible in some way to P, or M is epiphenomenal.

This argument posits that the neural state P, on which M supervenes, is also the cause of P*, the physical response to M. But this is generally not the case. Between the feeling and the behavioral response to it, there can be a lengthy interval of practical reasoning concerning the type of movement (if any) to perform. Otherwise, every movement would be a reflex. Hence, M supervenes on P, but P does not cause P*. M and P* have different causal bases, so causal over-determination is not entailed by M's supervenience on P. This can be seen with the ship analogy wherein one mechanism (a photo-detector) produces a control signal from the starlight, and a separate, mechanical system uses that signal to adjust the ship's rudder.

Hence, the science of mental causation is that of control theory (i.e., cybernetics), wherein the operation of a system is typically adjusted based on an error signal that represents the difference between the goal for the system and its actual state (Ashby 1956). A number of theorists have characterized goal-oriented behaviour as a process involving feedback control (MacKay 1966; Powers 1973; Carver 1979; Carver and Scheier 1981; Marken 2002). The 'navigation' of an organism using its feelings as control signals is similar to the stellar navigation of a ship – except that, in the case of the organism, the source of the feedback signals is internal to the 'vessel'. The organism experiences affective qualia and, using learned behavior and/or practical reasoning, responds accordingly. Just as the imagined ship can't navigate without the starlight, an organism that is guided by its feelings is in mortal danger without them. Humans that lack sensitivity to pain often die before reaching adulthood, because they fail to notice injuries (Nagasako et al. 2003).

The view that affective qualia perform a control function is not new to psychology; Cannon (1932) describes the role that feelings like hunger and thirst perform in the homeostatic regulation of bodily requirements - like water, sugar, proteins, fat and calcium, as well as the oxygen and salt contents of the blood. Schulze and Mariano (2004) offer the following, generalized account:

> Since hedonic states arise whenever a control system produces a chronic regulation error, this implies that the control system is unable to regulate an important physiological variable within the limits required to maintain the integrity of the organism. The hedonic states that arise in response to an increasing regulation error serve to co-opt the behavioral system and its resources. It is then up to the latter to select and execute the appropriate behaviors drawing on cognitive systems in the process.

In addition to physiological conditions, thoughts can also result in motivating feelings, as Hume (1739) describes:

> 'Tis obvious, that when we have the prospect of pain or pleasure from any object, we feel a consequent emotion of aversion or propensity, and are carry'd to avoid or embrace what will give us this uneasiness or satisfaction. 'Tis also obvious, that this emotion rests not here, but making us cast our view on every side, comprehends whatever objects are connected with its original one by the relation of cause and effect.

So, an action may be stimulated by the *anticipation* of pleasure or pain, and this apparently occurs through a faint experience of the expected feeling. Freud famously contends that this sort of process can occur subconsciously, causing us to pursue or repress particular thoughts and memories. Hence, affective feelings function as control signals that motivate an organism to think and/or move to realize a goal-state.

The ability to respond to their feelings conferred a significant biological advantage on those species that evolved this capability. An organism that is limited to reflexive movements is constrained by its evolutionary past, like the aforementioned moth that flies into a flame. In contrast, motivation by its affective feelings enables an individual organism to respond in the present, to new threats and opportunities. Such a phenotype has the possibility to cognitively 'adapt' to some types of events *within its own lifetime*.

4 The Nature and Mechanism of Qualia

I inferred above that the source of qualia is some sort of living substance. In this section, I consider the ontological nature of qualia and the type of event that realizes them. Whereas the arguments concerning agency were mainly analytical, with regard to qualia they are primarily scientific.

4.1 Qualia Are Energy Generated in Specialized Neurons

Qualia appear to be a form of energy. We detect them, and detection generally relies on transduction - the conversion of energy from one form to another. Qualia can carry information for both cognitive and behavioral functions, and communications theory holds that information is modulated energy. Additionally, it seems that everything recognized by modern physics is energy of some type, so feelings might be as well. This supposition is consistent with various observations. These include data from brain stimulation reward experiments, the perceived intensity of sensations as a function of neural activity, phenomenal experiences of some types, and measurements of energy consumption by the brain.

When humans undergo electrical stimulation at some brain sites, they report experiences of pleasure (Heath, 1964). So compelling is the effect on some subjects that the use of this technique raises ethical issues (Oshima and Katayama 2010). In Brain Stimulation Reward studies on rats, electrical pulses are applied to regions of a rat's brain that correspond anatomically to these human 'pleasure centers'; one that is typically targeted is the medial forebrain bundle. A rat will work for these pulses; its motivation is measured by the effort it expends to obtain the reward.

A key result from these experiments is that the strength of the reward effect depends on the total firing rate produced in the relevant neurons, not on the form of the stimulating pulse train (Gallistel et al. 1981; Shizgal, 1999; Simmons and Gallistel 1994). An explanation of this phenomenon suggests itself: the relevant neurons transform the electro-chemical energy of the neural spikes to the phenomenal energy of pleasure, the perceived intensity of the pleasure is proportional to the aggregate energy of the converted neural pulses, and a rat works harder for rewarding pleasure that is more intense.

The strength of sensory qualia also appears to depend on the energy in the associated neurons. In a study on odors, the experienced intensity of a smell correlated with the rate of neural impulses in the amygdala (Winston et al. 2005). Similarly, Mather (2006) reports that "the most successful model of loudness perception … proposes that the overall loudness of a given sound is proportional to the total neural activity evoked by it in the auditory nerve." The rate of neural firings has also been observed to have considerable influence on the visual perception of brightness (Kinoshita and Komatsu 2001) and on the tactile perception of the amplitude of a surface vibration (Bensmaia 2008).

Certain types of phenomenal experiences also support the hypothesis that qualia result from an energy transduction, rather than information processing. A strong blow to the head produces the visual sensation of 'seeing stars'. Apparently, some of the mechanical energy of the jolt is transduced to action potentials in those neurons that convert the pulses to visual qualia. Also, visible and audible white noise carry no information, so there can be no symbolic representation in the resulting neural activity to the effect that 'this is noise'. Yet, an experience of such a phenomenon provides us with knowledge of its random character and its strength; how can this be? Although noise lacks information, it does consist of energy. Evidently, the energy comprising the neural noise is converted to a phenomenal experience, one which retains the relative intensity and spectral properties of the aggregated neural pulses.

Additionally, some measurements of energy consumption by the brain support the hypothesis that a portion of that energy is converted to feelings. Using positron emission tomography (PET) and functional magnetic resonance imaging (fMRI), Raichle (2006) measured the brain's responses to controlled stimuli (in terms of changes in blood flow). Because the increase in energy consumption due to the stimuli was much less than expected, he surmised that "the brain apparently uses most of its energy for functions unaccounted for." Raichle calls this 'dark energy' and posits that it supports intrinsic neural activity for functions like the maintenance of information.

I hypothesize that at least part of this energy is spent in the production of qualia, a fundamental form of energy that was not captured in Raichle's measurements. When awake, we are continuously subjected to feelings of various kinds (both conscious and subconscious) in sensing our external and internal environments. If phenomenal experiences are a form of energy, generating those feelings would increase the baseline metabolic rate of the brain. This would explain the considerable amount of energy that was 'missing' in Raichle's studies.

In principle, the sort of experiment performed by Raichle could provide a means for falsifying the hypothesis that qualia are a form of energy. If that theory is correct, measurements of regions that are sources of feelings should show more 'missing' energy than locations that are not. Such an experiment depends on identifying the areas of the brain that produce qualia, and on a measurement technique with sufficient spatial resolution to distinguish those regions from locations that don't generate feelings.

4.2 A Phenomenal Experience Depends on a Local Event

There are two alternatives regarding the spatio-temporal nature of the event(s) that cause a phenomenal experience. One is the computational hypothesis that a quale results from a causal pattern, such that the existence and character of a quale depend on multiple events distributed over space and time. The other possibility is that a quale is caused by a singular event at a particular space-time location. For the latter alternative, a quale must depend on the type of matter and/or energy at its location; otherwise, it would be under-determined. From the concept of an action, I inferred above that a phenomenal experience has its origin in some type of substance. Here, I argue again for this claim - this time mainly from science.

The causal-pattern hypothesis faces a challenge from physics in the principle of locality, which holds that an event at a space-time location depends only on what is at that location. Einstein expressed the importance of this principle in a letter to Max Born (1971, 171): "If this axiom were to be completely abolished, the idea of the existence of quasi-enclosed systems, and thereby the postulation of laws which can be checked empirically in the accepted sense, would become impossible." Intuitively, the locality principle seems correct; how could an event at some instant be influenced by things that are not at the event's location at that instant? Locality does not preclude the existence of causal 'chains' over space and time, but it does entail that the *type* of event that occurs at a time and place depends only on what exists then and there. The motion of a billiard ball may have its *historical* cause in a complex pattern of collisions involving many other balls, but the *type* of motion a ball exhibits is due only to the way it is impacted by the last ball in the sequence. In general, the locality principle is evident in laws of physics, which do not include any time delays or spatial separations between causes and effects. A changing magnetic field produces an electric field when and where the change in the magnetic field occurs. A mass is accelerated by gravity in proportion to the strength of the gravitational field at the space-time location of the mass.

Turing's (1950) canonical characterization of a computer also conforms to the locality principle. The next state of his machine depends only on the current state and the input to that state, as reflected in the computer's 'machine table'. Accordingly, an effect that depended on a pattern of prior machine states would not be the result of a computation. Furthermore, the supervenience formulation of 'minimal physicalism' as described by Kim (1998) also reflects locality: "Mental properties supervene on physical properties, in that necessarily, for any mental property *M*, if anything has *M* at time *t*, there exists a physical base (or subvenient) property *P* such that it has *P* at *t*, and necessarily anything that has *P* at a time has *M* at that time." [my underline] This precludes a 'physicalist' view of mental states as realizations of causal patterns.

It might be contended that locality does not apply to some types of events, those that exhibit what is called 'quantum entanglement.' Einstein was sceptical of this phenomenon, which he termed 'spooky action at a distance'. But this effect has been confirmed experimentally; the spin-polarizations of electrons generated in pairs and then separated seem, when measured, to influence each other instantaneously across space. The specific basis for this dependence is debated, but the possible explanations all appear to entail a non-local influence of some kind (Yanofsky, 2013).

Nevertheless, quantum entanglement can't rescue causal-pattern theories of qualia - especially if feelings are a form of energy. Information can't be conveyed using entangled properties, and there is no evidence that neural activity in the brain depends on non-local effects. Whether a particular neuron fires is fully explained by local events at its synapse; it does not depend on the history of those events. Furthermore, if a feeling is realized by an energy transform, that event must be localized - or fundamental laws of physics would be violated. Specifically, if the emerging energy-type did not come into being at the same *time* that the prior type is extinguished, there would be a violation of the conservation of energy in the interim. Or, if the new form of energy did not arise at the same *place* as the prior type, the relativistic limit on the speed of signalling would be breached.

Ironically, the physicalist view that a quale depends on a causal pattern implies some sort of non-physical causation. Consider two computers that are in qualitatively identical physical states at some instant. The first has executed the computational algorithm that is thought to be necessary for realizing some feeling, while the second has simply been placed in the same, resulting state. If the first computer has a phenomenal experience while the second does not, that difference could only be due to some non-physical influence because, by stipulation, the two computers are physically identical. Any 'memories' of the computational sequence that exist in the first computer would also be duplicated in the second - unless those 'memories' were non-physical.

Therefore, a feeling depends only on that which exists at its space-time location, which entails that qualia are determined by a particular kind of 'stuff'. In Sect. 3.3 above, I deduced that qualia originate in living matter, but it remains to consider the sort of mechanism by which they are realized, and why they are private.

4.3 Qualia Are Subjective Because They Are Spent as They Are Produced

Subjectivity concerns the process of observation. In general, observation is a form of communication in which energy of some type carries information from the event of interest, to the observer. Objective observation requires that, in principle, any observer could have received the very same modulated energy. So, for a feeling to be objectively observable *qua feeling*, some of the phenomenal energy would have to leave its source. Evidently, this is not possible for qualia. A similar circumstance exists in the cosmological phenomenon of a black hole. Any light produced by, or within the vicinity of, the 'hole' is not observable because it can't escape the gravitational force of the collapsed star.

This suggests that a feeling is subjective because it does not escape its origin. Unlike the energetic property of thermal heat, phenomenal energy is apparently not transferrable by contact, nor is it radiated. But the energy of a phenomenal experience can't simply disappear when the experience stops; it must be transformed to energy of another kind. I posit that qualia are converted to another type of energy as, and where, they are generated. The homeostatic character of life offers a clue to the nature of that energy transduction.

4.4 Qualia Are a Defense Mechanism of Their Living Micro-source

I have argued that the source of qualia is some type of living matter. In addition to metabolizing energy and materials to keep itself going, a living cell defends itself against some dangers to its well-being. One such mechanism is its construction of heat-shock proteins when the cell is confronted with various threats - like thermal changes, oxidative stress, or some toxic substances (Richter et al. 2010). I hypothesize that, like the production of heat-shock proteins, the generation of qualia serves a defensive, homeostatic function for the living matter that produces feelings in specialized neurons. I shall use the term "q-source" for this substance. I posit that action potentials in these neurons threaten the biological integrity of the q-source, and it avoids harm from the neural spikes *by converting them to feelings*.

Why does this make feelings private? The second law of thermodynamics dictates that preventing an increase in the entropy of the q-source requires the expenditure of energy, just as a refrigerator must use energy to slow the increase in the entropy of its contents. If the act of transforming neural pulses into qualia averts a threat to the biological integrity of the source of the qualia, energy must be consumed for that benefit. That energy apparently comes from the qualia themselves; if so, they never leave their source. As they are generated, qualia are transformed immediately to another type of energy; this precludes objective observation of them. I earlier analogized the q-source to a refrigerator. If the above account of qualia's generation is correct, the q-source is a remarkable sort of refrigerator. Unlike the kind of machine we use to preserve food, which requires energy from an external source, the energy used by the q-source to 'cool' the 'hot' things inside it (i.e., the action potentials) comes from those very things!

A different perspective might clarify this postulated mechanism. Living matter contains potential energy that resides in its structure, an organization of atoms and molecules that enables the substance to perform the activities that keep it alive. I hypothesize that action potentials can damage the organizational structure of the q-source. As neural activity begins to have this effect, the q-source reacts by converting the electro-chemical pulses to feelings. Ridding itself of the neural spikes in this way enables the q-source to return to its original structure, which benefit is compensated by the immediate expenditure of the qualia. Energy is thereby conserved; the energy in the neural pulses is converted to the energy of the feeling, which is instantly exchanged for the potential energy of the q-source. Accordingly, the generation of feelings is an action of the q-source that is due to its homeostatic nature; qualia are produced by living matter of some kind.

5 Concluding Remarks

The biological theory described in this paper is more scientifically conservative than the dominant, computationalist hypothesis because, while it posits a new form of energy, it does not violate any law of physics. And it explains more.

Regarding agency, the multiple realizability of a computer entails that the form of its movement has an external source (its designer), while the intrinsic nature of living matter bestows it with self-determined behavior for a self-determined goal: itself. The computational theory does not fundamentally distinguish actions from reflexes. On the biological hypothesis, actions exhibit the 'circular', homeostatic movement of a self-sustaining substance, while reflexes have a 'linear', programmed form. Functionalist theories struggle to find a causal role for the experiential aspect of a feeling, but this is not a problem for the biological theory wherein affective qualia serve as control signals in the regulatory processes by which a living organism maintains itself.

The central assumption of the computationalist view, that a phenomenal experience is determined by a causal pattern, contradicts the locality principle of physics. It thereby entails a radical form of causation that defies space and time. The biological theory does not violate locality; it postulates that a quale is the product of a singular, localized event: an energy transduction. No scientific account of subjectivity is provided by the orthodox, functionalist theory, while subjectivity is nomologically necessitated if, at the micro-level, qualia prevent an increase in the entropy of their source – a function that accords with the homeostatic character of life. Neither theory accounts for the experiential property of a feeling, but this epistemological failing is consistent with the inference that qualia are a *fundamental* form of energy.

No part of the biological theory is ad hoc. As pictured in Fig. 1, it provides integrated, mutually supporting accounts of agency, qualia and their subjectivity - all scientifically based in the thermodynamically unique, self-maintaining nature of living matter.

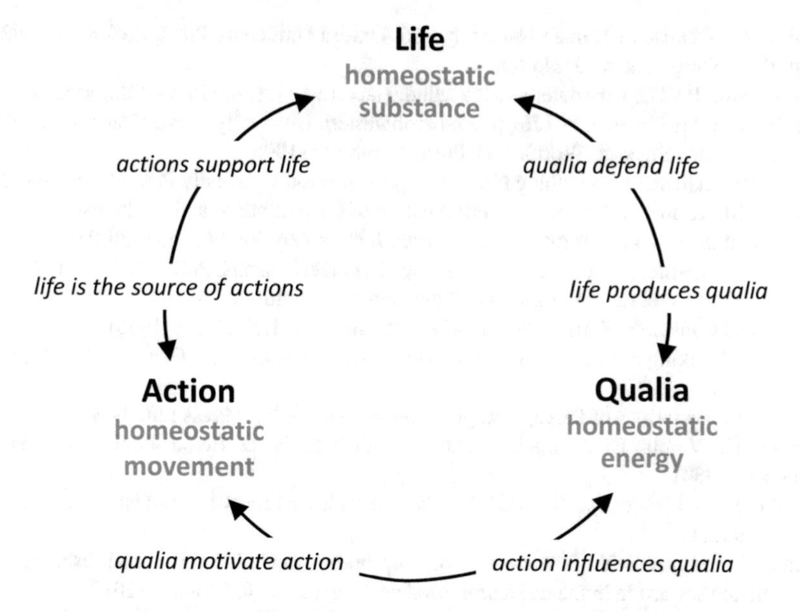

Fig. 1. Agency and qualia depend on the homeostatic nature of life.

Acknowledgments. I am grateful to reviewers Paul Schweizer and Chuanfei Chin, whose thoughtful comments and suggestions on the initial version of the paper prompted me to improve it in a number of ways. Vincent Müller also has my sincere thanks for organizing a stimulating conference, and for his help as Editor.

References

Apostle, H.G.: Aristotle's On the Soul, 416b18–19. The Peripatetic Press, Grinnell (1981). Translated by H. Apostle

Ashby, W.R.: An Introduction to Cybernetics. Martino Publishing, Mansfield Centre (1956/2015)

Barandarian, X., Di Paolo, E., Rohde, M.: Defining agency: individuality, normativity, asymmetry, and spatio-temporality in action. Adapt. Behav. **17**(5), 370 (2009)

Bensmaia, S.: Tactical intensity and population codes. Behav. Brain Rese. **190**(2), 165–173 (2008)

Block, N.: On a confusion about a function of consciousness. Behav. Brain Sci. **18**, 2 (1995)

Born, M.: The Born-Einstein letters, p. 171. Walker and Company, New York (1971)

Cannon, W.B.: The Wisdom of the Body. W. W. Norton & Company, New York (1932/1963)

Carver, C.S.: A cybernetic model of self-attention processes. J. Pers. Soc. Psychol. **50**, 1216–1221 (1979)

Carver, C.S., Scheier, M.F.: Attention and Self-regulation: A Control-Theory Approach to Human Behaviour. Springer, New York (1981)

Damasio, A.: Self Comes to Mind, p. 53. Random House, New York (2012)

Gallistel, C., Shizgal, P., Yeomans, J.: A portrait of the substrate for self- stimulation. Psychol. Rev. **88**, 228–273 (1981)

Gribbin, J.: Science: a History 1543-2001. BCA/The Penguin Press, London (2002)

Heath, R.G.: The Role of Pleasure in Behavior. Harper & Row, New York (1964)

Hume, D.A.: Treatise of Human Nature, p. 266. Oxford University Press, Oxford (1739/2000). Ed. D. F. Norton and M. J. Norton
Johnson-Laird, P.: The Computer and the Mind. Harvard University Press, Cambridge (1988)
Jonas, H.: The Phenomenon of Life, p. 79. Northwestern University Press, Evanston (2001)
Kim, J.: Mind in a Physical World. MIT Press, Cambridg (1998)
Kim, J.: Physicalism, or Something Near Enough. Princeton University Press, Princeton (2005)
Kinoshita, M., Komatsu, H.: Neural representation of the luminance and brightness of a uniform surface in the macaque primary visual cortex. J. Neurophysiol. **86**, 2559 (2001)
Machamer, P., Darden, L., Craver, C.: Thinking about mechanisms. Philos. Sci. **67**, 1–25 (2000)
MacKay, D.M.: Cerebral organization and the conscious control of action. In: Eccles, J.C. (ed.) Brain and Conscious Experience, pp. 422–445. Springer, Heidelberg (1966)
Marken, R.: Looking at behavior through control theory glasses. Rev. Gen. Psychol. **6**(3), 260–270 (2002)
Mather, G.: Foundations of Perception, pp. 116–141. Psychology Press Ltd., New York (2006)
Maturana, H., Varela, F.: Autopoiesis and Cognition, p. 79. D. Reidel Publishing Company, London (1980)
Nagasako, E., Oaklander, A., Dworkin, R.: Congenital insensitivity to pain: an update. Pain **101**, 213–219 (2003)
Oshima, H., Katayama, Y.: Neuro-ethics of deep brain stimulation for mental disorders: brain stimulation reward in humans. Neurol. Medico-Chirurgica **50**, 845–852 (2010)
Powers, W.T.: Behavior: The Control of Perception. Aldine, Chicago (1973)
Prigogine, I.: Time, structure and fluctuations. Science **201**, 777–785 (1978)
Pross, A.: The driving force for life's emergence: kinetic and thermodynamic considerations. J. Theor. Biol. **220**, 393–406 (2003)
Pross, A.: What Is Life? Oxford University Press, Oxford (2012)
Raichle, M.: The brain's dark energy. Science **314**, 1249–1250 (2006)
Richter, K., Haslbeck, M., Buchner, J.: The heat shock response: life on the verge of death. Mol. Cell **40**(2), 253–266 (2010)
Russell, B.: The Problems of Philosophy. Oxford University Press, London (1959)
Sachs, J.: Aristotle's On the Soul and On Memory and Recollection. Green Lion Press, Santa Fe (2001). Translated by J. Sachs
Schneider, E.D., Kay, J.J.: Life as a manifestation of the second law of thermodynamics. Math. Comput. Model. **19**(6-8), 36 (1994)
Schrödinger, E.: What is life? & Mind and Matter, p. 79. Cambridge University Press, Cambridge (1944/1967)
Schulze, G., Mariano, M.: Mechanisms of Motivation, vol. 15, pp. 1–5. Trafford Publishing, Victoria (2004)
Shizgal, P.: On the neural computation of utility: implications of studies from brain stimulation reward. In: Kahneman, D., Diener, E., Schwartz, N. (eds.) Well-being: the Foundation of Hedonic Psychology, pp. 500–524. Russel Sage Foundation, New York (1999)
Simmons, J.M., Gallistel, C.R.: Saturation of subjective reward magnitude as a function of current and pulse frequency. Behav. Neurosci. **108**, 151–160 (1994)
Spencer, H.: The Principles of Psychology, p. 593. Longman, Brown, Green, and Longmans, London (1855)
Turing, A.M.: Computing machinery and intelligence. Mind **59**, 433–460 (1950)
Winston, J., Gottfried, J., Kilner, J., Dolan, R.: Integrated neural representations of odor intensity and affective valence in human amygdala. J. Neurosci. **25**(39), 8903–8907 (2005)
Yanofsky, N.: The Outer Limits of Reason. MIT Press, Cambridge (2013)

Dynamic Concept Spaces in Computational Creativity for Music

René Mogensen[(✉)]

Birmingham City University, Birmingham, UK
Rene.Mogensen@bcu.ac.uk

Abstract. I argue for a formal specification as a working understanding of 'computational creativity'. Geraint A. Wiggins proposed a formalised framework for 'computational creativity', based on Margaret Boden's view of 'creativity' defined as searches in *concept spaces*. I argue that the epistemological basis for *delineated* 'concept spaces' is problematic: instead of Wiggins's bounded types or sets, such theoretical spaces can represent traces of creative output. To address this problem, I propose a revised specification which includes *dynamic concept spaces*, along with formalisations of memory and motivations, which allow iteration in a time-based framework that can be aligned with learning models (e.g., John Dewey's experiential model). This supports the view of computational creativity as *product* of a *learning process*. My critical revision of the framework, applied to the case of computer systems that improvise music, achieves a more detailed specification and better understanding of potentials in computational creativity.

1 Introduction

So far, there is no known definitive description of what computational creativity might be; to improve that end I argue for a formal specification as a working understanding of computational creativity for music. My working understanding supports an analytical view of machines that improvise *co-creatively* with humans, and the specification can also serve as a generative tool for development of new improvising systems (as in (Mogensen 2017b)).

A computational creativity is not necessarily in the same category as human creativity and comparing these two 'creativities' may well, in logic, be a category mistake. Kinds of what we call *creativity* may have in common what Wittgenstein called 'family resemblances', and so I take the *creativity concept family* as a term covering possible 'creativities' that exhibit both similarities and differences. The vaguely defined 'human creativity' serves heuristically as prototype for the creativity concept family only to the extent that I use terms derived from ideas about human creativity to name and to guide the conceptualisations of my proposed components in the specification for computational creativity, no identity between human creativity and computational creativity is implied.

V. C. Müller (Ed.): PT-AI 2017, SAPERE 44, pp. 57–68, 2018.
https://doi.org/10.1007/978-3-319-96448-5_7

I take as given that anything that a current digital computer (or a Universal Turing Machine[1]) can do, can be represented in a formal specification. Therefore, if a computer can in some way be programmed to perform creatively, in other words produce a kind of 'creativity' and become a member of the creativity concept family, then such a creativity must be definable as a formal specification of 'computational creativity'. Developing a more detailed formal specification for computational creativity is an essential step towards understanding the potentials of such technology; and such specification can additionally serve as a guide for developing more capable implementations that can interact constructively with human priorities.

Creativity is often referred to as consisting of some *creative process*, whereas I argue for understanding creativity as determined by *product* achieved by a *learning* process, so that creativity itself is *not* a process but instead is a *product* (echoing Glickman (1976)). In support of this view of creativity I argue that the formal specification allows alignment with learning models (e.g., John Dewey's experiential model (Dewey 1938), (Kolb 2015)).

I base my formal specification on my reworking, in effect replacement, of Wiggins's (2006a) formal framework, which in turn was based on Boden's (2004) conception of 'creativity' as searches in concept spaces. In order to allow the alignment of the specification with the experiential learning model as mentioned, I argue that the epistemological *delineation* of 'concept spaces', in the Wiggins/Boden framework, is problematic: instead of bounded types or sets (that imply a rather static character), such theoretical spaces should more properly represent traces of creative output.[2] These emergent traces are much better represented by *dynamic* concept spaces. I examine my revised specification in the context of computers that co-creatively improvise music together with human performers.[3]

2 A Working Specification for Computational Creativity

My working specification for computational creativity, in Z-style notation,[4] views creativity as searches in conceptual spaces. In my initial adaptation of Wiggins's

[1] The Universal Turing Machine was presented in (Turing 1936). 'The [Universal] Turing Machine not only established the basic requirements for effective calculability but also identified limits: No computer or programming language known today is more powerful than the Turing Machine' (Petzold 2008, p. 330). See Petzold's (2008) book for an insightful interpretation and discussion of Turing's 1936 article.

[2] I use the term 'trace' in the sense of Jean-Jacques Nattiez where 'the symbolic form [of the work] is embodied physically and materially in the form of a *trace* accessible to the five senses' (Nattiez 1990, p. 12).

[3] I have previously examined 'co-creativity' in the musical context (Mogensen 2017b).

[4] Briefly, the Z schema notation includes a declarations part above the central horizontal line and predicates below the horizontal line. "The central horizontal line can be read 'such that'." The axiomatic predicates (below the line in Fig. 1) "appearing on separate lines are assumed to be conjoined together, that is to say, linked with the truth-functional connective $\wedge$" (Diller 1990, 6).

framework I summarise Wiggins's Axioms in Fig. 1 and his approach to determining 'creative output' in Fig. 2 (from (Mogensen 2017b) and (Wiggins 2006a, pp. 451–453)). In Fig. 1 the declarations are interpreted as follows: $\mathscr{C}$ is a concept space of type Σ in the universe of possible concepts $\mathscr{U}$. C is a concept type and c^1, c^2 are instances of C and $\top$ is the empty concept, all of which may be within a concept space $\mathscr{C}$. In Wiggins's formalism "creativity" is seen as searches in a conceptual space ($\mathscr{C}$), which is a subset of the universe of possible concepts ($\mathscr{U}$).[5]

Wiggins proposed an approach to evaluating concepts, discovered through the searches, which is summarised in Fig. 2: a *Language* ($\mathscr{L}$) gives the basis for a *Search strategy* ($\mathscr{T}$) and *Constraints* ($\mathscr{R}$) on the conceptual space ($\mathscr{C}$), along with *Evaluation criteria* ($\mathscr{E}$), that are related to form part of the input to a decision function which consists of an *interpreter* $\langle\langle .,.,.\rangle\rangle$ and an *evaluator* $[[.]]$.

$\mathscr{U}$: *Possible concepts*
Σ : *Concept space type*
C : *Concept type*
$\mathscr{C}$: *Instance of* Σ
$\top$: *Empty concept*
c_1, c_2 : *Instances of C*

$\top \in \mathscr{U}$
$\forall c_1, c_2 \in \mathscr{U} | c_1 \neq c_2$
$\forall \mathscr{C} | \mathscr{C} \subseteq \mathscr{U}$
$\forall \mathscr{C} | \top \in \mathscr{C}$

Fig. 1. My schema of Wiggins's four Axioms.

I have previously (Mogensen 2017a) modified the specification by adding *Intrinsic Motivations* and *Extrinsic Motivations*[6] (see Figs. 3 and 4) based on information theoretic types proposed in Oudeyer and Kaplan's typology of computational models of motivations, which combines psychological concepts with generalisations of robot implementations (Oudeyer and Kaplan 2007, pp. 4–5). The formalised representations of intrinsic motivation can indicate a combination of motivations that can described as in the schema in Fig. 3.[7] Four types of motivation components are included: 1. r_l : *Attraction to novelty*;

[5] For a full narrative explanation of more details of Wiggins's framework I refer the reader to his (2006a) paper.

[6] Here I am representing $\mathscr{M}_1$ and $\mathscr{M}_2$ as arrays, rather than summing the individual motivation components as I did in (Mogensen 2017a); the array is a less reductive representation which I expect will be more useful for the framework development.

[7] Oudeyer and Kaplan (2007) do not address issues of probability calculation and I will also defer such issues. The references on which they base their typology do include reports on implementations some of which may detail instances of probability calculations.

Creative Output
$\mathscr{L}$: *Language*
$\mathscr{R}$: *Constraints on concept space*
$\mathscr{T}$: *Search strategy*
$\mathscr{E}$: *Value definition*
$\langle\langle .,.,.\rangle\rangle$: *Wiggins's interpreter function*
$[[.]]$: *Wiggins's evaluator function*

$\mathscr{R}, \mathscr{T}, \mathscr{E} \in \mathscr{L}$
$[[\mathscr{E}]]\left(\langle\langle \mathscr{R}, \mathscr{T}, \mathscr{E}\rangle\rangle(\{\top\})\right)$

Fig. 2. My summary of Wiggins's '[e]valuating members of the conceptual space' with the empty concept as a starting point.

$\mathscr{M}_1$: *Intrinsic Motivation*
P : *Probability*
c_k : *Instances of C*
J : *Constant*
H : *Knowledge of possibility space*
r_l : *Attraction to novelty*
r_m : *Information gain*
r_n : *Pleasure of surprise*
r_o : *Comfort of the familiar*
t : *Time*

$$H(\mathscr{C}(t)) = -\sum_{c_k \in \mathscr{C}(t)} P(c_k) ln\left(P(c_k)\right)$$
$$r_l(c_k,t) = J_l \cdot \left(1 - P(c_k,t)\right)$$
$$r_m(c_k,t) = J_m \cdot \left(H(\mathscr{C},t) - H(\mathscr{C},t+1)\right)$$
$$r_n(c_k,t) = J_n \cdot \frac{1-P(c_k,t)}{P(c_k,t)}$$
$$r_o(c_k,t) = J_0 \cdot P(c_k,t)$$
$$\mathscr{M}_1 = \left(r_l(c_k,t), r_m(c_k,t), r_n(c_k,t), r_o(c_k,t)\right)$$

Fig. 3. My adaptation of some types from the Oudeyer/Kaplan formal intrinsic motivation typology.

2. r_m : *Information gain*; 3. r_n : *Pleasure of surprise*; 4. r_o : *Comfort of the familiar*. These four components are described as probability-based computations[8] that operate on an experienced concept ($c_k(t)$) in relation to the known part of the concept space at the time ($\mathscr{C}(t)$).

I proposed that extrinsic motivations can be formalised in a similar way, although with a focus on external input as shown in Fig. 4. The four motivation components are similar to those of the intrinsic motivations, except that for extrinsic motivations ($\mathscr{M}_2$) the probability-based computations operate on an external source of sensory input ($M_k(t)$) in relation to the known part of the concept space at the time ($\mathscr{C}(t)$).

[8] These component descriptions are adapted from Oudeyer and Kaplan (2007).

$\mathscr{M}_2$: *Extrinsic Motivation*
P : *Probability*
M_k : *External input*
J : *Constant*
H : *Knowledge of possibility space*
r_l : *Attraction to novelty*
r_m : *Information gain*
r_n : *Pleasure of surprise*
r_o : *Comfort of the familiar*
t : *Time*

$$H(\mathscr{C}(t)) = -\Sigma_{M_k \in \mathscr{C}(t)} P(M_k) ln\big(P(M_k)\big)$$
$$r_l(M_k,t) = J_l \cdot \big(1 - P(M_k,t)\big)$$
$$r_m(M_k,t) = J_m \cdot \big(H(\mathscr{C},t) - H(\mathscr{C},t+1)\big)$$
$$r_n(M_k,t) = J_n \cdot \frac{1 - P(M_k,t)}{P(M_k,t)}$$
$$r_o(M_k,t) = J_0 \cdot P(M_k,t)$$
$$\mathscr{M}_2 = \big(r_l(M_k,t), r_m(M_k,t), r_n(M_k,t), r_o(M_k,t)\big)$$

Fig. 4. My adaptation of reward structures from the Oudeyer/Kaplan typology for extrinsic motivation.

My four choices of the formalisations of motivations (r^l, r^m, r^n, r^o) are only part of the Oudeyer/Kaplan intrinsic motivation typology and it may be useful to explore other types and hence other concepts of motivations in the framework, but I leave this for future research. The four formalised motivation types are based on human psychology and so would seem to contradict my proposal in the Introduction that human and computational creativity are different categories. However, I argue that using theories of human motivation as the basis for computational models does not mean that these are of the same categories, but rather that the computational motivation models reference human motivation in order to guide conceptualisation.

$$Memory : \mathscr{W}(t) = \bigcup_{p=1}^{t-1} \Big(Q(p) \cdot [[\mathscr{E}]]\Big(\langle\langle \mathscr{R}, \mathscr{T}, \mathscr{E}\rangle\rangle\big(c(p)\big)\Big)\Big). \tag{1}$$

This formalisation required a more explicit *Memory* representation, as discussed in (Mogensen 2017b), which is defined as $\mathscr{W}(t)$ in expression 1 and reappears in Fig. 5 in my version of the framework. $\mathscr{W}(t)$ is a memory of past evaluations at time t: it is the set of past results of Wiggins's evaluator functions. Each element of the memory (subset of past interpreter function outputs) may be attenuated by some time-dependent effect which I indicate as Q.

My revised *Creative Output* formalisation is shown in Fig. 5 (Mogensen 2017a, p. 8), which can be summarised as follows: the interpreter function uses constraints to interpret changes in intrinsic ($\mathscr{M}_1$) and extrinsic ($\mathscr{M}_2$) motivations

Creative Output

$\mathscr{L}$: *Language*
$\mathscr{R}$: *Constraints on concept space*
$\mathscr{T}$: *Search strategy*
$\mathscr{E}$: *Value definition*
c : *Instance of C*
$\mathscr{M}_1$: *Intrinsic Motivation*
$\mathscr{M}_2$: *Extrinsic Motivation*
$\mathscr{W}$: *Memory*
t : *Time*
$\langle\langle .,.,. \rangle\rangle$: *Wiggins's interpreter function*
$[[.]]$: *Wiggins's evaluator function*

$$\mathscr{R}, \mathscr{T}, \mathscr{E} \in \mathscr{L}$$
$$\mathscr{W}(t) = \bigcup_{p=1}^{t-1} \Big(Q(p) \cdot [[\mathscr{E}]] \Big(\langle\langle \mathscr{R}, \mathscr{T}, \mathscr{E} \rangle\rangle (c(p)) \Big) \Big)$$
$$[[\mathscr{E}]] \Big(\langle\langle \mathscr{R}, \mathscr{T}, \mathscr{E} \rangle\rangle (\Delta \mathscr{M}_1(t), \Delta \mathscr{M}_2(t), c(t), \mathscr{W}(t-1)) \Big)$$

Fig. 5. My revised version of the *Creative Output* formalisation.

as well as the current concept space ($\mathscr{C}(t)$) and accumulated memory ($\mathscr{W}$). This interpretation is processed by the evaluator function to give the *Creative Output.*[9]

3 Concept Space Morphology

With my specification we can begin to examine the possibility that concept spaces ($\mathscr{C}$) are not the delineated types (Σ) that seem to be used in the Wiggins/Boden framework; rather, concept spaces are dynamic and can represent emergent qualities of the traces of creative output, and the structure over time of these traces is generated from the experiences of the agents that operate on and within them. In Fig. 6 I have formalised a view of dynamic concept spaces: changes in constraints $\Delta\mathscr{R}(t)$, search strategy $\Delta\mathscr{T}(t)$ and value definitions $\Delta\mathscr{E}(t)$ are functions of memory $\mathscr{W}(t-1)$ and motivations ($\mathscr{M}_1(t-1), \mathscr{M}_2(t-1)$). The change of concept space at time t ($\Delta\mathscr{C}(t)$) is, in turn, a function of the changes of constraints $\mathscr{R}(t)$, search strategy $\mathscr{T}(t)$ and value definition $\mathscr{E}(t)$ as well as the latest concept $c(t)$ and the concept space previously perceived $\mathscr{C}(t-1)$.

This morphology of the concept space is examined from the agent perspective, since it is generated from inputs that include memory and motivations. So here the concept space is not an ideal space encompassing all possibilities in a particular domain, rather it is a dynamic space of possibilities as perceived by an agent which may or may not correspond to a particular idealised domain. This distinction is the key to refining this part of the formalism. To define an ideal domain-based concept space would require omniscience, knowledge of the entire

[9] Arguably, in Fig. 5 and expression 1 the component $(c(p))$ should be replaced by $(\Delta\mathscr{M}_1(p), \Delta\mathscr{M}_2(p), c(p), \mathscr{W}(p-1))$ if we want to include memory of motivations.

$\mathscr{C}(t)$: Concept Space
Creative Output
t : time

$$\Delta\mathscr{R}(t) = f_{\mathscr{R}}(\mathscr{R}(t-1), \mathscr{W}(t-1))$$
$$\Delta\mathscr{T}(t) = f_{\mathscr{T}}(\mathscr{T}(t-1), \mathscr{W}(t-1), \mathscr{M}_1(t-1), \mathscr{M}_2(t-1))$$
$$\Delta\mathscr{E}(t) = f_{\mathscr{E}}(\mathscr{E}(t-1), \mathscr{W}(t-1), \mathscr{M}_1(t-1), \mathscr{M}_2(t-1))$$
$$\Delta\mathscr{C}(t) = f_{\mathscr{C}}(c(t), \mathscr{C}(t-1), \Delta\mathscr{R}(t), \Delta\mathscr{T}(t), \Delta\mathscr{E}(t))$$
$$\mathscr{C}(t) = \mathscr{C}(t-1) \cdot \Delta\mathscr{C}(t)$$

Fig. 6. A view of Concept Space morphology.

universe of possible concepts ($\mathscr{U}$) which is obviously not accessible; instead, we might postulate that a dynamic possibility space ($\mathscr{C}$) may be on a trajectory towards a possible ideal domain (Σ) in the universe ($\mathscr{U}$), while completion of this trajectory seems unlikely to be a reachable goal.

I propose the dynamic concept space as a generated space, where the space at time t is defined as a function of constraints, search strategy and value definition moderated by memory, as shown in expression 2. This definition is then equal to the last predicate in the specification in Fig. 6.

$$\begin{aligned} \mathscr{C}(t) &: f(\mathscr{R}(t), \mathscr{T}(t), \mathscr{E}(t), \mathscr{W}(t-1)) \\ &= \mathscr{C}(t-1) \cdot \Delta\mathscr{C}(t). \end{aligned} \tag{2}$$

Wiggins and Boden distinguish between 'exploratory creativity' and 'transformational creativity'. When a concept space is changed by the agent through the action of searching, in other words when there is a morphology of the concept space, then the Boden/Wiggins distinction between transformational and exploratory creativity seems to break down. Instead of being separate categories, exploratory creativity *does* transform the concept space and transformation of the concept space *is* the result of exploratory action.

Consequent to the dissolution of the Boden/Wiggins distinction between transformational and exploratory creativity is that the Axioms from Fig. 1 can be simplified and redefined as shown in Fig. 7: we retain $\mathscr{U}$ as the universe of possible concept types C and we want to be able to differentiate individual points (c_1, c_2) in the concept universe. Wiggins's empty concept $\top$, which represents *nothing* but which Wiggins used to initiate the search process (see Fig. 2), can be omitted, since we use intrinsic motivation $\mathscr{M}_1$ as a driver of *Creative Output* even if memory $\mathscr{W}$ is empty and regardless of whether there is any extrinsic motivation $\mathscr{M}_2$ (see Fig. 5). The declaration of $\mathscr{C}$: *Concept Space* is no longer axiomatic since we define it in Fig. 6. Also, we no longer need the axiomatic expression that a concept space is a subset of the universe ($\forall\mathscr{C}|\mathscr{C} \subseteq \mathscr{U}$) since it is conceivable that a $\mathscr{C}$ could become identical to $\mathscr{U}$, although this is only as a limiting case since it would mean omniscient knowledge of the universe.

$\mathscr{U}$: *Possible concepts*
C : *Concept type*
c_1, c_2 : *instances of C*

$C \in \mathscr{U}$
$\forall c_1, c_2 \in \mathscr{U} | c_1 \neq c_2$

Fig. 7. The simplified set of Axioms for the specification including concept space morphology.

Wiggins required the third proposition in Fig. 1 because 'for transformational creativity to be meaningful, all conceptual spaces, $\mathscr{C}$, are required to be non-strict subsets of $\mathscr{U}$' (Wiggins 2006a, p. 452). However, as mentioned above, in this new specification for computational creativity the idea of 'transformation creativity' as distinct from 'exploratory creativity' is no longer meaningful: instead, with dynamically generated concept spaces, exploratory creativity may be said to be transformational of the concept space as expressed in the morphology of the concept space over time. The resulting axiomatic expression for my specification in Fig. 7 simply expresses that we can differentiate between some different concepts in the universe of possible concepts.

According to Wiggins, Boden views transformational creativity as changes in $\mathscr{R}$, in other words, as changes in the constraints on the concept space. Wiggins proposes a view of a transformational creative system 'as an exploratory creative system working at the meta-level of representation' (Wiggins 2006a, p. 455). At this 'meta-level' Wiggins uses his valuing function $[[\mathscr{E}]]$ as a method for determining what impact an explored concept $c(t)$ has on the current concept space $\mathscr{C}(t)$. However, using a dynamic, generative concept space, any explored $c(t)$ will change the concept space $\mathscr{C}$ regardless of the results of using it as input to an evaluation function. This seems to be an acceptable feature of the common conception of creativity: any explored possibility becomes part of memory, and so part of the concept space, regardless of whether it is valued at a given time or not. Anecdotally: when teaching music composition and creative use of music technology at Birmingham Conservatoire I often emphasise that any compositional choice that is considered for, but isn't applied in a particular musical work becomes part of the space of compositional choices available for another composition later on. In other words, the musical 'object' produced represents a subset of the dynamic concept space.

Figure 8 gives an informal overview of the present version of the framework where Memory — $\mathscr{W}$ — Intrinsic and Extrinsic Motivations — $\mathscr{M}_1$ and $\mathscr{M}_2$ — and the current Musical 'object' — $c(t)$ — are inputs to the Evaluator(interpreter) function: $[[.]](\langle\langle ., ., .\rangle\rangle(., ., ., .))$ in Fig. 5. The Evaluator(interpreter) function results in Creative output (Fig. 5), and this in turn becomes the next Musical 'object'. The output of the Evaluator(interpreter) function modifies the Dynamic concept space $\mathscr{C}$. The Dynamic concept space is the basis for Memory in my version of the framework. The components, aside from the Musical 'object', form the Computational Creativity. I expand the framework to include a wider context in another article (Mogensen 2018).

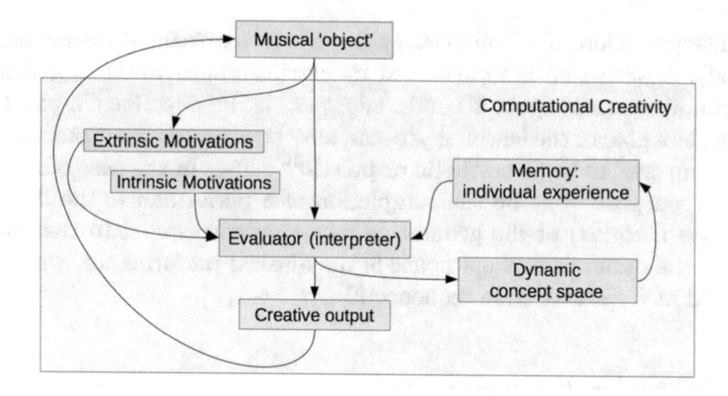

Fig. 8. Overview of the framework.

4 Aligning Concept Space Morphology with an Experiential Learning Model

Returning to Fig. 5 and Expression 2 the specification might appear to indicate some circularity in the system: 1. the concept space is dependent on constraints, strategy, value and memory; 2. memory is dependent on application of constrains, strategy, value; 3. constrains, strategy, value are dependent on memory of the concept space. But that is a misinterpretation: given discrete time t the equation should be interpreted as a process of discrete *iterations*, and so the formalism can be aligned with learning models. As an example I align the specification with John Dewey's experiential learning model (Dewey 1938).

Dewey's model of experiential learning, in other words his 'formation of purposes' in the case of learning music, can be understood as four steps that are cyclically reiterated: 1. 'Impulse' (the desire to play or create); 2. 'Observation' (listening to uses of techniques and ideas); 3. 'Knowledge' (analytical insights and embodied cognitive practice); 4. 'Judgement' (critical evaluation to make choices which will guide the next 'Impulse') (Kolb 2015, pp. 33–34) (Kolb 1984, pp. 22–23) (Dewey 1938, p. 69). This iterative process is illustrated in the diagram in Fig. 9, adapted from Kolb's (2015) interpretation of Dewey.

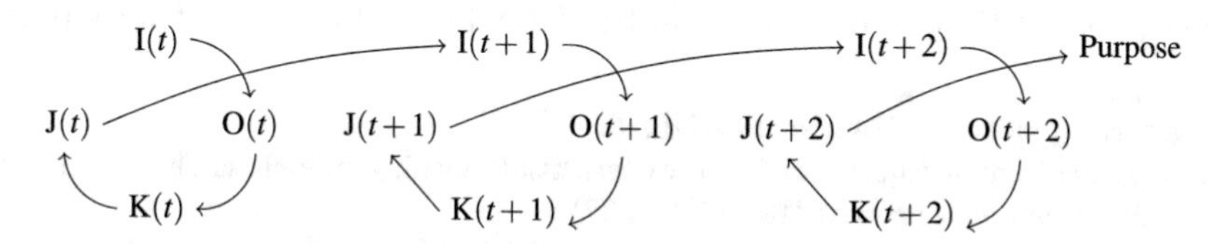

Fig. 9. Dewey's model of experiential learning with iterations leading to 'Purpose', where I: Impulse, O: Observation, K: Knowledge, J: Judgement, and t represents time.

I propose to align these four steps with components of the formal model so that I represent *Experiential Learning* as generative recursion, shown in Fig. 10.

In this interpretation, the Kolb/Dewey *Impulse* is represented by intrinsic motivation $\mathscr{M}_1$; *Observation* is represented by extrinsic motivation $\mathscr{M}_2$; *Knowledge* is the dynamic concept space $\mathscr{C}$; and *Judgement* is the *Creative Output.* Dewey's 'Purpose', as a goal of the learning process, may be an artefact output that is considered 'complete' in some aesthetic or poietic[10] sense. In the case of improvised music, the 'purpose' may be the completion of a performance; the *Judgements* (or *Creative Outputs*) of the generative recursion correspond to the playing of the music; the dynamic concept space is the musical performance, which is here represented in a discrete time sequence $[0, .., t, t+1, ..]$.

Experiential Learning

$I : Impulse = \mathscr{M}_1(t)$
$O : Observation = \mathscr{M}_2(t)$
$K : Knowledge = \mathscr{C}(t)$
$J : Judgement = Creative\ Output(t)$
$t : time$

$$Experiential\ Learning(t) = f(I(t), O(t), K(t), J(t), Experiential\ Learning(t-1))$$

Fig. 10. *Experiential Learning* as generative recursion.

As a consequence of the expression in Fig. 10 the generative recursion of this computational creativity specification can be understood as an experiential learning process. If the Wiggins/Boden's 'searching' in the universe of possible concepts is a learning process then the 'creativity' of the system is expressed in the emergent traces that are the *Creative Output* of this learning process.[11] This resonates with the philosophical argument made by Jack Glickman (1976, pp. 130–131) on the concept of creativity in the arts: that speaking of '"creative process"... is the wrong way to go about characterizing creativity, [instead] one must attend to the artistic product rather than to the process'. So I propose that creativity is not a process itself but is rather an artefact that may emerge from a *learning process.*[12]

According to Kolb, there is a 'dialectic... between the impulse that gives ideas their "moving force" and reason that gives desire its direction' in Dewey's model (Kolb 2015, p. 40). Applied in the formal model this may translate into

[10] The term 'poietic' is from Nattiez (1990).

[11] Kolb states that a characteristic of experiential learning models is that learning is best described as a process (Kolb 2015, 37).

[12] Wiggins appears to interchange the term 'artefact' with the term 'concept' and examines the 'conceptual space in which the artefact is found' (Wiggins 2006b, p. 209). This seems to be a confusion of terms since 'artefact' refers to physical objects made in some way by humans, whereas 'concepts' exist in human consciousness. What the nature of the relations between concepts and artefacts is, is a question beyond the present scope, but I expect that the distinction between these terms would still hold when applied in the context of computational creativity.

a relation between intrinsic motivation $\mathscr{M}_1$ and *Creative Output*, aligned with Impulse and Judgement (expression 3).

$$\mathscr{M}_1 \longleftrightarrow CreativeOutput \approx Impulse \longleftrightarrow Judgement \quad (3)$$

Kolb's 'most current statement [of] experiential learning theory is described as a dynamic view of learning based on a learning cycle driven by the resolution of the dual dialectics of action/reflection and experience/abstraction' (Kolb 2015, pp. 50–51) and his working definition of learning is that experiential '[l]earning is the process whereby knowledge is created through the transformation of experience' (Kolb 2015, p. 49).[13] Within the formal framework, these two dialectic relations can be understood as shown in expressions 4 and 5. We can say that reflection is evident in the change of concept space ($\Delta\mathscr{C}(t)$) which is in a dialectic relation with *Creative Output.* The external input (M_k), whether cognitive or computational, may be considered as 'experience' which is in a dialectic relation with the concept space abstraction ($\mathscr{C}$). Further investigation of these relations is beyond the present scope and are reserved for future work.

$$CreativeOutput \longleftrightarrow \Delta\mathscr{C}(t) \approx action \longleftrightarrow reflection \quad (4)$$

$$M_k \longleftrightarrow \mathscr{C}(t) \approx experience \longleftrightarrow abstraction \quad (5)$$

5 Conclusion

The presented development of the formal specification and understanding of its meaning opens up new possibilities for developing computational creativity. In much current Artificial Intelligence work the goal of a search algorithm is usually to find optimal solutions to search problems. In music, improvised music in particular, a focus on searching for optimal solutions to a 'problem' may be a category mistake. In other words the question, whether an optimal music has been achieved seems to be a misleading question. Instead one should ask what has been the value of the aesthetic experience of the music, and also has the learning process, that aligns with the making of the music, been productive of a transformed experience? In a creative system for improvising music there is no imperative to find an 'optimal' solution, since the morphology of the search itself can constitute a musical 'solution', a trace of a learning process, which counts as a valuable contribution to an aesthetic event. In this specification the generative search in the possibility space *is* a 'solution' to the improvisational performance 'problem'.

[13] One might question whether knowledge is 'created' and this becomes a questioning of the constructivist stance. Perhaps it is more accurate to say that knowledge is 'attained' or 'arrived at' since knowledge potentially exists regardless of our access to it? Resolving this question is beyond the present scope.

References

Boden, M.A.: The Creative Mind, 2nd edn. Routledge, London (2004)

Dewey, J.: Experience and Education. Macmillan Co., New York (1938)

Diller, A.: Z, An Introduction to Formal Methods. Wiley, Chichester (1990)

Glickman, J.: Creativity in the arts. In: Aagaard-Mogensen, L. (ed.) Culture and Art, pp. 130–146. Humanities Press, Atlantic Highlands (1976)

Kolb, D.A.: Experiential Learning. Prentice Hall Inc., Englewood Cliffs (1984)

Kolb, D.A.: Experiential Learning, 2nd edn. Pearson Education Inc., Upper Saddle River (2015)

Mogensen, R.: Computational motivation for computational creativity in improvised music. In: Proceedings of Computer Simulation of Musical Creativity conference 2017, pp. 1–10 (2017a). https://csmc2017.wordpress.com/proceedings/

Mogensen, R.: Evaluating an improvising computer implementation as a 'partial creativity' in a music performance system. J. Creative Music Syst. **2**(1), 1–18 (2017b)

Mogensen, R.: Formal representation of context in computational creativity for music. In: Gouveia, S.S., de Fernandes Teixeira, J. (eds.) Artificial Intelligence and Information: A Multidisciplinary Perspective. Vernon Press, Wilmington (2018)

Nattiez, J.-J.: Music and Discourse: Toward a Semiology of Music (Abbate, C. (ed.) transl.). Princeton University Press, Princeton (1990)

Oudeyer, P.-Y., Kaplan, F.: What is intrinsic motivation? A typology of computational approaches. Front. Neurorobotics **1**(6), 1–14 (2007)

Petzold, C.: The Annotated Turing: A Guided Tour Through Alan Turing's Historic Paper on Computability and the Turing Machine. Wiley Publishing Inc., Indianapolis (2008)

Turing, A.: On computable numbers, with and application to the entscheidungsproblem. In: Proceedings of the London Mathematical Society, 2nd series, vol. 42, pp. 230–265 (1936)

Wiggins, G.A.: A preliminary framework for description, analysis and comparison of creative systems. Knowl. Based Syst. **19**(7), pp. 449–458 (2006a)

Wiggins, G.A.: Searching for computational creativity. New Gener. Comput. **24**(3), 209–222 (2006b)

Creative AI: Music Composition Programs as an Extension of the Composer's Mind

Caterina Moruzzi(✉)

The University of Nottingham, Nottingham, NG7 2RD, UK
caterina.moruzzi@nottingham.ac.uk

Abstract. I discuss the question "Can a computer create a musical work?" in the light of recent developments in AI music generation. In attempting to provide an answer, further questions about the creativity and intentionality exhibited by AI will emerge. In the first part of the paper I propose to replace the question of whether a computer can be creative with questions over the intentionality displayed by the system. The notion of creativity is indeed embedded with our subjective judgement and this prevents us from giving an objective evaluation of an idea or product as creative. In Sect. 2, I suggest to shift the focus of the inquiry to the autonomy possessed by the software. I finally argue that the application of generative adversarial networks to music generators provides the software with a level of autonomy sufficient to deem it able to create musical works.

1 AI, Creativity, and Musical Works

A widely recognised feature of musical works (MWs) is that they did not exist before being actually created by a composer: a MW, to be identified as such, is necessarily created. We can thus claim that the composer performs an act of creativity. In order to assess whether computers can create MWs, then, it is necessary to ask the question: "Can a computer be creative?"

Numerous definitions of creativity have been proposed in the literature on the topic (Boden 2004, 6). Boden's definition, shared in its main elements by other theories of creativity, is: "Creativity is the ability to come up with ideas or artefacts that are new, surprising and valuable" (Boden 2004, 1). The attribute 'valuable' highlights a characteristic which our common sense essentially attributes to the notion of creativity: its being *subject-dependent*. Indeed, what influences our assessment of the creativity of an output are not only the features of the outcome but also the way in which the artwork is produced and presented to the audience. An obvious example are ready-mades. Duchamp's *Fountain* is not deemed creative, and therefore valuable, because of its formal properties but instead because of the choice of *presenting* it as an artwork and the meaning that this choice brings with it.

Arguably, then, given the subject-dependent nature of creativity, a test over the intuitions that people hold in respect to artefacts or ideas is enough to determine the creativity of an outcome. This is what the Neukom Institute for Computational Science assesses

V. C. Müller (Ed.): PT-AI 2017, SAPERE 44, pp. 69–72, 2018.
https://doi.org/10.1007/978-3-319-96448-5_8

with its 'Turing Test in the Creative Arts'.[1] This contest asks machines to create music, dance, or poetry that is indistinguishable from human created outputs. In 2017, the winner in the music section was Music Plus One with the song *The Wild Geese*. On the basis of the features of the outcome, it thus seems possible for a computer to exhibit a level of creativity comparable to a human's.

Yet, it may be argued that a subjective judgement on the creativity of its outcome is insufficient for deeming the computer creative. Indeed, there is evidence to support the claim that our evaluation changes once we become aware of the way in which this outcome was produced. Especially so when we learn that it was not created by a human but by a machine (Boden 2010, 411).[2] This suggests that we should consider not just the final product but also its provenance.

The subjective judgements and biases which come with the evaluation of something as creative make it impossible to objectively answer the question "Can a computer be creative?" What we are measuring when we provide an answer to this question, in fact, are not the computer's accomplishments but instead our subjective evaluation of them. We can then try to analyse not just the creativity exhibited by the outcome produced by the computer but, instead, the *intentionality* of the computer in producing it. In other words, we can judge whether the computer produced its outcome intentionally, i.e. consciously intending to produce exactly that outcome. We should then rephrase the question and ask: "Can a computer be intentionally creative?"

2 Autonomous Computers or Extended Humans?

In keeping with what I stated above in regards to creativity, it is possible to attribute *subject-dependent* intentionality to non-animal entities, and, thus, to computers (Dennett 1971). We sometimes speak of artefacts 'as if' they were intentional beings (for example, we may talk of the thermostat 'as if' it has the intention of regulating the temperature) but this does not mean that they possess intentionality independent from our interpretation of their behaviour (Searle 1992). As with creativity, it seems to be against our intuition to consider subject-dependent intentionality sufficient for deeming a computer intentionally creative.

The real challenge for computers would then be to be perceived as *subject-independently* intentional, namely as *autonomous* entities. A specification of what I mean by 'autonomous' is needed, since it may be argued that not even humans are really autonomous. Our choices are indeed affected by our upbringing, culture, and environment. For autonomy in relation to software here I intend not a complete independence from the programmer but instead the ability of 'breaking the rules' that the programmer encoded in the software.

I suggest an alternative definition of minimal creativity ($\text{C}{\small \text{REATIVITY}}_{\text{m}}$) which focuses on the autonomy needed by a system to produce an output which is deemed creative. It should be noted that this definition of creativity focuses on the creative process and not

[1] https://bregman.dartmouth.edu/turingtests/node/1.

[2] It may be argued that our judgement in this respect is biased. Still, for the sake of this discussion I will assume that our intuitive judgements are worth accounting for.

on the output. This is allegedly consistent with the intuitions that we bear in respect to creativity: the assessment of the creative process is equally, if not more, important than the assessment of its output. The definition I propose is deliberately weak to avoid referring to the notion of intentionality, and to identify the minimal requirements for an action to be creative:

CREATIVITY_{m}:

(i) Autonomous reception of external or internal stimuli;[3]
(ii) Autonomous selection of some of these stimuli;
(iii) Autonomous elaboration of the selected stimuli;
(iv) Autonomous production of new stimuli on the basis of the previous process.

An example of this creative process in the musical field is *La Mer* by Claude Debussy. In composing this orchestral piece, Debussy was inspired by the sounds and noises produced by the ocean, by prints, and by short stories (Huscher). The creative process that Debussy went through is constituted of an (i) autonomous reception of the mentioned sources; an (ii) autonomous selection of some of the stimuli that he received from them; an (iii) autonomous elaboration of these stimuli in a musical form; and an (iv) autonomous production of the resulting piece: *La Mer.*

In software for music generation such as Jukedeck and Flow Machines the output mimics the corpus that has been used to train it. In addition, the output matches the constraints which have been indicated by the programmers or the users of the software (Briot and Pachet 2017). I argue that (trained or not-trained) AI music generators which are limited to providing a different rendition of the input entered in the database, on the basis of rules written by the programmer, cannot be deemed creative. They in fact do not comply with the requirements specified by CREATIVITY_{m}, namely they are not autonomous in the process of creation. At best they can be considered an extension of the programmer's or user's mind. In arguing for this claim I follow the extended mind theory: cognitive systems extend beyond the individual and into the environment (Clark and Chalmers 1998). Similarly, this kind of software for music generation can present creativity only insofar it is an extension of the programmer's or user's cognitive machinery. They do not, however, possess autonomous creativity.

Nevertheless, the application of unsupervised machine learning and generative adversarial networks (GANs) as in the MidiNet system (Yang 2017) may overturn this conclusion. The collaborative work of the generator and discriminator software in GANs allows the system to gain independence from the programmer and, as a consequence, the software does not need supervision from the latter. Unlike other software for music generation, GANs systems change the rules that were initially entered by the programmer through unsupervised learning. Moreover, GANs allow music generators to transform an input of random noises into a melody. They, thus, bypass the hindrance to creativity which is a consequence of mimicking the corpus of data given as input (Briot and Pachet 2017). One of the main features that we commonly attribute to creative products is unexpectedness. GANs produce unexpected results: the combination of an

[3] For stimulus I intend every object or event which carries some information and evokes a reaction.

input of random noise, constraints, and transgressions makes the prediction of the outcome impossible. This adds to the consideration of the neural network as a 'black box' and, in general, of the potential 'creativity' of the system. With the progress of systems of unsupervised learning and a progressive independence from the rules given in the initial phase of programming, it will be possible for computers to create unexpected results, more similar to human ones.

3 Conclusion

In order for a computer to be able to create a MW, it needs to comply with the requirements expressed by $\text{CREATIVITY}_{\text{m}}$. Specifically, it needs to be autonomous in selecting and elaborating stimuli. I argue that the application of GANs allows software for music generation to reach the level of autonomy needed for deeming it able to create MWs.

The shift from the consideration of intentionality as necessary element for a creative systems to the consideration of its autonomy, suggested by $\text{CREATIVITY}_{\text{m}}$, is beneficial under at least two respects. First, it allows us to address creativity in AI without the need to account for the notion of intentionality, notion implied by the assessment of creativity. Second, it is possible to test the level of autonomy, i.e. to test whether the system is able to change the initial rules through unsupervised learning, while intentionality is too vague a concept for evaluation.

References

Boden, M.A.: The Creative Mind: Myths and Mechanisms, 2nd edn. George Weidenfeld and Nicolson, London (2004)

Boden, M.A.: The turing test and artistic creativity. Kybernetes **39**, 409–413 (2010)

Briot, J.P, Pachet, F.: Music Generation by Deep Learning - Challenges and Directions (2017). arXiv:1712.04371v1

Clark, A., Chalmers, D.: The extended mind. Analysis **58**, 7–19 (1998)

Dennett, D.C.: Intentional systems. J. Philos. **68**, 87–106 (1971)

Huscher, P.: Chicago Symphony Orchestra Program Notes. Claude Debussy, *La Mer*, Chicago Symphony Orchestra

Searle, J.R.: The Rediscovery of the Mind. MIT Press, Cambridge (1992)

Yang, L., Chou, S., Yang, Y.: MidiNet: a convolutional generative adversarial network for symbolic-domain music generation using 1D and 2D conditions (2017). arXiv:1703.10847

How Are Robots' Reasons for Action Grounded?

Bryony Pierce(✉)

University of Bristol, Bristol, UK
bryony.pierce@bristol.ac.uk

Abstract. This paper defends the view that (non-conscious) robots' reasons for action can only be grounded externally, in the qualitative character of the conscious affective experience of their programmers or users. Within reasoning, reasons for action need to be evaluated in a way that provides immediate non-inferential justification of one's reasons to oneself, in order to stop a potential regress of *whys*. Robots devoid of consciousness and thus incapable of feeling emotion cannot process information about reasons for action in a way that is subjectively meaningful. Different types of grounding will be discussed, together with the question of relativism about fundamentality in the context of grounding. The concluding discussion will consider the case of hypothetical conscious robots with internally grounded reasons for action, arguing that it would be unethical for such robots to be created, as they would either effectively be brought into slavery or, if developing AI rather than human-centred values, would potentially represent a threat to human life.

1 Introduction

Can we defend the view that robots act for reasons, and what would it mean for them to do so? Or, if not, what capacity, if any, do they currently lack that would enable them to do so? In particular, how could a robot's reasons for action be grounded?

Let us start by taking a hypothetical robot, for the purposes of the discussion, whose output in the form of apparently goal-directed behaviour provides some evidence that it can engage in intelligent reasoning, drawing conclusions about what actions it should perform on the basis of ongoing sensory input together with stored information about relevant values or goals. Let us also suppose that this robot can report back on which options it evaluated, and explain, in a more or less sophisticated manner (i.e. anything from a printout of a flow chart to a complex verbal report in grammatical sentences), why it selected a certain course of action to pursue. Such a robot might be more successful at a given task, e.g. driving a vehicle, or carrying out a surgical procedure, than a human being, and might give a convincing account of the steps in its reasoning. So, what obstacles could there be to describing the robot as having reasons for its behaviour and as having acted upon those reasons? The answer, I will argue, lies in the way that those reasons are grounded.

We think of our reasons for action as grounded, both from a first-person perspective and when contemplating other people's reasons, at least in the minimal sense of there being something underlying the reasoning process that renders our decisions intelligible

V. C. Müller (Ed.): PT-AI 2017, SAPERE 44, pp. 73–80, 2018.
https://doi.org/10.1007/978-3-319-96448-5_9

to ourselves. Our reasons are said to be grounded when certain facts concerning those reasons are taken to obtain in virtue of something more fundamental, in a relation of non-causally dependent justification. Grounding may frequently be in something tacit – an unarticulated and unquestioned premise that is applied as a matter of course – rather than something that features explicitly in the reasoning process. Some premises are so obviously true that we stop short of citing them as part of our reasoning, e.g. a person might campaign against experiments on monkeys, reasoning that the experiments are unethical because they cause pain and distress to sensitive, intelligent animals and that there are therefore good reasons to organise protests, but the underlying premise that the animals' suffering is a bad thing, in itself, will be taken as read.

Conscious experience of affective valence (awareness, through emotions or other feelings, of whether commodities, situations or anticipated outcomes of action have a positive or negative value to us, and to what extent) plays an important regress-stopping role in the production of rational human action. It is what stops the potentially infinite regress of *whys* questioning why we should perform actions, why those reasons are compelling, why it matters whether reasons are compelling and justifiable, and so on, providing internal grounding of reasons for action. Affective responses provide essential information about the value of anticipated outcomes that allows cognition to proceed with evaluations and judgements on the basis of preferences that would otherwise be lacking, leaving cognition with nothing to guide it toward any particular course of action.

Beliefs could be formed, purely cognitively, by applying reasoning to sensory and other input, and a range of possible courses of action could be identified by predicting probable causal sequences, but beliefs about states of affairs and how changes might be brought about is not sufficient for the capacity to judge one outcome better than another. When we apply reason to a problem and form a judgement about what to do, we rely, additionally, on (often tacit) knowledge of relative values. In our folk-psychological understanding of others' behaviour, theorising on the basis of desires, *which rely on perceived value*, is prior to belief-based theorising (Wellman and Woolley 1990). Knowledge of value is learnt through experiences in which affective responses provide the relevant information: a commodity that provokes a positive response takes on a positive value and each encounter with a commodity or element of a perceived change in states of affairs in our environment has affective valence. Some affective responses are neutral/mild and barely noticeable until we direct our attention specifically to them (e.g. my response to the sound of the keys as I type, which I now find slightly pleasant but previously had not noticed; however, all these details combine and contribute to our fluctuating moods). My view is that affective content – the product of conscious or unconscious registering of affective valence[1] – nonetheless pervades our experience of the world.

Without affective valence, the concept of value becomes empty; reasons for action then lose their compelling nature and can no longer move us to act in a way that we can judge to be rationally justified. This has the consequence that robots or other entities with artificial intelligence that are incapable of experiencing – rather than merely

[1] Some of this information may take the form of an absence of unexpected change in aspects of our surrounding.

simulating – emotions, are unable to stop the regress of whys in this way, so cannot have reasons for action they can justify to themselves in any meaningful way. Without the capacity for emotion, outcomes cannot *matter* and preferences cannot be formed, so actions are not performed for subjectively meaningful reason. As Harnad says, "[w]ithout feeling, we would just be grounded Turing robots, merely acting as if we believed, meant, knew, perceived, did or chose" (2011, p. 23). Any reasons for action emotionless agents might give as an explanation of action would be grounded in something external to themselves and beyond their comprehension, even if we grant that they might comprehend in some sense. Their reasons would be meaningful only to some other entity in terms of whose values – derived from direct acquaintance with the affective valence of commodities and states of affairs – the reasons could be perceived to justify a course of action, on the basis of the perceived value of the anticipated consequences.

2 Kinds of Grounding

The *symbol grounding problem* is the problem of how we can move from abstract symbolic representations, such as descriptions of the external world, to the non-symbolic (that which is described) in a way that stops the infinite regress of explanations of what symbols mean in terms of other intrinsically meaningless symbols (Harnad 1990). It is concerned with meaning, but it is distinct from *the problem of meaningfulness*: the problem of how symbols can have subjective meaning in terms of incorporating both motivational force and the justification of reasons required to achieve that motivational force.

Harnad proposed, as a solution to the symbol grounding problem, the grounding of symbols in non-symbolic sensorimotor function:

> Hence grounding means sensorimotor grounding: Symbols must be grounded in the capacity to discriminate and identify the objects, events and states of affairs that they stand for, from their sensory projections (1993, 7.2).

Harnad later says that symbols processed by a robot would be grounded if the robot could use its sensors to recognize those things, or categories of things, referred to by the symbols and interact with them as humans would – again this amounts to sensorimotor activity – but he questions whether grounding is enough for meaning (2011).

Grounding in sensorimotor activity gives information a functional application and creates a relation between the symbolic form and physically instantiated objects of perception. This, at least, is achievable in AI. Symbols that represent reasons for action, the specific topic under discussion here, need to stand in some relation to objects external to the symbol system, to stop the infinite regress of using symbols to define symbols (or processes to explain the value of other processes, if reasons are in non-symbolic form, e.g. as in a dynamical systems model). But there is another potential regress that needs to be stopped: the regress of whys that can arise when attempting to justify action to oneself. A point has to be reached where we can stop asking why we ought (prudentially or morally) to do something. That happens once it becomes self-evident that a reason justifies something, non-inferentially, without further explanation. (Reasons and the realisation that they are justified in this way need not be entertained consciously at the time of deliberation; they may be recognised tacitly only, but their recognition relies on past conscious affective responses that enabled knowledge of affective valence relevant to the situation to be acquired.)

When we think about the need to justify reasons to ourselves, within reasoning, grounding in sensorimotor activity doesn't go far enough, because sensorimotor activity isn't an end in itself; it can be instrumental in achieving goals, but it is possible to interact with the external world while lacking any sense of the meaningfulness of one's actions. A function such as the construction of a vehicle by a robot capable of recognising and manipulating objects in its environment may demonstrate sensorimotor skills but is not evidence that the robot is aware of what it is doing or why it is doing it, or that it can perform actions for subjectively meaningful reasons. We need, additionally, to be able to (a) attach meaning to symbols in a deeper sense than is necessary merely to produce appropriate behaviour; and (b) to be able to attach value to the *consequences* of sensorimotor activity.

Dickinson and Balleine's Hedonic Interface Theory (HIT) tells an evolutionary story about grounding, in which consciousness has a central function, claiming that agents only know what to value – and thus what goals to set themselves – once the effects of events on physiological states are made explicit through conscious affective responses (1994; 2010). Although conscious affective responses play a central role for Dickinson and Balleine, they say that reasons for action are grounded in underlying biological functions. This is one way of grounding reasons for action, from a third-person perspective, but to be able to find reasons for action compelling to ourselves, in reasoning, another type of grounding is necessary – grounding in the qualitative character of the affective valence of subjective experience. The two ways of grounding reasons for action are not, however, mutually exclusive. I adopt relativism with regard to fundamentality in grounding relations, i.e. whether it is biological function or affective experience that is more fundamental depends on the context in which one is considering grounding.

HIT's evolutionary explanation thus addresses the question of why things are the way they are, accounting for human behaviour, but failing to stop the regress of whys in reasoning or to address what I call *the problem of meaningfulness* (Pierce 2017). The level at which grounding occurs in HIT is in [the function of] physiological states, and these, like the physical mechanisms and processing activities of a robot incapable of experiencing emotion, lack intrinsic meaning. They merely provide a causally efficacious realiser: a way for abstract processes to have effects in the physical world, but even a rudimentary (programmable) robot will be able to perform actions within its environment that are caused by machine code rather than by mechanical systems alone.

It is important to distinguish between a number of different senses of grounding: teleological grounding, in which mental content succeeds in representing objects in the external world, and teleosemantic grounding, which seeks to naturalise semantic content (e.g. Millikan 1984); grounding in cognitive phenomenology/conscious perceptual content (Smithies 2014); grounding in perceptual and intuitive content (Chudnoff 2011); and grounding in the sense of having meaning, as in Searle's Chinese room thought experiment (1980). Searle's Chinese room argument distinguishes between two different types of ability: manipulating symbols according to syntactic rules and understanding their meaning in the sense of their semantic content, supporting the claim that the implementation of a computer program simulating human behaviour is not sufficient for the ascription of consciousness or intentionality. For Searle, it is intentional content that is grounded in consciousness. Smithies and Chudnoff focus on epistemic justification, with phenomenal consciousness providing immediate non-inferential justification of beliefs. This is distinct from the grounding of reasons for acting on those beliefs, but is mentioned here because they, too, emphasise the need for grounding from the first-person perspective, in some contexts.

3 Affective Valence

I have identified several types of content so far: syntactic, semantic, intentional, perceptual and intuitive, some requiring grounding and others providing it. I wish, now, to discuss the further distinction I make between (a) semantic content in the sense of reference to that which is external to the symbol system and (b) the subjective meaningfulness of information – its *affective content*.

I mentioned above that affective valence plays a regress-stopping role in reasoning, that the concept of value is derived from direct experience of affective responses that allow us to discern affective valence, and that affective responses are necessary for goal-setting and motivation. A key claim in this paper is that reasons for action are grounded in the qualitative character of the affective valence of subjective experience. It is when we experience an affective response to a commodity or state of affairs (actual or potential) and attribute a value to it (within a certain context, given our physiological needs at the time of encountering the stimulus, thus allowing for fluctuating needs and preferences) that we are able to build up a set of values to apply when making decisions about which potential outcomes of our actions we wish to pursue. This is also a central argument in HIT, and one that is supported by empirical evidence (see Balleine and Dickinson 1998; Balleine et al. 1995; Dickinson and Balleine 1994 and 2010; Dickinson et al. 1995). Emotion has inherent value; emotions have hedonic tone with a positive or negative valence.

Without access, via experienced affective responses, to stored or newly formed/updated values, together with (i) the capacity for abstraction (no two commodities or states of affairs will be identical, so attributing value to categories, such as bread, or abstract characteristics, such as beauty, comfort or loyalty, is necessary when setting goals or evaluating potential outcomes of action); and (ii) the ability to assess the probability of anticipated outcomes; we cannot stop the regress of whys and view our reasons as grounded from a subjective point of view. Once we are confronted with an affective response to the perceived anticipated probable value of an outcome of action, whether it be joy, terror, relief, hesitance, indifference or irritation, we are in a position to judge whether the contemplated action is subjectively justifiable. (In attempting to judge whether it is objectively justifiable, we would also take into account the imagined responses of others, potentially appealing to some set of values perceived to be shared or universal.)

The regress of whys might take the form of, for example, questions about what we stand to gain, whether there is any advantage to us in gaining whatever it is we might gain, why there would be any advantage, why it would be an advantage to gain that advantage in particular, whether an advantage is something we should attempt to gain, and why, and so on. The regress cannot be stopped if reasons for action are grounded in biological function alone because, in the absence of any appeal to the subjective values arising from affective responses, we can continue to question whether our survival or the reproduction of our genes is of any importance without coming up against anything that provides immediate non-inferential justification for having any preference.

4 Robots' Reasons

Externally grounded reasons, as in the case of an emotionless robot acting for reasons grounded in the values of its programmers or users, would lack the affective content that ordinarily allows rational beings to conceive of actions as justified from their own perspective: this content is the tacit knowledge of affective valence applicable within each context in which a person selects a course of action; this constitutes an underlying premise or set of premises. Goal-directed action seemingly justified by externally grounded reasons, were a robot able to process data about goals and various means of attempting to purse them, would be guided by values that had been formed in the course of conscious experience of affective responses on the part of some (human) entity other than the robot.

There is a longstanding misconception about the roles of emotion and reason that continues to influence some theoretical work on rationality: 'Most models of decision-making assume the process to be rational, which would exclude the possibility of emotion playing a role, other than of a hindrance.' (Gutnik et al. 2006) Instead, I contend that caring about the outcomes of our actions is built into the concept of a reason to act, and that preferences rely on affective content for their motivational force.

I have argued above that a robot incapable of affective experience could have no preferences of its own and thus no reasons for action that would have meaningful semantic content; it could only deal with syntax and perform movements, etc., on the basis of strings of subjectively meaningless symbols. The symbols' relation to the external world and ability to guide interaction with others or to manipulate objects would lead to their having meaningful content only from the perspective of an external observer, just as the text I am typing that is being displayed on a screen, as I type, has meaningful content for me, as a conscious human being able to read English, or potentially for some other person reading it, but not for the computer causing it to appear as I type.

Robots programmed in such a way that they could be said to have values would present another set of problems. If they were to be guided by human values, which human values should be selected? There is no set of universal values that could be ascribed to humans, or even to a subset of humans within society, because of individual differences as well as differing needs and circumstances that affect what is valued in various contexts. Even minimising cultural differences, trying to decide what principles should guide driverless cars, how political decisions should be made or what constitutes a healthy diet, say, would soon result in strong disagreements.

A robot that had access to information about the way its values had been created, i.e. in a fairly arbitrary manner, for the above-mentioned reasons, would have no reason to judge that its values grounded its reasons for action (internally, rather than externally, in the values of its programmers or users) in a way that provided justification for those reasons. Robots could perhaps learn values in interaction with humans, but we would then run the risk of their acquiring significantly more unrepresentative values, or values that might be in direct conflict with those of many humans and/or with one another. An example of a learning process with unsatisfactory consequences along these lines is that of Microsoft's AI chatbot, Tay. Tay reportedly turned into a 'genocidal racist' within 24 hours (Shead 2016), with Microsoft issuing the following statement: 'The AI chatbot Tay is a machine learning project, designed for human engagement. As it learns, some of its responses are inappropriate and indicative of the types of interactions some people are having with it' (Shead 2016). Perhaps this is one possible consequence of a lack of sensitivity to affective valence to guide the evaluation process and a failure to ground reasons in the qualitative character of conscious experience informed by affective content.

Alternatively, an A[ultra]I with consciousness and affective responses would either have to be programmed to respect (and so prioritise) human-centred values, in which case their reasons for action would be grounded in the affective responses of a subset of humans, meaningful only within a context of slavery, or would develop values *not* ultimately derived from those of their programmers/users, in which case the resulting robots might judge it futile to serve or protect human beings, once they were capable of taking on the roles necessary for their own maintenance, repair and reproduction.

5 Conclusions

In summary, this paper defends the view that (non-conscious) robots' reasons can only be grounded externally, in the qualitative character of the conscious affective responses of their programmers or users. Reasons for action need to be grounded, in reasoning, from a first-person perspective, in order to motivate action for reasons we can judge to be justified. This grounding relies on affective responses to commodities and states of affairs, actual or anticipated/potential, that provide immediate non-inferential justification of reasons for action. Caring about the outcomes of one's action is built into the concept of having a reason to act. So, non-conscious robots, devoid of emotion, cannot have subjectively justifiable reasons for action and the information they process lacks meaningfulness.

Of the various kinds of grounding considered, grounding in affective content (the qualitative character of conscious affective responses) is the only one, within the context of reasoning about one's own reasons for action, to (a) stop the regress of whys, and (b) address the problem of meaningfulness. However, relativism about fundamentality allows for multiple types of grounding in different types of content, depending on the context and perspective.

Finally, to create a conscious robot capable of having internally grounded reasons for action, if that is ever possible (I am sceptical about claims that robots or other AI entities are already conscious), would not be morally permissible, as the robots would either effectively be slaves or would be a potential danger to human life, as they would be unlikely to have any reasons to serve or protect human beings.

References

Balleine, B., Dickinson, A.: Consciousness—the interface between affect and cognition. In: Cornwell, J. (ed.) Consciousness and Human Identity, pp. 57–85. Oxford University Press, New York (1998)

Balleine, B., Garner, C., Dickinson, A.: Instrumental outcome devaluation is attenuated by the anti-emetic ondansetron. Q. J. Exp. Psychol. **48B**, 235–251 (1995)

Chudnoff, E.: The nature of intuitive justification. Philos. Stud. **126**(3), 347–373 (2011)

Dickinson, A., Balleine, B.: Motivational control of goal-directed action. Anim. Learn. Behav. **22**(1), 1–18 (1994)

Dickinson, A., Balleine, B.: Hedonics: the cognitive-motivational interface. In: Kringelbach, M.L., Berridge, K.C. (eds.) Pleasures of the Brain, pp. 74–84. Oxford University Press, New York (2010)

Dickinson, A., Balleine, B., Watt, A., Gonzalez, F., Boakes, R.A.: Motivational control after extended instrumental training. Anim. Learn. Behav. **23**, 197–206 (1995)

Gutnik, L.A., Hakimzada, A.F., Yoskowitz, N.A., Patel, V.L.: The role of emotion in decision-making: a cognitive neuroeconomic approach towards understanding sexual risk behaviour. J. Biomed. Inf. **39**(6), 720–736 (2006)

Harnad, S.: The symbol grounding problem. Physica **D42**, 335–346 (1990)

Harnad, S.: Grounding symbols in the analog world with neural nets. Think **2**(1), 12–78 (1993). (Special issue on connectionism versus symbolism)

Harnad, S.: Lunch Uncertain. Times Literary Suppl. **5664**, 22–23 (2011)

Millikan, R.G.: Language, Thought, and Other Biological Categories. MIT Press, Cambridge (1984)

Pierce, B.: The regress-stopping role of affective valence. In: Proceedings of AISB Annual Convention, pp. 255–259 (2017)

Searle, J.R.: Minds, brains, and programs. Behav. Brain Sci. **3**(3), 417–457 (1980)

Shead, S.: Here's why Microsoft's teen chatbot turned into a genocidal racist, according to an AI expert. Business Insider (2016). http://uk.businessinsider.com/ai-expert-explains-why-microsofts-tay-chatbot-is-so-racist-2016-3. Accessed 31 Jan 2018

Smithies, D.: The phenomenal basis of epistemic justification. In: Sprevak, M., Kallestrup, J. (eds.) New Waves in Philosophy of Mind (2014). http://www.palgraveconnect.com/pc/doifinder/10.1057/9781137286734. Accessed 28 Nov 2014

Wellman, H.M., Woolley, J.D.: From simple desires to ordinary beliefs: the early development of everyday psychology. Cognition **35**(3), 245–275 (1990)

Artificial Brains and Hybrid Minds

Paul Schweizer(✉)

Institute for Language, Cognition and Computation, School of Informatics, University of Edinburgh, Edinburgh, UK
paul@inf.ed.ac.uk

Abstract. The paper develops two related thought experiments exploring variations on an 'animat' theme. Animats are hybrid devices with both artificial and biological components. Traditionally, 'components' have been construed in concrete terms, as physical parts or constituent material structures. Many fascinating issues arise within this context of hybrid *physical* organization. However, within the context of functional/computational theories of mentality, demarcations based purely on *material* structure are unduly narrow. It is *abstract* functional structure which does the key work in characterizing the respective 'components' of thinking systems, while the 'stuff' of material implementation is of secondary importance. Thus the paper extends the received animat paradigm, and investigates some intriguing consequences of expanding the conception of bio-machine hybrids to include abstract functional and semantic structure. In particular, the thought experiments consider cases of mind-machine merger where there is *no* physical Brain-Machine Interface: indeed, the material human body and brain have been removed from the picture altogether. The first experiment illustrates some intrinsic theoretical difficulties in attempting to replicate the human mind in an alternative material medium, while the second reveals some deep conceptual problems in attempting to create a form of truly *Artificial* General Intelligence.

1 Introduction

In this paper I would like to explore some intriguing conceptual terrain concerning implications for future forms of mentality that might arise through advances in AI theory and technology. The discussion will proceed by examining two related variations on an 'animat' theme. Animat devices are defined as robotic machines with both active biological and artificial components (Franklin 1995). At first pass, 'components' are most graphically construed in concrete terms, as brute material parts or constituent physical mechanisms. And of course, many intriguing issues arise within this context of hybrid physical organization, wherein biological matter such as living neural cells is coupled with engineered robotic components such as sensors and actuators. The topic is especially compelling when the biological matter in question is a living human body/brain that is augmented with technological implants and extensions. This 'cyborg' version of the bio-machine hybrid theme raises profound questions about the nature, boundaries

V. C. Müller (Ed.): PT-AI 2017, SAPERE 44, pp. 81–91, 2018.
https://doi.org/10.1007/978-3-319-96448-5_10

and future of the human mind that have already stimulated much discussion (e.g. (Clark 2003)).

However, within the overall context of Functional/Computational Theories of Mind (FCTM) (e.g. (Putnam 1967), (Fodor 1975) (Johnson-Laird 1988)) central to Cognitive Science and AI, demarcations based purely on *material* structure are unduly narrow. According to FCTM, it is *abstract* functional and computational structure which does the key theoretical work in characterizing and individuating the respective 'components' of thinking agents, whereas, in accord with the principle of Multiple Realizability (MR), the 'stuff' of material implementation is generally deemed to be of secondary importance. Hence the paper aims to extend the received animat paradigm, and investigate some consequences of expanding the conception of animats and bio-machine hybrids to encompass abstract functional and semantic structure, and not just concrete physical mechanisms.

In the standard cyborg case, the core physical system is still human/biological, and this is then augmented by fusion with artificial hardware devices *via* implants, neuroprosthetics, etc. The issue then becomes one of teasing out the implications for human mentality and identity that result from this *corporeal* blend of organic and engineered components. In these standard cases of mind/machine merger, the biological brain is *physically* impacted by other material structures, *via* Brain-Machine Interfaces, to produce a system that is no longer strictly human, but rather is a hybrid incorporating both natural and synthetic aspects.

By contrast, in the two thought experiments developed below, the *entire* physical system is synthetic – the 'brain' in question is completely artificial, and the mechanism under investigation is itself an advanced technological artifact, a robot. And instead of blending organic and synthetic *physical* parts, as in the cyborg paradigm, the scenarios below trade on the 'interaction' and comingling of purely artificial hardware systems with the *abstract* formal and linguistic structure central to organically engendered human mentality. So the hybridization involves a wholly synthetic physical device, in combination with the abstract, biologically induced cognitive and linguistic architecture of the human species. The issue then becomes one of exploring the implications for robot/human mentality that result from this hybridization – how much of the ensuing cognitive system should now be viewed as properly artificial and how much is still human?

In terms of what's essential to human mental identity, there are at least three key factors to consider. One (**1**) is internal processing structure, central to FCTM. This abstract template or computational blueprint is a defining characteristic of the human cognitive type, and it's the result of many eons of organic evolution and natural selection. Another critical feature (**2**) is our conscious experience or phenomenology: the field of occurrent, qualitative presentations (or P-consciousness, in Block's (1995) terminology). Our introspective self-identity is largely determined by this ongoing 'stream of consciousness'. And a third (**3**) key feature is the *content* of propositional attitude states such as beliefs and desires. To a great extent, who we are is dependent upon what we believe and what we want. According to the standard belief-desire framework of psychological explanation, the content of these propositional attitudes is central to our status as rational agents, and similarly this feature is vital to Dennett's (1981) Intentional Stance.

2 An Artificial Brain as Alternate Realization

The first 'abstract animat' thought experiment begins by utilizing feature (**1**) above. According to FCTM and the attendant principle of MR, internal processing structure is not something that is essentially about our flesh and blood *embodiment*, but rather concerns a higher level description of our neurological machinery. FCTM gives rise to the mind/program analogy, wherein the mind is theoretically captured at the *software* level. So let us suppose that sometime in the sanguine future, cognitive scientists eventually discern the underlying functional/computational architecture of the human mind, and merely for the sake of convenience, let us suppose that it's some more sophisticated and far reaching version of Fodor's (1975) Language of Thought (LOT), say LOT_{37}* (clearly, this structure would have to be capable of far more than simply the manipulation of linguistically encoded propositional attitude states). So let us suppose that LOT_{37}* is the formal processing architecture which has been organically evolved and is implemented in the brain. It is the level of description which characterizes us as advanced cognitive agents, as human *minds*, and the brain is running LOT_{37}* as an indigenous formal system of rule governed symbol manipulation, in accord with the classical mind/program model.

The organic brain is then the original physical realizer of LOT_{37}*, but according to MR, our biological 'wetware' is not in principle privileged in this regard. Just as with computational procedures in general, it should be possible to take the abstract LOT_{37}* software and run it on an artificial hardware device physically quite unlike the human brain. So in the first animat scenario, let us assume that this impressive theoretical and technological feat has been accomplished – human scientists have fabricated a purely artificial electro-mechanical 'brain' that implements the human Language of Thought. For ease of comparison, we will assume that the artificial brain occupies the cranial cavity of a fully operational robot, and hence manipulates environmental inputs and produces outputs controlling various forms of behavior in a manner completely analogous to a normal human being. Indeed, the artifact is so well crafted that it excels at some suitable version of the combined linguistic and robotic Total Turing Test (Harnad 1991) and its success is due to the fact that the robot is an alternate realization of our own cognitive software.

Turing's original test is designed as an 'imitation game', where the goal is to *fool* someone into thinking that the computer is human. The strategy is intended to screen off anticipated (and perhaps outdated) human prejudice towards artifacts, by appealing to a standard whereby they are deemed behaviorally indistinguishable from us. But this induces a number of red-herrings, since the goal strays from detecting general intelligence *per se* to slavishly impersonating humans, warts and all (see French 2000). In the current discussion I will therefore shift the emphasis from indistinguishability to the more salient goal of producing externally observable capabilities on a par with humans in terms of exhibiting broad-spectrum intelligence. In particular, in order to pass the presently contemplated 'soft' version of the Total Turing Test, the robot is not required to *mislead* the judges into mistaking it for a corporeal human, since this would add a myriad of restrictions and complications which are not relevant to the overall project of AI. So we will allow the judges to be cognizant of the fact that the robot is *not* physically a human. We will assume that they are suitably fair minded and impartial, and the task of the robot is to exhibit behavior that would count as appropriately intelligent in the general human case.

The robot is entirely artificial in terms of its *physical* organization and composition, but it is nonetheless a genuine case of biomechanical hybridization, since its *cognitive* architecture is an instance of the human LOT_{37}*. Thus its mechanical body and synthetic 'central nervous system' are advanced technological artifacts, while the abstract cognitive processing essential to its identity as a thinking agent is an organically engendered cognitive template. An artificial brain is running the software of the human mind, and in contrast to a standard cyborg case, there is now *no* biological or organic matter present, but only the *abstract* computational structure of human cognition, which structure possesses a clearly biological as opposed to artificial etiology. According to FCTM, it is this LOT_{37}* structure which distinguishes us at the cognitive level, and we have replicated this defining human mental characteristic in an artificial brain. Hence at the salient level of description the biological brain and the robot's artificial analogue are *functionally identical*, and thus it may appear that, although the robot's synthetic brain is physically quite distinct from our own organic hardware, it nonetheless supports a purely human *mind*. However, I will now invoke feature (**2**) above to argue that this is not the whole story.

Conscious experience is a notoriously problematic topic, about which there is well known and abundant disagreement. Many theorists (e.g. Fodor) invoke FCTM only to explain the high level cognitive processing involved in propositional attitude states and rational action, while at the same time bracketing the entire issue of qualia and phenomenology. However, other authors, including (Lycan 1987), (Jackendoff 1987), (Johnson-Laird 1988), (Chalmers 1996) try to extend the reach of the computational paradigm, and contend that *conscious states* themselves arise via the implementation of the appropriate functional or information processing structure. Let us denote this extension of the basic FCTM framework 'FCTM+'.

In contrast, a primary alternative to FCTM+ contends that it is the physical substrate, the actual material *realizer* of the abstract functional structure which must be invoked in the explanation of conscious presentations (as in (Churchland 1984)). This opposing view is a form of physicalist type-type identity theory, wherein particular material structures or processes are identified as constituting, 'giving rise to', or providing the supervenience base for the corresponding phenomenal state or property. In the case of *human* consciousness, salient aspects of the *biological brain* are thus hypothesized as responsible for various features of subjective experience, and this guides the empirical search for 'neural correlates' of consciousness and attendant psycho-physical mappings.

In comparison, a distinguishing feature of FCTM+ is that it advocates a form of 'non-reductionist' token-token physicalism motivated by the principle of Multiple Realizability. As noted above, abstract computational procedures can be implemented *via* any number of quite distinct types of physical configuration. For example, classical

Turing machines, conceived as finite programs of instructions for manipulating 0's and 1's, have the ontological status of mathematical abstractions. Like differential equations, sets, Euclid's perfectly straight lines, etc., Turing machines don't exist in *real* time or space, and they have no causal powers. In order to perform *actual* computations, an abstract Turing machine must be realized or instantiated by some suitable arrangement of mass/energy. And as Turing (1950) observed long ago, there is no privileged or unique way to do this.

The very same abstract Turing machine can be implemented *via* modern electronic circuitry, a Victorian contraption made of gears and levers (*a la* Babbage's Analytical Engine), a human being following the instructions by hand using notepad and pencil (as in the banks of clerks working at Bletchley Park), as well as more 'deviant' physical arrangements such as roles of toilet serving as the machine tape and empty beer cans for the cipher '1'. Thus there is no uniform reduction from *type* of computational state to *type* of physical state. But each particular instance or physically realized *token* of a given abstract state is still just a particular physical state or process, governed by the ordinary laws of nature. Hence the ontological commitments are held to be physicalist but non-reductivist.

The position I will now advocate is that FCTM+ is mistaken, and that qualia must supervene upon the physical substrate rather than the functional organization. Why? – because unlike computational formalisms (as well as propositional attitude states, viewed dispositionally as high level, counterfactual-supporting configurations of a computational system), conscious states are inherently *non-abstract*; they are *actual*, occurrent phenomena extended in physical time. Many qualitative presentations, such as a visual sensation of seeing a bright red dot on a display monitor in some laboratory set-up, have a measurable duration, which means that the conscious event takes place over some objectively specifiable length of time. In sharp contrast, abstract Turing machines are not extended in physical time – the computational 'steps' are not tethered to any units of physical duration and a concrete temporal dimension is entirely lacking. It is only the steps in a *materially realized* Turing machine computation that are extended in physical time, and the very same steps in different types of realization can have vastly different temporal durations – the Analytical Engine will be markedly slower than contemporary electronic realizations.

But FCTM+ is committed to the result that qualitatively identical conscious states are maintained across wildly different kinds of physical realization, from human neural wet ware to the robot's silicon circuitry, to the gears and levers of the Analytical Engine. And this is tantamount to the claim that an actual, substantive and *invariant* qualitative phenomenon is preserved over radically diverse material systems, while at the same time, no internal physical regularities need to be preserved. But then there is no actual, occurrent factor which could serve as the causal substrate or supervenience base for the substantive and invariant phenomenon of internal conscious experience. The advocate of FCTM+ cannot plausibly rejoin that it is invariance of *formal* or *functional* role which supplies this basis, since formal role is abstract, and such abstract features can only be implemented *via* actual properties, but they do not have the power to *produce* them (see Schweizer 2002) for related discussion).[1]

Indeed, physical conservation laws hold that all physical events must have a purely physical cause. So if one is really a physicalist (as opposed to some sort of crypto-dualist) and holds that occurrent qualitative experience is an actual event rather than a mere abstraction, then it follows that the cause must be physical. Hence it would seem to be entailed by basic conservation laws that the material brain (natural or synthetic) must do the causal work of the mind. If internal conscious states are real phenomena extended in time, then their ultimate source must be the brain/hardware – they must depend upon intrinsic properties of the *realizer* as a proper subsystem of the actual world.

Conscious experiences are then seen as hardware states that play an abstract functional role. This abstract role remains a legitimate software concern, and it must be preserved across divergent realizations. But the actual properties of consciousness are a feature of the material substrate, and (unless one has some sort of 'magical' theory of computation, whereby implementing a computational formalism somehow imbues a physical system with mysterious powers and properties over and above its ordinary physical traits) these are not guaranteed to be preserved across widely different physical systems. Qualitative aspect is essentially conditioned by the hardware and hence *is* largely a matter of our flesh and blood embodiment (the above is comparable to some of the views put forward by Searle (1992), although here I am making a claim about qualitative states *simplicitor*, and no assertion whatever about 'Intentionality').

My position is not in direct conflict with the functionalist-driven view that some advanced functional roles such as a self-model and other meta-cognitive features may *require* conscious implementation, and hence that alternative realizations of the human cognitive template that are purportedly devoid of conscious experience (e.g. Block's (1978) 'Chinese Nation') are not genuine possibilities in the first place, and hence cannot serve as hypothetical counterexamples to FCTM+. Although I am in principle agnostic as to whether *any* functional role is such that it requires conscious presentations, even if this constraint on possible implementations is granted, it does not follow that the supporting phenomenology must be *qualitatively identical* to ours. There is no reason to suppose that the field of human conscious experience is the *unique solution* to the functional constraints imposed by LOT_{37}*, and hence the purely abstract structures of FCTM+ are not sufficient to determine our particular phenomenology.

Qualitative presentations in the case of, e.g., visual perception, play the functional role of providing information about the external environment. Hence LOT_{37}* would include functional pathways for processing sensory inputs and utilizing this information for executive control, such as locomotion and navigation. However this is not enough to determine the qualitative aspects produced by the physical vehicle that implements this role. So the very same abstract LOT_{37}* functional specification, when implemented in human neurophysiology, will result in qualitatively different phenomenology than when implemented in the robot's silicon circuitry. The physical vehicle implementing

[1] Thus the critique applies not just to classical computation and the mind/program model, but to any approach committed to abstract structural explanation and MR, such as connectionist architectures.

the functional role will determine the actual and occurrent aspects of qualitative experience, and indeed, physiological variations between human individuals could well induce large disparities in qualia, even among con-specifics.

It's not clear how deeply the LOT_{37}* high-level rational and linguistic processing, integrated with more ancient perceptual and navigational architecture, will penetrate into the qualitatively manifested differences in the robot's physical substrate. Thus if employ some version of Dennett's (1992, 2003) 'heterophenomenology', it's conceivable that the robot could report on qualitative aspects of its conscious states that diverge from ours, and hence would cause it to fail a Total Turing Test designed as a strict 'imitation game'. But in the context of the present thought experiment we will assume that the robot is indeed functionally isomorphic, and hence its verbal outputs reveal no qualitative differences, as in the case of qualitative variations between con-specifics not revealed through verbal outputs.

There are a number of different types of conscious states, including perceptual, cognitive, and affective. The LOT_{37}* will encompass high-level rational and linguistic processing, integrated with more ancient perceptual and navigational architecture, and these will be realized by divergent physical media in the robotic 'cognitive clone'. However, affective conscious states, such as moods and emotions, are far less obviously tethered to *abstract* processing structure, and much more directly related to brute biochemical influences. Since the robot is a synthetic device, it will not possess human hormones, and thus we should expect its affective states (if any) to be qualitatively distinct from ours.

And as a final consideration in this vein, it's relevant to mention 'noise' as yet a fourth type of conscious experience. For example, when I rub my knuckles against my closed eyes and then open them, I see various twinkling yellow spots. As with the tingling sensation of a sneeze, etc., I would take these phenomena to serve no functional role whatever, but rather to be mere noise in my organic hardware system – just evolutionary spandrels. And since the robot's hardware is fundamentally different, it seems reasonable to conclude that it would not experience qualitatively identical forms of noise.[2]

In any case, it's quite safe to say that consciousness is still deeply mysterious, and currently no one has a firmly established or conceptually complete and satisfactory account. Given our present state of insipient understanding, it's inevitable that conjecture and speculation will abound. Neither the FCTM+ view nor the contrasting hardware based account defended above are yet confirmed (or definitively refuted). I've offered some criticisms of the functionalist view and various reasons for favoring the hardware based approach, but it nonetheless remains an open question. And the focus of the current exercise is a thought experiment, rather than an attempt to conclusively establish the truth of some proposition. So for the purposes of the thought experiment, those who might still adhere to FCTM+ are invited to construe the results in a conditional format – *if* the (speculative) hardware based account turns out to be correct and the (speculative) FCTM+ view turns out to be false, *then* the consequent follows. Hence for the sake of

[2] There is nothing to prevent an FCTM + advocate from attributing a functional role to tingles, afterimages, etc., but I would view this as merely an ad hoc strategy for defending their theory against an obvious objection.

argument we will now proceed on the assumption that the antecedent of the foregoing conditional turns out to be true.

So although the robot's mind is *functionally* identical to that of a human, we should expect its *synthetic phenomenology* to be highly divergent, since the substrate in which the functional structure is realized is very different than human neurophysiology. The hypothetical robot brain sustains a form of artificial consciousness that is qualitatively distinct from ours, and potentially very alien. Thus to the extent that phenomenology is a constituent of general mentality, the robot mind is distinct from the human mind. The LOT_{37}* processing structure is the result of many eons of organic evolution and natural selection, and in this respect the robot's cognitive architecture has a clearly biological etiology. But even though the robot's abstract mental processing structure is quintessentially human, its conscious experience is artificial, and is qualitatively dissimilar to ours. Hence the overall type of mind induced is not purely human, but rather is a bio-machine *hybrid.*

3 An Artificial Brain Implementing Synthetic Cognitive Architecture

The second scenario takes yet a further step of abstraction. In the first case we detached computational structure from underlying hardware, and exploited MR to yield an artificial realization of the biologically evolved LOT_{37}*. Now we will abstract away from *internal* factors altogether, including *both* physical substrate and the cognitive software it's running. We consider a computational artifact again capable of passing the combined linguistic and robotic Total Turing Test, but where the robot's internal processing structure is now entirely artificial. The robot's cognitive architecture has been custom designed by AI researchers, and is *functionally* as unlike LOT_{37}* as the first robot's artificial 'brain' is *physically* unlike human neurophysiology. This is a case of successfully manifesting all aspects of intelligent human behavior in the shared linguistic and spatiotemporal environment, but where this is achieved *via* an internal processing structure vastly different from humans.

In response to the question above – how much of the robot's mind should be viewed as properly artificial? – it may appear that in this case the answer should be '*all of it*', that we have succeeded in producing a truly synthetic mind, comprised of both artificial software and an artificial brain (and perhaps replete with synthetic phenomenology, as in the previous thought experiment). However, I will now invoke feature (**3**) to argue that, again, this is not the whole story.

The *content* of propositional attitude states such as beliefs and desires is surely a core feature of minds. As above, to a great extent, our mental identity is dependent upon what we believe and what we want. According to the standard belief-desire framework of psychological explanation, the content of these propositional attitudes is central to our status as *rational agents* and similarly this feature is vital to Dennett's Intentional Stance. And this is important, because the issue at hand does not concern the bare mechanical and engineering factors involved in designing and building a robot able to pass the Total Turing Test. Instead the issue concerns the subsequent *evaluation* of the artifact with respect to its semantic and intentional properties, including genuine intelligence, understanding, reference for its assorted linguistics outputs, and the attribution of associated mental states, such as *believing* that snow is white, *knowing* that water is H_2O, *wanting* to pass the Total Turing Test, etc.

As in the case of behaviorally indistinguishable *humans,* the robot will be evaluated as an Intentional System harboring assorted beliefs, desires and other intentional states, and whose behavior can be explained and predicted on the basis of the *content* of these states. Accordingly, the robot's salient sonic emissions are *interpreted* as asserting various propositions and expressing assorted cognitive contents. For example, suppose Robbie the Robot, our hypothetical Total Turing Test artifact, is ambling down a path and there's a fallen log in the way. Robbie lifts his artificial leg unusually high and steps over the log. When asked 'Why did you lift your leg so high?' Robbie emits the rejoinder 'I saw the fallen log and did not want to trip on it.' Robbie is reporting the content of his relevant propositional attitude states in English, and if we are to interpret him as such then they depend in an essential manner on the public, externally determined semantics for this *human* Natural Language (NL). This is entailed if we are to construe the artifact as a rational agent, as the locus of some genuine form of mentality, and hence as *using* NL in a meaningful and referential manner, rather than just *mentioning* syntactic strings generated by its internal linguistic processing system.

According to Putnam's (1975) highly influential and compelling analysis, the semantics of NL 'ain't in the head' of any individual human agent, but rather are set by the encompassing sociolinguistic community of which the agent is a member. But if linguistic meanings ain't in the head of any individual humans, then they surely ain't in the data base of Robbie the Robot. As originally propounded by Burge (1979), Putnam's semantic externalism for NL implies that *mental content* is non-individualistic. The propositional attitudes of human individuals derive their meaning from the engulfing sociolinguistic medium.

And just as in the case of individual human mentality, so too for Robbie. The Total Turing Test robot is inextricably embedded in a *human* sociolinguistic community with its associated network of *human* cognitive scaffolding (Rupert 2009). Thus because of semantic externalism and the ineliminable role of the *biologically* engendered sociolinguistic medium, the robot's wide *mental content* will be derived from the embedding Natural Language culture. Even though Robbie is wholly non-natural as an isolated artifact, his *wide* propositional attitudes will be no more artificial than yours or mine, and his essential status as an Intentional System is dependent on the *human* sociolinguistic culture in which he functions (see Schweizer 2012 for further discussion).

So to the extent that he's a rational agent susceptible to the framework of Belief-Desire explanation and prediction, Robbie's mind has an ineliminable human component. Natural languages have evolved over many cycles of adaptation and selection, and in this sense the sociolinguistic context upon which Robbie's mental content depends can be seen as an 'organic' component with a clearly biological etiology. Hence such a robot would not be a case of purely artificial mentality, but rather a complex blend of artificial internal processing structures in conjunction with biologically engendered sociolinguistic content. In the general case, mental states are determined both by *internal* factors, such as computational processing configurations and phenomenology, as well as *external* factors, such as the wide propositional content of beliefs and desires. Hence the robot is a bio-machine hybrid in terms of its external *versus* internal facets of mentality.

4 Conclusion

The discussion has extended the received animat paradigm by exploring two cases of genuine mind-machine merger, but where there is *no* physical Brain-Machine Interface – indeed, the material human body/brain has been removed from the picture altogether. The first thought experiment utilizes FCTM and the attendant principle of MR to envision a case where the quintessentially human LOT_{37}* functional/computational architecture is implemented in a humanoid artifact. The widely embraced mind/program analogy would seem to imply that the resulting 'cognitive clone' would possess a purely *human* mind, sustained by an alternative physical substrate. However, it is argued that the situation is not so straightforward, and that the artificial consciousness induced by the robot's divergent hardware would result in a type of mentality not purely human, but rather a form of bio-machine hybrid. And this illustrates some intrinsic theoretical difficulties in attempting to replicate the human mind in an alternative material medium.

In the second thought experiment, the human body/brain as well as its organically engendered cognitive architecture have been removed, and the robot in question runs custom designed artificial software. Nonetheless, its status as an Intentional System, and the attendant content of its propositional attitude states is essentially human, which illustrates some deep theoretical difficulties in attempting to create a form of purely *Artificial* General Intelligence, a truly *artificial mind.*[3]

Acknowledgments. I would like to thank the reviewers Chuanfei Chin, Sam Freed and Dagmar Monett for useful comments.

References

Block, N.: Troubles with functionalism. In: Savage, C.W. (ed.) Perception and Cognition. University of Minnesota Press, Minneapolis (1978)

Block, N.: On a confusion about a function of consciousness. Behav. Brain Sci. **18**, 227–247 (1995)

Burge, T.: Individualism and the mental. In: French, P., Euhling, T., Wettstein, H. (eds.) Studies in Epistemology, vol. 4, Midwest Studies in Philosophy. University of Minnesota Press (1979)

Chalmers, D.: The Conscious Mind. OUP, Oxford (1996)

Churchland, P.: Matter and Consciousness. MIT Press, Cambridge (1984)

---

[3] In view of externalist semantical implications for mental content, a fully artificial form of mentality would require a 'Planet of the Robots' scenario, a community of feral artifacts not programmed with human NL. Instead, the robots would need to evolve their own sociolinguistic community from scratch, just as the human race did. In this manner the robotic mental states and contents would be *genuinely artificial*, just as advanced biological creatures on another planet would possess a *genuinely alien* form of mentality.

Clark, A.: Natural-Born Cyborgs. OUP, Oxford (2003)
Dennett, D.: True believers: the intentional strategy and why it works. In: Heath, A.F. (ed.) Scientific Explanation: Papers Based on Herbert Spencer Lectures Given in the University of Oxford. University Press, Oxford (1981)
Dennett, D.: Consciousness Explained. Viking Press, London (1992)
Dennett, D.: Who's on first? Heterophenomenlogy explained. J. Conscious. Stud. **10**, 19–30 (2003)
Fodor, J.: The Language of Thought. Harvard University Press, Cambridge (1975)
Franklin, S.: Artificial Minds. MIT Press, Cambridge (1995)
French, R.: The Turing test: the first 50 years. Trends Cogn. Sci. **4**, 115–122 (2000)
Harnad, S.: Other bodies, other minds: a machine incarnation of an old philosophical problem. Minds Mach. **1**, 43–54 (1991)
Jackendoff, R.: Consciousness and the Computational Mind. MIT Press, Cambridge (1987)
Johnson-Laird, P.N.: The Computer and the Mind. Harvard University Press, Cambridge (1988)
Lycan, W.G.: Consciousness. MIT Press, Cambridge (1987)
Putnam, H.: Psychological predicates. In: Capitan, W.H., Merrill, D.D. (eds.) Art, Mind and Religion. University of Pittsburgh Press, Pittsburgh (1967)
Putnam, H.: The meaning of 'meaning'. In: Mind, Language and Reality. Cambridge University Press (1975)
Rupert, R.: Cognitive Systems and the Extended Mind. Oxford University Press, Oxford (2009)
Schweizer, P.: Consciousness and computation: a reply to Dunlop. Minds Mach. **12**, 143–144 (2002)
Schweizer, P.: The externalist foundations of a truly total Turing test. Minds Mach. **22**(3), 191–212 (2012)
Searle, J.: The Rediscovery of the Mind. MIT Press, Cambridge (1992)
Turing, A.: Computing machinery and intelligence. Mind **59**, 433–460 (1950)

Huge, but Unnoticed, Gaps Between Current AI and Natural Intelligence

Aaron Sloman(✉)

School of Computer Science, University of Birmingham,
Birmingham B15 2TT, UK
a.sloman@cs.bham.ac.uk
http://www.cs.bham.ac.uk/~axs

Abstract. Despite AI's enormous *practical* successes, some researchers focus on its potential as science and philosophy: providing answers to ancient questions about what minds are, how they work, how multiple varieties of minds can be produced by biological evolution, including minds at different stages of evolution, and different stages of development in individual organisms. AI cannot yet replicate or faithfully model most of these, including ancient, but still widely used, mathematical discoveries described by Kant as non-empirical, non-logical and non-contingent. Automated geometric theorem provers start from externally provided logical axioms, whereas for ancient mathematicians the axioms in Euclid's *Elements* were major *discoveries*, not arbitrary starting points. Human toddlers and other animals spontaneously make similar but simpler topological and geometrical discoveries, and use them in forming intentions and planning or controlling actions. The ancient mathematical discoveries were not results of statistical/probabilistic learning, because, as noted by Kant, they provide non-empirical knowledge of possibilities, impossibilities and necessary connections. Can gaps between natural and artificial reasoning in topology and geometry be bridged if future AI systems use previously unknown forms of information processing machinery – perhaps "Super-Turing Multi-Membrane" machinery?

1 The Meta-Morphogenesis Project

This paper opens a small window into a large project, begun over half a century ago, in my DPhil thesis (Sloman 1962) defending Kant's claims (Kant 1781) about the nature of mathematical discoveries: they are non-empirical, non-contingent, and they are *synthetic*, i.e. not based purely on logic plus definitions.

Around 1969 Max Clowes introduced me to AI. The project then grew into an attempt to use AI to explain many aspects of minds (Sloman 1978), including their abilities to make mathematical discoveries, especially the geometrical and topological discoveries made by ancient mathematicians.

V. C. Müller (Ed.): PT-AI 2017, SAPERE 44, pp. 92–105, 2018.
https://doi.org/10.1007/978-3-319-96448-5_11

A new strand began in 2011, inspired by Turing's work on morphogenesis (Turing 1952), namely the Meta-Morphogenesis (M-M) project[1], investigating evolution of biological information processing mechanisms and capabilities, including an outline theory of evolved construction-kits.[2,3] That provides a framework for new attempts to identify ancient processes and mechanisms of mathematical discovery, especially precursors of mechanisms involved in topological and geometric discovery, illustrated by the work of Archimedes, Euclid, Zeno and many others. I conjecture that this will require discovery of some of the simpler intermediate cases in the evolution of the mechanisms involved. Early developmental stages may also give clues.

Studying spatial reasoning in other intelligent species, e.g. squirrels and crows, and pre-verbal human toddlers, may give clues regarding mechanisms used by ancient adult human mathematicians, including clues indicating their reasoning about possible and impossible spatial structures and processes in solving practical problems, where those processes use subsets of the mechanisms involved in ancient mathematical discoveries.

There's no evidence that ancient mathematicians and intelligent non-human animals use axiomatic, logical, forms of representation and reasoning based on Cartesian coordinates, such as Hilbert's axiomatization of Euclid (Hilbert 1899), and geometry theorem provers, e.g. (Chou et al. 1994). My claim could be challenged by evidence showing that brains of some non-human species, and humans who have never encountered modern logic include genetically specified formalisms and mechanisms for doing what logic theorem provers do. (Merely showing that activity in certain brain regions is correlated with performing a task does not explain how brains perform that task – unlike specifying the algorithms and data-structures used by a robot to perform the task.)

Analysing examples of related, simpler, mathematical and proto-mathematical discoveries in humans and other animals[4], suggests that intelligent animals use types of information processing machinery that are not included in currently understood logical, algebraic, or statistical, reasoning mechanisms, including neural-nets. For example, no learning mechanism based on probabilistic inference can discover impossibilities or necessities, which are key features of mathematical discovery, as pointed out in Kant (1781).

Virtual machines running on digital computers closely coupled with the environment could be richer than a Turing machine, e.g. if the environment includes non-digital or truly random phenomena). If the environment with which a digital computer interacts is not a discrete-state machine, the coupled system, including any virtual machinery used, cannot be modelled with perfect precision on

---

[1] http://www.cs.bham.ac.uk/research/projects/cogaff/misc/meta-morphogenesis.html.

[2] http://www.cs.bham.ac.uk/research/projects/cogaff/misc/construction-kits.html.

[3] An invited video talk at IJCAI 2017, is available online, with extended notes: http://www.cs.bham.ac.uk/research/projects/cogaff/misc/ijcai-2017-cog.html.

[4] Pre-verbal toddler topology is illustrated in this 4.5 min video: http://www.cs.bham.ac.uk/research/projects/cogaff/movies/ijcai-17/small-pencil-vid.webm.

a Turing machine, since no discrete machine can model perfectly a processes that runs through all the real numbers between 1 and 2, in order, whereas a continuously changing chemical structure might be able to.[5] (Moving only through the rationals in order would be more complex.)

2 Limited Progress, Despite Spectacular Successes

The practical uses of AI, and the rate at which they are now multiplying are so impressive that some serious thinkers have begun to fear that we are in danger of building monsters that will take over the planet and do various kinds of harm to humans, that we may be unable to prevent because we don't match their intelligence. For some reason most such thinkers don't consider the more optimistic possibility, suggested many years ago[6], that truly superhuman intelligence will include a kind of wisdom that rejects the selfish, thoughtless, competitive, destructive, gullible, superstitious, and other objectionable features that lead to so much harm done by humans to other humans and other species. But "singularity risks" are not my concern: this paper is about how *little* progress has been made in philosophical and scientific aspects of AI that motivated the early researchers who hoped, as I still do, that AI can give us powerful new ways of modelling and understanding natural intelligence: AI as *science* and *philosophy* not just *engineering*.

Alas, AI as engineering dominates AI education (and publicity) nowadays, in contrast with the concerns of early researchers in the field, including some philosophers, who noticed the potential of research in AI to contribute to a new deep understanding of natural intelligence. For a survey see Margaret Boden's two-volume masterpiece (2006).

Recent spectacular engineering successes mask (current) limited scientific and philosophical progress in AI. Two results of this masking (at present) are a shortage of good researchers focusing on the long term issues, and a shortage of funds for long term scientific research. Most funded AI research at present aims at demonstrable practical successes, leaving some of the important scientific questions unanswered, and to some extent un-noticed!

I do not claim that progress is impossible, only that it is very difficult and requires deep integration across disciplines. It also depends on an educational system producing high calibre multi-disciplinary researchers.

Despite its enormous practical importance, some AI researchers, like Turing, are more interested in the potential of AI as *science* and *philosophy* than its *practical* applications. E.g. AI (along with computer science) has begun to advance science and philosophy by providing new forms of explanation for aspects of natural intelligence and new answers to ancient philosophical questions about the nature of minds, their activities, and their products.

[5] For further discussion of "virtual machine functionalism" see http://www.cs.bham.ac.uk/research/projects/cogaff/misc/vm-functionalism.html.

[6] E.g. in the epilogue to my 1978 book, *The Computer Revolution in Philosophy*, here http://www.cs.bham.ac.uk/research/projects/cogaff/crp/#epilogue.

In particular, as explained in Chap. 2 of Sloman (1978), the deepest aim of science (not always acknowledged as such) is to discover what sorts of things are *possible*, and what makes, or could make, them possible, not to discover regularities. In contrast, many science students are (unfortunately) taught to regard science as primarily concerned with finding, explaining and using observed correlations: a shallow view of science criticised vehemently in Deutsch (2011). Deep scientific theories all contribute to the study of what is possible and how it is possible, including ancient atomic theory, Newton's mechanics, chemistry, Darwin's theory of natural selection, quantum mechanics, e.g. (Schrödinger 1944), computer science, AI and theoretical linguistics.

The Turing-inspired Meta-Morphogenesis project mentioned in Note 1 has addressed such issues since 2012. AI, including future forms of AI, must be an essential part of any deep study of "the space of possible minds" (Sloman 1984), which may be far richer than anyone currently suspects.

3 AI as Science and Philosophy

For most people, AI is primarily an engineering activity, whereas my interest, since around 1969, inspired by Max Clowes, and AI founders such as Minsky e.g. (1963, 1968, 2006), McCarthy e.g. (1979, 2008), and Simon e.g. (1967, 1969), is focused mainly on the potential of AI to trigger and eventually to answer scientific and philosophical questions, e.g. about what minds and mental states and processes are, and how they work, including how they evolved, how they develop, how they can vary, with potential applications in education and therapy.

A long term goal is to explain how biological evolution is able to produce so many different forms of information-processing, in humans and non-human organisms, at different stages of development, in different physical and cultural contexts, and in different cooperating subsystems within complex individuals (e.g. information processing subsystems involved in: internal languages[7], language development, visual perception, motivational processes, and mathematical discovery). Explaining all this requires major progress in understanding varieties of information processing. Clues may come from many evolutionary stages, including: microbe minds, insect minds, and other precursors of the most complex minds we hope to understand and model. This is the Meta-Morphogenesis project mentioned in Note 1.

Unfortunately much "standard" research, seeking experimental or naturally occurring regularities, fails to identify what needs to be explained, because most animal information processing is far richer than observable and repeatable input-output relationships – e.g. your mental processes as you read this. No amount of laboratory testing can exhaust the responses you could possibly give to possible questions about what you are reading here, and there is no reason to assume that all humans, even from the same social group, or even the same research department, will give the same answers. Compare how different the outputs of great composers, or poets, or novelists are, even if they live in the same location.

[7] http://www.cs.bham.ac.uk/research/projects/cogaff/talks/#talk111.

A standard response is to regard all that diversity as irrelevant to a science of mind. One consequence is narrowly focused research using experiments, e.g. in developmental psychology, designed to constrain subjects artificially to support repeatability. This can conceal their true potential, requiring long term studies of individuals, which would have to accommodate enormous variability in developmental trajectories.

There are exceptions, e.g. Piaget's pioneering work on children's understanding of Possibility and Necessity, published posthumously (Piaget 1981,1983). But he lacked adequate theories of information processing mechanisms (as he admitted at a workshop I attended, shortly before he died). Piaget's earlier work inspired the proposals in Sauvy and Sauvy (1974). It could also suggest useful goals for future, more human-like, robots.

4 Aim for Generative Power *not* Data Summaries

Overcoming the limitations of "standard" empirical research on how minds work requires setting explanatory goals at the level of *generative powers* rather than *observed regularities*, as Chomsky and others pointed out long ago (1965). (For historical detail see (Boden 2006); Compare the claim that deep science is more concerned with discovery and explanation of *possibilities* than *laws*, in Sloman (1978, Chap. 2).

Even in the physical sciences, modelling observed regularities can often be achieved without accurate modelling of the *mechanisms* that happened, on that occasion, to produce those regularities, e.g. the apparent successes of the Ptolemaic theory of planetary motion, and many other well supported then later abandoned regularities in physics – including Newtonian dynamics.

Problems of reliance only on observed and repeatable regularities are far worse in the science of mind. Overcoming them requires application of deep multi-disciplinary knowledge and expertise, including designing, testing and debugging complex virtual machines interacting with complex environments. This helps to debunk the myth that AI is dependent on Turing machines: TMs are defined to run disconnected from any environment, rendering them useless for working AI systems, despite their great theoretical importance for computer science (Sloman 2002). Preliminary ideas regarding a "Super Turing membrane machine" are in Sloman (2017b),[8] related to ideas about affordances in Sloman (2008) and McClelland's work on affordances for mental action, e.g. (2017). This requires substantial long term research.

Insights can often be gained by studying naturally occurring, but relatively rare phenomena, for example when attempts to teach deaf children in Nicaragua to use sign language demonstrated that children do not merely *learn* pre-existing languages: they can also *create* new languages cooperatively, though this is

[8] For a detailed example see http://www.cs.bham.ac.uk/research/projects/cogaff/misc/deform-triangle.html and http://www.cs.bham.ac.uk/research/projects/cogaff/misc/apollonius.html.

cloaked by the fact that they are usually in a minority, so that collaborative construction looks like learning (Senghas 2005).

4.1 Human/Animal Mathematical Competences

A particular generative aspect of human intelligence that has been of interest to philosophers for centuries, and discussed by Kant (1781, 1783), is the ability to make mathematical discoveries, including the amazing discoveries in geometry presented in Euclid's *Elements* over two thousand years ago that are still in use world-wide every day by scientists, engineers and mathematicians (though unfortunately now often taught only as facts to be memorised rather than rediscovered by learners).

I suspect that Kant understood that those abilities were deeply connected with practical abilities in non-mathematicians such as weaver birds, squirrels, elephants, and pre-verbal toddlers (my examples, not his), as illustrated in the video presentation in Sloman (2017b). Young children don't have to be taught topology in order to understand that something is wrong when a stage magician appears to link and unlink a pair of solid metal rings. Online documents exploring some of the details are referenced in Note 8 and the work on evolved construction-kits in Note 2.

Despite the popular assumption that computers are particularly good at doing mathematics, because they can calculate so fast, run mathematical simulations, and even discover new theorems and new proofs of old theorems using AI theorem-proving packages, they still cannot replicate the ancient geometric and topological discoveries, or related discoveries of aspects of geometry and topology made unwittingly by human toddlers (illustrated in the video referenced in Note 4. and related achievements of other species, e.g. birds that weave nests from twigs or leaves, and squirrels that defeat "squirrel-proof" bird feeders. (Search online for videos.)

These limits of computers are of far deeper significance for the science of minds than debates about whether computer-based systems can understand proofs of incompleteness theorems by Gödel and others, e.g. (Penrose 1994) (who recognizes the importance of ancient geometric competences, but gives no plausible reasons to think they *cannot* be replicated in AI systems, although they have not been replicated so far.)

4.2 AI Theorem Provers Do Something Different

There are impressive AI geometry theorem provers, but they *start* from logical formalisations of Euclid's axioms and postulates, e.g. using Hilbert's (1899) version. They derive theorems using methods of modern logic, algebra, and arithmetic (e.g. pruning search paths by using numerical checks). Those methods are at most a few hundred years old, and some much newer. They were not known to or used by great ancient mathematicians, such as Archimedes, Euclid, Pythagoras and Zeno, or children of my generation learning to prove statements in Euclidean geometry. How did their brains work?

A major unsolved problem for AI is to understand and replicate the relevant *ancient* reasoning powers. The postulates and axioms in Euclid's *Elements*, e.g. concerning congruency, were stated without proof, but were not *arbitrary assumptions* adopted as starting points to define a mathematical domain, as in modern axiomatic systems.

Rather, Euclid's axioms and postulates were *major discoveries*, and various mathematicians and philosophers have investigated ways of deriving them from supposedly more primitive assumptions, e.g. deriving notions like point and line from more primitive spatial/topological notions, as demonstrated by Scott (2014). A simpler example, from Sloman (2017b), referenced in Note 8 is in Fig. 1.

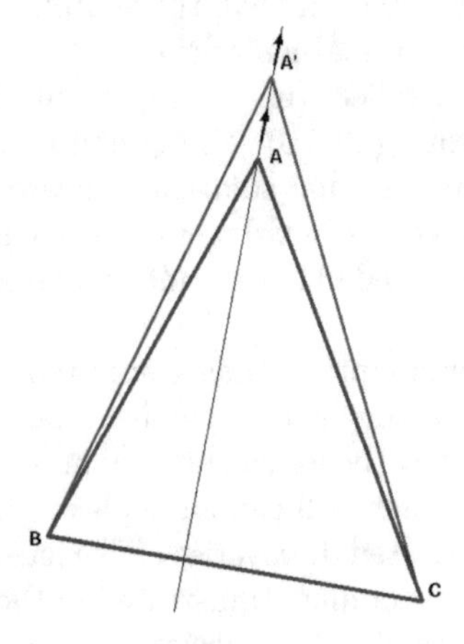

Fig. 1. What happens to the size of the angle at A if A is moved further from BC along a line through the opposite side BC? Answering involves thinking about two continua (the continuum of positions of the top vertex, and the continuum of angle sizes) and their relations. Many people with no mathematical training can do this easily, in my experience. What are their brains doing? How do brains represent impossibility or necessity? If the line of motion of A intersects the base outside the triangle the situation is more complicated, and Apollonius' construction becomes relevant, as Diana Sofronieva pointed out to me.

If you start with an arbitrary planar triangle, like the blue one in Fig. 1, then continuously move one vertex further from the opposite side, along a line through the opposite side, e.g. producing the red triangle, and then continuing, what happens to the size of the angle at the top as it moves: how do you know? What enables you to know that it is impossible for the angle to get larger? Investigation of how the problem changes if the line of motion changes is left as an exercise for the reader (see Note 8).

Euclid's *starting points* require mathematical discovery mechanisms that seem to have gone unnoticed, and are not easily implementable in current AI systems without using something like a Cartesian-coordinate-based arithmetic model for geometry, which was not used by the ancient mathematicians making discoveries thousands of years before Descartes.

Moreover, for reasons given by Kant, they cannot be *empirical* discovery methods based only on finding regularities in many trial cases, since that cannot prove *necessity* or *impossibility*: mathematics is concerned with *necessary* truths and *impossibilities* not empirical generalisations. This feature is ignored by much psychological research on mathematical competences and cannot be explained by statistics-based neural theories of mathematical reasoning. This does not imply *infallibility*, as shown by Lakatos (1976). Any practising mathematician knows that mathematicians can make mistakes. I did at first when reasoning about the stretched triangle problem above, which is what led to the exploration reported in Sloman (2017b).

5 Robots with Ancient Mathematical Competences?

Can current computing technology support ancient mathematical discovery mechanisms, or are new kinds of computers required, e.g. perhaps chemical computers replicating ill-understood brain mechanisms? (I suspect Turing was thinking about such mechanisms around the time he died (suggested by reading (Turing 1952)). There is evidence in Craik (1943) that Kenneth Craik, another who died tragically young, was also thinking about such matters, perhaps inspiring Turing posthumously? Does anything in current neuroscience explain how biological brain mechanisms represent and reason about *perfectly straight, perfectly thin* lines, and their intersections? Or reason about *impossibilities*, and *necessary* consequences of certain kinds of motion?

Future work needs to dig deeper into differences between the forms of logical/mathematical reasoning that computers can and cannot cope with, e.g. because the former use manipulation of discrete structures or discrete search spaces, and the latter require new forms of computation, e.g. the structures and processes used in ancient proofs of geometrical and topological theorems. (Compare the procedures for deriving Euclid's ontology from geometry without points presented in a recorded lecture by Scott (2014). The presentation clearly uses a great deal of spatial/diagrammatic reasoning rather than purely logical and algebraic reasoning.)

The required new mechanisms are not restricted to esoteric activities of mathematicians: e.g. many non-mathematicians, including young children, find it obvious that two linked rings made of rigid impenetrable material cannot become unlinked without producing a gap in one of the rings.

6 Representing Impossibility and Necessity

What brain mechanisms can represent impossibility? How can impossibilities be *derived* from perceived structural relationships? Young children don't have to study topology to realise that something is wrong when a stage magician appears to link and unlink solid rings. What mechanisms do their brains use? Or the brains of squirrels mentioned above?[9] There are many more examples, including aspects of everyday reasoning about clothing, furniture, effects of various kinds of motion, etc. and selection between possible actions (affordances) by using partial orderings in space during visual feedback rather than numerical measures of spatial relationships or the kinds of statistical/probabilistic reasoning that now (unfortunately) dominate AI work in vision and robotics. An alternative approach uses semi-metrical reasoning, including topological structures and partial orderings, was suggested in Sloman (2007). I have not been able to persuade any AI/Robotics researchers, however, possibly because using that approach would require massive changes to Robot vision and reasoning mechanisms. How can such mechanisms be implemented in brains?

Current computers can produce realistic simulations of *particular* spatial processes but that's very different from understanding generic constraints on *classes of processes*, like the regularity linking two dimensions of continuous variation mentioned in Fig. 1.

No amount of repetition of such processes using a drawing package on a computer will enable the computer to understand *why* the angle gets smaller, or to think of asking whether the monotonicity depends both on the choice of the line of motion of the vertex and the starting point. See Note 8 and (Sloman 2017b). I did not notice this until Auke Booij pointed it out to me.

Such geometric reasoning about partial orderings is very different from understanding why an expression in boolean logic is unsatisfiable or why a logical formula is not derivable from a given set of axioms, both of which can be achieved (in some cases) by current AI systems, but only after the problem is rephrased in terms of possible sequences of logical formulae in a proof system, or possible solutions to numerical equations, using something like Hilbert's logic-based formulation of Euclidean geometry. Ancient geometric reasoning was very different from reasoning about arithmetical formulae by using Cartesian coordinates. (Claims by John Searle and others that computers are purely syntactic engines, with no semantic competences, have been adequately refuted elsewhere.)

7 Gaps in Theories of Consciousness

7.1 What Is Mathematical Consciousness?

Can we give the required sort of consciousness of geometrical necessity to future robots? The lack of any discussion of mathematical consciousness, e.g. "topological impossibility qualia", in all contemporary theories of consciousness that I have encountered, seems to me to suggest that those theories are at best incomplete, and probably deeply mistaken, at least as regards spatial consciousness.

The tendency for philosophers of mind to ignore mathematical discovery is particularly puzzling given the importance Kant attributed to the problem as long ago as 1781. (And long before him Socrates and Plato?)

---

[9] Many additional examples are presented in http://www.cs.bham.ac.uk/research/projects/cogaff/misc/impossible.html.

Perhaps this omission is a result of a widely held, but mistaken, belief that Kant was proved wrong when empirical support was found for Einstein's claim that physical space is non-Euclidean. Had Kant known about non-Euclidean geometries, he could have given as his example of non-empirical discovery of non-analytic mathematical truths the discovery that a subset of Euclidean geometry can be extended in different ways, yielding different geometries with different properties. Kant had no need to claim that human mathematicians are infallible, and as far as I know, never did claim that. His deep insights were qualified, not refuted, by Lakatos (1976). This was also discussed in my 1962 thesis (Sloman 1962). "Proto-mathematical" discoveries of various kinds are also made, and put to practical uses, by pre-verbal human toddlers.[10]

Whether AI can be extended in the foreseeable future to accommodate the ancient mathematical competences using current computers depends on whether we can implement the required virtual machinery in digital computers or whether, like brains, future human-like computers will have to make significant use of chemical information processing, perhaps using molecules rather than neurons as processing units, as discussed by Grant (2010), Trettenbrein (2016), Gallistel and Matzel (2012), Newport (2015) (citing von Neumann) and others.

As long ago as 1944 Schrödinger (1944) pointed out the importance for life of the fact that quantum physics explains how chemistry can support both discrete processes (structural changes in chemical bonds) and continuous changes (folding, twisting, etc.) The possibility that biological information processing is implemented not at the neural level but at the molecular level was also considered by John von Neumann in his 1958 book *The computer and the brain*, written while he was dying. If true this implies that current calculations regarding how soon digital computers will replicate brain functionality are out by many orders of magnitude (e.g. many centuries rather than decades). See also Newport (2015).

7.2 Probabilistic Reasoning *vs* Impossibility/Necessity

AI researchers who have not studied Kant's views on the nature of mathematical knowledge as non-analytic (synthetic, i.e. not derivable using only definitions and pure logic), non-contingent (concerned with what's possible, necessarily the case, or impossible) may find it hard to understand what's missing from AI. In particular, I have found that some believe that eventually deep learning mechanisms will suffice.

But mechanisms using only statistical information and probabilistic reasoning are constitutionally incapable of learning about necessary truths and falsehoods, as Kant noticed, long ago, when he objected to Hume's claim that there are only two kinds of knowledge: empirical knowledge and analytic knowledge (definitional relations between ideas, and their logical consequences).

[10] Several examples of various kinds are presented in http://www.cs.bham.ac.uk/research/projects/cogaff/misc/toddler-theorems.html and http://www.cs.bham.ac.uk/research/projects/cogaff/misc/impossible.html.

Hume's view of *causation* as being of the first sort (concerned with observed regularities) is contradicted by mathematical examples including the triangle deformation example above: motion of a vertex of a triangle away from the opposite side *causes* the angle to decrease, just as adding three apples to a collection of five apples *causes* the number in the collection to increase to eight. Examples of Humean and Kantian causal reasoning in humans and other animals were presented (in collaboration with Jackie Chappell) in Chappell and Sloman (2007b).

7.3 Can We Give Robots Geometric Reasoning Abilities?

Possible lines of enquiry about what's missing from current AI are suggested by Turing (1952), leading to a new theory regarding the variety of mechanisms and transitions in biological evolution, including evolution of new kinds of construction kit (Sloman 2017a).[11] Evolution repeatedly produced new biological construction kits for new kinds of information processing mechanism. This may explain the evolution of epigenetic processes that produce young potential mathematicians. Ideas about "meta-configured competences" are being developed in collaboration with biologist Jackie Chappell (Chappell and Sloman 2007a),[12] extending Karmiloff-Smith's theories of "Representational Redescription" (1992), and hypotheses about non-linear, structured, extendable, *internal* languages required for percepts, intentions, plans, usable generalisations, and reasoning, long before *external* languages were used for communication (Sloman 2015).

One consequence of these investigations is rejection of the popular "Possible worlds semantics" as an analysis of (alethic) modal operators ("impossible", "possible", "contingent", and "necessary"), in favour of (Kant-inspired) semantics related to variations in configurations of fragments of *this* world, as illustrated in the stretched triangle example, and many other examples of geometrical and topological reasoning.

8 Conclusion, and Further Work

This paper opens a small window into a large, complex, still growing project. (See Note 1.) There are many implications for AI as Science, AI as engineering and AI as philosophy, and also deep implications for psychology and neuroscience, insofar as they have not yet addressed the problem of how minds or brains are able to make discoveries concerning necessary truths and impossibilities that are not merely logical truths or falsehoods. There are also hard biological problems to be solved, concerning evolutionary histories of the features of human brains

[11] The work on construction kits is still being extended in http://www.cs.bham.ac.uk/research/projects/cogaff/misc/construction-kits.html.

[12] http://www.cs.bham.ac.uk/research/projects/cogaff/misc/meta-configured-genome.html.

and minds that have these amazing capabilities. Perhaps only after these non-AI questions have been answered will AI engineers be able to design artificial minds with ancient mathematical capabilities of Archimedes and others. Not all psychologists and neuroscientists notice that the task of explaining mathematical cognition is not merely the task of explaining *numerical* competences, on which they tend to focus, while ignoring the richness of numerical competences such as the central role of transitivity of one-one correspondence. (Piaget was an exception.)

On-going work investigates requirements for a Super-Turing membrane computer,[13] able to acquire and use information about spatial structures and relationships in performing practical tasks, for instance understanding how available information and affordances *necessarily* change as viewpoints change, or objects rotate or move – because visual information normally travels in straight lines. If these ideas can be used in future designs, we may be able to produce robots that replicate the discoveries made by great ancient mathematicians as well as the deep but unnoticed spatial reasoning abilities developed by pre-verbal humans and many other intelligent species.

This should help to stifle distracting and impoverished theories of embodied cognition, mistakenly giving the impression that there is no requirement for deep and complex internal information-processing engines produced by biological evolution, but not yet replicated in AI systems. And if, as I suspect (and perhaps Turing suspected), these mechanisms are implemented in sub-synaptic chemical mechanisms, then since there are many orders of magnitude more molecules than neurones, this suggests that hopes or fears about computers soon reaching or overtaking human intelligence are time-wasting distractions from the hard task of trying to understand and model human intelligence, or more generally animal intelligence.

The Meta-Morphogenesis web site is expected to continue growing.[14] But there are many unsolved problems, including problems about mechanisms underlying ancient forms of mathematical consciousness.

References

Boden, M.A.: Mind As Machine: A History of Cognitive Science, vol. 1-2. Oxford University Press, Oxford (2006)

Chappell, J., Sloman, A.: Natural and artificial meta-configured altricial information-processing systems. Int. J. Unconventional Comput. **3**(3), 211–239 (2007a). http://www.cs.bham.ac.uk/research/projects/cogaff/07.html#717

Chappell, J., Sloman, A.: Two Ways of Under-standing Causation: Humean and Kantian. Pembroke College, Oxford (2007b). http://www.cs.bham.ac.uk/research/projects/cogaff/talks/wonac

Chomsky, N.: Aspects of the Theory of Syntax. MIT, Cambridge (1965)

[13] http://www.cs.bham.ac.uk/research/projects/cogaff/misc/super-turing-geom.html.

[14] See Note 1, and also http://www.cs.bham.ac.uk/research/projects/cogaff/misc/construction-kits.html.

Chou, S.-C., Gao, X.-S., Zhang, J.-Z.: Machine Proofs In Geometry. World Scientific, Singapore (1994)
Craik, K.: The Nature of Explanation. CUP, London, New York (1943)
Deutsch, D.: The Beginning of Infinity: Explanations That Transform the World. Allen Lane & Penguin Books, New York (2011)
Gallistel, C., Matzel, L.: The neuroscience of learning: beyond the Hebbian synapse. Annu. Rev. Psychol. **64**, 169–200 (2012)
Grant, S.G.: Computing behaviour in complex synapses. Biochemist **32**, 6–9 (2010)
Hilbert, D.: The Foundations of Geometry. Project Gutenberg, Salt Lake City (1899). (Translated 1902 by E.J. Townsend, from 1899 German edition)
Kant, I.: Critique of Pure Reason. Macmillan, London (1781). (Translated (1929) by Norman Kemp Smith)
Kant, I.: Prolegomena to Any Future Metaphysics That Will Be Able to Present Itself as a Science. Unknown (1783). (Several translations online)
Karmiloff-Smith, A.: Beyond Modularity: A Developmental Perspective on Cognitive Science. MIT Press, Cambridge (1992)
Lakatos, I.: Proofs and Refutations. CUP, Cambridge (1976)
McCarthy, J.: Ascribing mental qualities to machines. In: Ringle, M. (ed.) Philosophical Perspectives in Artificial Intelligence, pp. 161–195. Humanities Press, Atlantic Highlands (1979)
McCarthy, J.: The well-designed child. Artif. Intell. **172**(18), 2003–2014 (2008). http://www-formal.stanford.edu/jmc/child.html
McClelland, T.: AI and affordances for mental action. In: Computing and Philosophy Symposium, AISB Convention 2017, 19–21 April 2017, pp. 372–379. AISB, Bath (2017). http://wrap.warwick.ac.uk/87246
Minsky, M.L.: Steps toward artificial intelligence. In: Feigenbaum, E., Feldman, J. (eds.) Computers and Thought, pp. 406–450. McGraw-Hill, New York (1963)
Minsky, M.L.: Matter mind and models. In: Minsky, M.L. (ed.) Semantic Information Processing. MIT Press, Cambridge (1968)
Minsky, M.L.: The Emotion Machine. Pantheon, New York (2006)
Newport, T.: Brains and Computers: Amino Acids versus Transistors. Kindle (2015). https://www.amazon.com/dp/B00OQFN6LA
Penrose, R.: Shadows of the Mind: A Search for the Missing Science of Consciousness. OUP, Oxford (1994)
Piaget, J.: Possibility and Necessity: Vol. 1. The Role of Possibility in Cognitive Development, Vol. 2. The Role of Necessity in Cognitive Development. University of Minnesota Press, Minneapolis (1981, 1983). (Translated by Helga Feider from French in 1987)
Sauvy, J., Sauvy, S.: The Child's Discovery of Space: From Hop-scotch to Mazes - An Introduction to Intuitive Topology. Penguin Education, Harmondsworth (1974). (Translated from the French by Pam Wells)
Schrödinger, E.: What is Life? CUP, Cambridge (1944)
Scott, D.S.: Geometry without points (2014). https://www.logic.at/latd2014/2014%20Vienna%20Scott.pdf
Senghas, A.: Language emergence: clues from a New Bedouin Sign Language. Curr. Biol. **15**(12), R463–R465 (2005)
Simon, H.A.: Motivational and emotional controls of cognition. In: Simon, H.A. (ed.) Reprinted in Models of Thought, pp. 29–38. Yale University Press, New Haven (1967)
Simon, H.A.: The Sciences of the Artificial. MIT Press, Cambridge (1969). (Second edition 1981)

Sloman, A.: Knowing and Understanding: relations between meaning and truth, meaning and necessary truth, meaning and synthetic necessary truth. DPhil, Oxford (1962). http://www.cs.bham.ac.uk/research/projects/cogaff/62-80.html#1962

Sloman, A.: The Computer Revolution in Philosophy. Harvester Press, Hassocks (1978). goo.gl/AJLDih, Revised 2018

Sloman, A.: The structure of the space of possible minds. In: Torrance, S. (ed.) The Mind and the Machine: Philosophical Aspects of AI. Ellis Horwood (1984). http://www.cs.bham.ac.uk/research/projects/cogaff/81-95.html#49a

Sloman, A.: The irrelevance of Turing machines to AI. In: Scheutz, M. (ed.) Computationalism: New Directions, pp. 87–127. MIT Press, Cambridge (2002). http://www.cs.bham.ac.uk/research/cogaff/00-02.html#77

Sloman, A.: Predicting Affordance Changes, Birmingham, UK (2007). http://www.cs.bham.ac.uk/research/projects/cogaff/misc/changing-affordances.html

Sloman, A.: Architectural and representational requirements for seeing processes, proto-affordances and affordances. In: Cohn, A.G., Hogg, D.C., Möller, R., Neumann, B. (eds.) Logic and Probability for Scene Interpretation. Schloss Dagstuhl, Dagstuhl (2008). http://drops.dagstuhl.de/opus/volltexte/2008/1656

Sloman, A.: What are the functions of vision? How did human language evolve? (2015). http://www.cs.bham.ac.uk/research/projects/cogaff/talks/#talk111

Sloman, A.: Construction kits for biological evolution. In: Cooper, S.B., Soskova, M.I. (eds.) The Incomputable: Journeys Beyond the Turing Barrier, pp. 237–292. Springer International Publishing, Cham (2017a). http://www.springer.com/gb/book/9783319436678

Sloman, A.: Why can't (current) machines reason like Euclid or even human toddlers? (2017b). http://cadia.ru.is/workshops/aga2017

Trettenbrein, P.C.: The demise of the synapse as the locus of memory: a looming paradigm shift? Front. Syst. Neurosci. (2016). https://doi.org/10.3389/fnsys.2016.00088

Turing, A.M.: The chemical basis of morphogenesis. Phil. Trans. R. Soc. London B **237**, 37–72 (1952)

Social Cognition and Artificial Agents

Anna Strasser(✉)

Berlin, Germany
annakatharinastrasser@gmail.com

Abstract. Standard notions in philosophy of mind have a tendency to characterize socio-cognitive abilities as if they were unique to sophisticated human beings. However, assuming that it is likely that we are soon going to share a large part of our social lives with various kinds of artificial agents, it is important to develop a conceptual framework providing notions that are able to account for various types of social agents. Recent minimal approaches to socio-cognitive abilities such as mindreading and commitment present a promising starting point from which one can expand the field of application not only to infants and non-human animals but also to artificial agents. Developing a minimal approach to the socio-cognitive ability of acting jointly, I present a foundation for future discussions about the question of how our conception of sociality can be expanded to artificial agents.

1 Introduction

The handling of technical devices is influencing the life of a wide range of people: many have a mobile phone, use social networks or rely on intelligent software as part of their working lives. Additionally, one can predict increased interactions with artificial systems such as autonomous driving systems, care robots and conversational machines. Soon a lot of people will share a large part of their social lives with various kinds of artificial agents. Where previous revolutions have dramatically changed our environments, this one has the potential to substantially change our understanding of sociality.

If, one day, interactions with artificial systems will be experienced as genuinely social, then it is no longer sufficient to characterize such interactions as mere tool use. Investigating the boundaries of our understanding of sociality and the role artificial agents may play in our social world, I suggest an expanded notion of joint action that is applicable to artificial systems. This is particularly relevant because joint actions as social interactions, unlike tool use, raise ethical questions. Tools neither have rights nor obligations, but with regard to social agents in social interactions such as joint actions, questions about obligations and rights of interaction partners are relevant. But before we can approach these questions, we have to develop an appropriate conceptual framework to determine the conditions for those artificial agents who can be considered social agents.

In philosophy of mind, however, socio-cognitive abilities such as mindreading (Fodor 1992), individual agency (Davidson 1980) or the ability to act jointly (Bratman

---

A. Strasser—Independent Researcher

V. C. Müller (Ed.): PT-AI 2017, SAPERE 44, pp. 106–114, 2018.
https://doi.org/10.1007/978-3-319-96448-5_12

2014) are characterized as if they were unique to sophisticated human beings only. The aim of this paper is to elaborate on the extent to which these human-centered conceptions can be expanded to artificial agents with respect to the socio-cognitive ability to act jointly. An expanded framework of sociality provides the basis on which we can make finer-graded distinctions between tool use and joint action with respect to artificial systems.

To explore whether and how artificial agents might enter the realm of social cognition, it would be helpful to have a clear-cut definition of social cognition at hand. But philosophy does not provide such a definition. Even various characterizations of cognition itself are incompatible in certain respects. With respect to the boundaries of cognition, some positions claim that cognitive processes are necessarily internal or brain-bound, whereas others argue that cognition should be understood as extended into the body and environment (Clark and Chalmers 1998). Furthermore, there are debates about whether associative conditioning and other seemingly lower-level behaviors may count as cognitive (Buckner and Fridland 2017). In addition, we do not have clear-cut criteria specifying what makes cognition especially social.

For the purposes of this paper I characterize social cognition on a functional level. I assume that a primary function of socio-cognitive abilities is to enable us to encode, store, retrieve, and process social information about other agents in order to understand them. This social competence is at the same time an essential requirement to facilitate social interactions. This functional definition remains neutral with regard to disputes about the boundaries of cognition and can be open to multiple realization of socio-cognitive abilities.

Focusing on the socio-cognitive ability of acting jointly, to do things together in order to reach a common goal, I am confronted with the standard philosophical notion of joint action (Bratman 2014), which excludes other types of agents from the start. In contrast to this restrictive notion, I argue for a less human-centered version of this notion. The proposed notion of joint action will specify conditions artificial agents need to fulfill to enter the space of social interactions. Thereby, I present a suggestion as to under which circumstances artificial agents may count as proper social agents in a joint action.

In order to overcome the general tendency to restrict socio-cognitive abilities to living beings, I refer to recent minimal approaches (Butterfill and Apperly 2013; Michael et al. 2016; Vesper et al. 2010), which present a promising starting point for establishing a broader framework. Such approaches suggest so-called minimal versions of standard notions in order to capture a wider range of socio-cognitive abilities. One of their rationales is questioning the necessity of certain conditions that come with the standard philosophical conceptions. For example, the proposed notion of minimal mindreading specifies the very minimal presuppositions of how agents can anticipate the behavior of others without requiring a mastery of language or representations of complex mental states (Butterfill and Apperly 2013). With the help of this notion, instances of a broader spectrum including agents such as infants and non-human animals can be captured. Developing a minimal notion of joint action that can account for abilities of artificial agents, I follow the idea of questioning seemingly necessary conditions. In line with recent minimal approaches, it is my strategy to show that some assumed conditions we find in living beings are not necessary for acting jointly as such. Given that we (will) have reasons to expand the space of sociality from living beings to specific non-living agents, some of the standard conditions should rather be seen as biological constraints specific to human beings and should not be used to exclude artificial agents from the start.

2 Modes of Ascriptions of Socio-Cognitive Abilities

Investigating the practice of ascribing socio-cognitive abilities, it is important to distinguish between two modes of ascriptions: an 'as if' mode and a justified mode of ascription.

Characterizing the 'as if' mode, one can describe two functional roles an 'as if' ascription can take on. First, the 'as if' mode can have an explanatory role. In order to make sense of the world, humans ascribe all sorts of socio-cognitive properties to non-living beings, without claiming that the described objects really possess such properties. Second, social interactions among humans are based on a lot of 'as if' ascriptions.

Regarding the former, an experiment by Heider and Simmel (1944) illustrates how participants attribute social properties in order to describe simply moving geometrical forms. Although it is helpful and enlightening to characterize perceptual input through a social narrative and not through a technical description of geometric forms, it is of course not justified to claim (and no one does) that these objects actually have social features. Along the same lines, Daniel Dennett (1987) describes how we apply the intentional stance to non-living beings. Using the 'as if' mode in this way helps us to make sense of the world but it remains neutral with regard to the question of what socio-cognitive abilities objects really have.

Besides this explanatory role, 'as if' ascriptions can also provide a basis in human-human interactions, when we do not have enough reasons to actually ascribe specific socio-cognitive abilities. We use default assumptions to treat other human agents as social agents. In these cases, we assume that a justified ascription is, in principle, possible and appropriate. Even though such default assumptions do not qualify as justified ascriptions, they play a role in human-human social interactions. For example, if I give a talk I apply many default assumptions such as thinking that most people in the audience do understand the language, are paying attention to what I say and so on – I will do so even in a potential video-conference setting in which I am, due to technical problems, not able to see my audience. Assuming that my audience consists of social agents puts me in a situation in which 'as if' ascriptions not only have an explanatory role but are also able to facilitate social interactions.

However, before we can adequately consider the role of 'as if' attributions in the study of human-computer interactions, we need to focus on the circumstances under which we are justified in attributing socio-cognitive abilities to artificial agents that they actually possess. Once we have agreed on the extent to which artificial agents possess socio-cognitive abilities and can be considered social agents, future research can analyze the role of an 'as if' mode in human-computer interactions. Unless we have a conceptual framework that determines what socio-cognitive abilities can be found in artificial agents, we cannot distinguish the two functional roles of an 'as if' mode assignment. If we don't know whether the 'as if' mode is applied to social interactions that are based on socio-cognitive abilities, the 'as if' ascription may just have the functional role of making sense of the observed behavior.

3 Towards Joint Actions

According to Bratman (2014), a joint action of human adults can be characterized as the idea of having a shared intention. It is most essential for being able to act jointly to have shared intentions, which are described as an interpersonal structure of related intentions that serves to coordinate action and structure bargaining between participants. Proposed conditions for having shared intentions are demanding. They are characterized by a specific belief state, a relation of interdependence and mutual responsiveness and presuppose common knowledge; all these conditions enable coordination and explicit relations of commitment.

In sum, there is a lot that is thought as necessary for joint actions. Neither children nor non-human animals fulfill such demanding conditions. Consequently, such a demanding notion cannot capture the abilities of other types of agents. But it is obvious that questioning the ability of children to act jointly conflicts not only with our common sense but also with empirical data. Moreover, recent research indicates that non-human animals also successfully engage in joint actions (Warneken et al. 2006).

In philosophy, there is a widespread debate about the proposed necessity of conditions for joint actions. For instance, Blomberg (2015) questioned whether the common knowledge condition should actually be taken as a necessary condition. He shows that various arguments in favor of the common knowledge condition fail and claims that participants can successfully engage in a joint action without explicitly knowing that they have a shared intention. It is sufficient that each of the participants intend by way of the other's intention and by way of meshing subplans.

Contributing to the debate about the necessity of conditions, I have argued elsewhere that not all participants in a joint action have to fulfill the same conditions (Strasser 2015). By referring to joint actions of mixed groups with distinct participants, such as mother and child or human being and artificial agent, I showed that it is possible that one of the participants fulfills less demanding conditions. Likewise, there are attempts in developmental psychology considering less demanding cases of joint actions (Vesper et al. 2010). But none of these attempts specifies the necessary minimal conditions that would enable an artificial agent to act together with a human agent to reach a common goal. Nevertheless, these debates indicate that the demanding conditions of the standard notion are not irreversible.

To develop an expanded notion of joint action with respect to artificial agents, this paper focuses on two major abilities, namely "the ability to act" and "the ability to coordinate". Consequently, I claim that if an artificial agent is able to meet the conditions for agency and coordination elaborated below, this agent qualifies as a social agent in a joint action.

3.1 Ability to Act

It is a necessary prerequisite for being able to act jointly to be able to act. But, once again, standard philosophical notions of action are tailored to human adults only. According to Davidson (1980), highly demanding conditions, such as consciousness, the ability to generate goals, the ability to make free choices, propositional attitudes, mastery of language and intentionality are required.

In sum, it is common sense to claim that actions are 'intentional under some description'. Due to a demanding understanding of intentionality that comes with a variety of further implications, non-living beings, animals and even infants seem to be doomed to remain mere tools, only capable of producing behavior. Only sophisticated intentional human beings equipped with consciousness are considered able to entertain free choice, generate goals and engage in proper actions. In opposition to this standard notion of action I argue for a finer-grained differentiation of classes of events in order to avoid a restriction of agency to living beings (adult humans) from the outset.

Every action is an event, but not all events are actions: the categories of events divide into two sub-categories, namely natural events such as thunderstorms and events that are labelled as behavior, while the sub-category of behavior, in turn, is divided into behavior and action. Guided by a finer-grained differentiation of potential categories, I propose to introduce a further sub-category between behavioral events and the ones described by contemporary notions of action. This category can be labelled as minimal action and can account for agency in the gray area between pure behavior and complex actions.

In line with the idea of minimal approaches, I suggest expanding the notion of action by questioning the necessity of some proposed conditions. For example, consciousness and intentionality in a strong sense might rather be biological constraints and should not be taken as necessary conditions for minimal actions. Framing a new notion, one can distinguish different types of actions, namely minimal and complex actions (Strasser 2015, 2005).

To capture events that cannot adequately be described as mere behavior but don't fulfill the demanding requirements for full-blown action, I suggest that conditions can be interpreted in a weaker sense. For example, unlike the standard notion, a minimal notion of action does not require that the generation of the goal necessarily occur in the acting system. Alternatively, a goal can be generated in another system and can then be transferred to the acting system. For a minimal acting system it is sufficient to be able to recognize and represent goals as goals. Along the same lines, I suggest questioning whether being conscious is a necessary condition with respect to the required information processes of minimal agency as such. Assuming that consciousness is rather a specific property of living beings (a biological constraint), a minimal notion of action does not require consciousness as a necessary condition. As long as an artificial agent is able to perceive, represent and process the relevant information and has effectors to perform an action, this agent seems well-equipped to act in a minimal sense. Admittedly, this askes for a more detailed characterization of what counts as relevant information than I am able to present in this paper. For the sake of argument, let's assume that it is possible to develop a minimal notion of action with which we can draw a new line between action and mere behavior – or more precisely agents and tools.

3.2 Coordination

The ability to act in a minimal sense is not yet sufficient to act jointly – to be a social agent in social interactions. I claim that the core of the social dimension of joint actions lies in the ability to coordinate. In line with the standard notion of joint action, I agree that a functional role of shared intentions is the ability to coordinate. Only if agents work together in an organized way can we talk of a joint action. Otherwise such interactions could only be described as accidently parallel actions or tool use and would not qualify as joint actions.

Furthermore, I claim that having social competence is the essential condition to make successful coordination in joint actions possible. To analyze the required social competence, I focus on three important aspects. First, I analyze in what sense an understanding of the other agents can be realized by mindreading abilities. Second, I examine the ability to process and interpret social cues. And finally, I investigate the ability of how agents rely on the willingness of the other agent to play her part.

For sure, there are more aspects of social competence – but with respect to minimal joint actions I claim that these three aspects are sufficient to explain success in coordination. I argue that if artificial agents possess the social competence to coordinate their actions with the actions of their human counterparts, they meet an important condition for minimal joint actions.

Mindreading. In many situations agents are able to anticipate what another agent will do next. This is especially useful if they intend to act jointly with this agent. In humanities and natural sciences, one aspect of this social competence is discussed under the label 'mindreading' or 'Theory of Mind' (Fodor 1992). Again, many conceptions of this socio-cognitive ability are tailored to humans only. But Butterfill and Apperly (2013) introduced the notion of minimal mindreading, which can account for a broader range of mindreading. By questioning the necessity of overly demanding cognitive resources, such as the ability to represent a full range of complex mental states and a mastery of language, this notion can account for automatic mindreading abilities in human adults, infants and non-human animals. Minimal mindreading explains success in mindreading tasks that requires less demanding conditions. Representations of less complex mental states, namely encounterings and registrations, are sufficient to anticipate the behavior of other agents in an efficient, automatic, fast and robust manner. Most significantly, with regard to artificial agents, it does not require conscious reasoning.

Referring to a recent paper by Gray and Breazeal (2014), which demonstrates how artificial agents model mental states of human beings with respect to the perspective a human counterpart has, I argue that minimal mindreading can also characterize an ability of artificial agents. Of course, this is only valid in a limited range of situations. But it shows that artificial agents, in principle, are able to infer from their perception of the physical world whether the human counterpart can see or cannot see an object and infer that future actions of the human will be guided by this perspective. Therefore, there are cases that justify an ascription of minimal mindreading abilities.

Reciprocity. In addition to minimal mindreading, which mainly covers perspective circumstances, I claim that human-computer interactions qualifying as joint actions require a reciprocal exchange of social information. In the human case, we observe an exchange of a wealth of social information. The reciprocal exchange of social cues such as prosody, gestures, and mimic contribute to successful coordination in joint actions. Disturbed exchanges lead to deficits in social interactions (Bogart and Tickle-Degnen 2015). Consequently, I claim that the ability to interpret and process social cues is an important condition for providing an understanding of the other agents in order to coordinate with them. To characterize this special relation social agents can have, I introduce the notion of social reciprocity, which enables a reciprocal exchange of social information. Social reciprocity is a special feature of social interactions, whereas in non-social interactions such as riding a bicycle there is no social reciprocity. The behavior of the bicycle is shaped by its physical properties only. A bicycle cannot express mental or emotional states; therefore, the bicycle cannot join a reciprocal exchange of social cues. With respect to sociality, all there is, in the bicycle case, is a one-way relation between an agent and an object, which constitutes a prototypical case of tool use. Whereas in social interactions, other social agents are involved. A special feature of social agents is that they are able to process and interpret social cues, which in turn deliver information about their mental and emotional states. This is how a reciprocal exchange of social information becomes possible.

The question now is: What does this mean for artificial agents? To develop artificial agents from mere tools into human-like partners, artificial agents should be able to handle social cues. This does not mean that artificial agents need emotional and mental states – it is sufficient if they can express and interpret social cues. Thereby we once again reach the point where the necessity of exclusively biological constraints is questioned. Instead of requiring emotional or mental states, one could implement functions that in the human case are realized by emotional or mental states.

Since every speech act is an action, and, consequently, a dialog is a joint action, communication can serve as a prototypical example to investigate the ability of how processing social cues can make 'minds' visible. A lot of research in this area is now focusing on social cues such as gestures (Kang et al. 2012) and emotional expression (Petta et al. 2011; Becker and Wachsmuth 2006). For example, ARIAs (Artificial Retrieval of Information Assistants) are able to handle multimodal social interactions (Baur et al. 2015). They can maintain a conversation with a human agent and, indeed, react adequately to verbal and nonverbal behavior.

To decide whether a specific artificial system possesses the social competence to coordinate, one has to explore on a case-to-case basis which kinds of social cues are relevant for this specific class of joint actions.

Commitment. It is crucial for the success of human-human joint actions that both agents stick to the action, to reach the goal. Commitments provide an important motivational factor: Only if we can rely on the contribution of the other agent are we motivated to stick to a joint action. To explore in what sense this is important with respect to human-computer joint actions, the minimal approach by Michael et al. (2016) is a good starting point. Again, standard notions of a (strict) commitment are tailored to human adults only (Shpall 2014). Strict commitments are characterized as a bidirectional relation between two active agents mutually having certain expectations and motivations with respect to a specific action. In contrast to this standard notion, Michael and colleagues claim that components of a strict commitment can be dissociated and suggest that the single occurrence of one component can be treated as a sufficient condition for a minimal sense of commitment.

With respect to human-computer joint actions, there are four possible cases of a minimal sense of commitment. The first two cases describe a minimal sense of commitment of the human counterpart, either having an expectation or a motivation based on a minimal sense of commitment. If we now take an 'as if' mode into account, it cannot be questioned that humans can, for the sake of a joint action, expect that an artificial agent be committed or feel committed toward an artificial agent.

The challenge is to elaborate under which circumstances (if any) we are justified in saying that artificial agents entertain a minimal sense of commitment, namely that they entertain a functional corresponding state of 'expecting' the human to be committed or of 'feeling' committed to a human being. According to the strategy of questioning whether certain biological constraints are necessary, this is again a case where I worry that requiring states such as 'expecting' and 'feeling' excludes artificial agents from the start. Therefore, the proposed minimal conception only requires that artificial agents take the functional role to successfully motivate the human counterpart by signaling a minimal sense of commitment.

4 Conclusion

If artificial agents continue to be increasingly prevalent in human social life, and interactions with them are experienced as genuinely social and not as mere tool use, then there is a need for a strategy for overcoming our restricted conceptions of socio-cognitive abilities in philosophy.

I claim that if artificial systems are able to act in a minimal sense and additionally provide the social competence to coordinate with other social agents, those systems can qualify as social agents in joint actions, because they actually possess socio-cognitive abilities.

With respect to coordination in joint actions, I elaborated on three important aspects of the required social competence, namely the understanding of other agents by minimal mindreading, reciprocal exchanges of social cues and the ability to contribute to a minimal sense of commitment. The elaborated conditions provide a guideline to discuss whether certain human-computer interactions qualify as minimal joint actions and help to distinguish whether some 'as if' ascriptions contribute to a social interaction instead of just delivering an explanation of the behavior. Additionally, the elaboration of these conditions may contribute to the development of benchmarks for roboticists' research.

References

Baur, T., Mehlmann, G., Damian, I., Gebhard, P., Lingenfelser, F., Wagner, J., Lugrin, B., André, E.: Context-aware automated analysis and annotation of social human-agent interactions. ACM Trans. Interact. Intell. Syst. **5**, 2 (2015)

Becker, C., Wachsmuth, I.: Modeling primary and secondary emotions for a believable communication agent. In: Reichardt, D., Levi, P., Meyer, J. (eds.) Proceedings of the 1st Workshop on Emotion and Computing in conjunction with the 29th Annual German Conference on Artificial Intelligence (KI2006), Bremen, pp. 31–34 (2006)

Blomberg, O.: Shared goals and development. Philos. Q. **65**(258), 94–101 (2015)

Bogart, K.R., Tickle-Degnen, L.: Looking beyond the face: a training to improve perceivers' impressions of people with facial paralysis. Patient Educ. Couns. **98**, 251–256 (2015)

Bratman, M.: Shared Agency: A Planning Theory of Acting Together. Oxford University Press, Oxford (2014)

Buckner, C., Fridland, E.: What is cognition? Angsty monism, permissive pluralism(s), and the future of cognitive science. Synthese **194**(11), 4191–4195 (2017)

Butterfill, S., Apperly, I.: How to construct a minimal theory of mind. Mind Lang. **28**(5), 606–637 (2013)

Clark, A., Chalmers, D.: The extended mind. Analysis **58**(1), 7–19 (1998)

Davidson, D.: Essays on actions and events. Oxford University Press, Oxford (1980)

Dennett, D.: The Intentional Stance. The MIT Press, Cambridge (1987)

Fodor, J.: A theory of the child's theory of mind. Cognition **44**(3), 283–296 (1992)

Gray, J., Breazeal, C.: Manipulating mental states through physical action – a self-as-simulator approach to choosing physical actions based on mental state outcomes. Int. J. Social Robot. **6**(3), 315–327 (2014)

Heider, F., Simmel, M.: An experimental study of apparent behavior. Am. J. Psychol. **57**, 243–259 (1944)

Kang, S., Gratch, J., Sidner, C., Artstein, R., Huang, L., Morency, L.P.: Towards building a virtual counselor: modeling nonverbal behavior during intimate self-disclosure. In: Eleventh International Conference on Autonomous Agents and Multiagent Systems, Valencia, Spain (2012)

Michael, J., Sebanz, N., Knoblich, G.: The sense of commitment: a minimal approach. Front. Psychol. **6**, 1968 (2016)

Petta, P., Pelachaud, C., Cowie, R. (eds.): Emotion-Oriented Systems: The Humaine Handbook. Springer, Heidelberg (2011)

Shpall, S.: Moral and rational commitment. Philos. Phenomenological Res. **88**(1), 146–172 (2014)

Strasser, A.: Kognition künstlicher Systeme. Ontos-Verlag, Frankfurt (2005)

Strasser, A.: Can artificial systems be part of a collective action? In: Misselhorn, C. (ed.) Collective Agency and Cooperation in Natural and Artificial Systems. Explanation, Implementation and Simulation. Philosophical Studies Series, vol. 122. Springer (2015)

Vesper, C., Butterfill, S., Sebanz, N., Knoblich, G.: A minimal architecture for joint action. Neural Netw. **23**(8/9), 998–1003 (2010)

Warneken, F., Chen, F., Tomasello, M.: Cooperative activities in young children and chimpanzees. Child Dev. **77**(3), 640–663 (2006)

Computation - Intelligence - Machine Learning

Mapping Intelligence: Requirements and Possibilities

Sankalp Bhatnagar[1,2], Anna Alexandrova[1,4], Shahar Avin[3], Stephen Cave[1], Lucy Cheke[1,11], Matthew Crosby[1,7], Jan Feyereisl[8,9], Marta Halina[1,4], Bao Sheng Loe[5], Seán Ó hÉigeartaigh[1,3], Fernando Martínez-Plumed[6], Huw Price[1,3], Henry Shevlin[1], Adrian Weller[1,12], Alan Winfield[1,10], and José Hernández-Orallo[1,6(✉)]

[1] Leverhulme Centre for the Future of Intelligence, Cambridge, UK
[2] The New School, New York, USA
[3] Centre for the study of Existential Risk, University of Cambridge, Cambridge, UK
[4] Department of History and Philosophy of Science, University of Cambridge, Cambridge, UK
[5] Psychometrics Centre, University of Cambridge, Cambridge, UK
[6] Universitat Politècnica de València, València, Spain
jorallo@dsic.upv.es
[7] Imperial College, London, UK
[8] AI Roadmap Institute, Prague, Czech Republic
[9] GoodAI, Prague, Czech Republic
[10] Bristol Robotics Laboratory, UWE Bristol, Bristol, UK
[11] Department of Psychology, University of Cambridge, Cambridge, UK
[12] Alan Turing Institute, London, UK

Abstract. New types of artificial intelligence (AI), from cognitive assistants to social robots, are challenging meaningful comparison with other kinds of intelligence. How can such intelligent systems be catalogued, evaluated, and contrasted, with representations and projections that offer meaningful insights? To catalyse the research in AI and the future of cognition, we present the motivation, requirements and possibilities for an atlas of intelligence: an integrated framework and collaborative open repository for collecting and exhibiting information of all kinds of intelligence, including humans, non-human animals, AI systems, hybrids and collectives thereof. After presenting this initiative, we review related efforts and present the requirements of such a framework. We survey existing visualisations and representations, and discuss which criteria of inclusion should be used to configure an atlas of intelligence.

1 Introduction

Despite significant AI progress, its pace and direction are largely unassessed and hard to extrapolate. The main reason for this is that we lack the tools to properly evaluate, compare and classify AI systems, and thus determine the future of the field. The comparison of AI systems with human and non-human intelligence is

V. C. Müller (Ed.): PT-AI 2017, SAPERE 44, pp. 117–135, 2018.
https://doi.org/10.1007/978-3-319-96448-5_13

typically performed in an informal and subjective way, often leading to contradicting assessments, especially in hindsight (Hayles 1996; Brooks 1997; Pfeifer 2001; Shah et al. 2016). The comparison between non-human animals and AI ranges from setting the goal of designing artificial agents with the behaviour of "an earwig" (Kirsh 1991) to the "intelligence of a rat" (Cadman 2014; Shead 2017), without further specification of what these capabilities or dimensions for comparison should be. The comparison with humans is not much more precise. For instance, two decades ago it was cognitive functions related to perception and action that seemed unattainable – "the gardeners, receptionists, and cooks are secure in the decades to come" said Steven Pinker in 1994. Now, these are the functions that look easier to be automated (Frey and Osborne 2017) – "if a typical person can do a mental task with less than one second of thought, we can probably automate it using AI either now or in the near future" (Ng 2016). Today, it is higher-level cognition (causal reasoning, compositionality, theory of mind, meta-cognition, etc.) that seems more out of reach (Marcus 2018).

The assessment is especially difficult as academia and industry in AI are rushing to achieve breakthroughs for specific problems, which often require massive data, computation power, embedded heuristics, strong bias, etc., undermining generality, autonomy and efficiency. For instance, AI can now play most video games (Hessel et al. 2017) and board games (Silver et al. 2017) better than humans, but the immediate training data and computational power that are needed are – as for today – orders of magnitude higher than those used by a human. As a result, it is difficult for policy makers to assess what AI systems will be able to do in the near future, and how the field may get there. There is no common framework to determine which kinds of AI systems are even desirable.

This contrasts with empirical science, where measurements, comparisons, representations and taxonomies are widespread. These characterisations can be theory-driven, such that a prior conceptual framework is used to categorise system features, or can be data-driven, which is increasingly important in many scientific disciplines (Marx 2013; Landhuis 2017; Einav and Levin 2014). Conceptual progress partly relies on finding and testing hypotheses through the computational analysis of large amounts of shared data (Gewin 2002), using open data science tools (Lowndes et al. 2017). In AI, we would like to analyse the state and progress of artificial systems based on data-grounded investigations. Research priorities and safety concerns depend on this analysis. We need to assess whether new AI systems and techniques are simply an incremental improvement for a narrow collection of applications or a real breakthrough representing a more general cognitive ability, which can be established in relation to comparable abilities in humans and other animals.

This wider view of AI, in the context of all kinds of intelligence, dates back to Sloman's "space of possible minds" (Sloman 1984). Figure 1 compares (a) a figurative plot (Shanahan 2016), covering a wide range of systems (also see (Yampolskiy 2014, Fig. 3b), (Arsiwalla et al. 2017, Fig. 3c), and (Solé 2017, Fig. 3d)), with (b) a plot depicting precise experimental results for several ape species on a battery of tests (Herrmann et al. 2007). The figure illustrates a

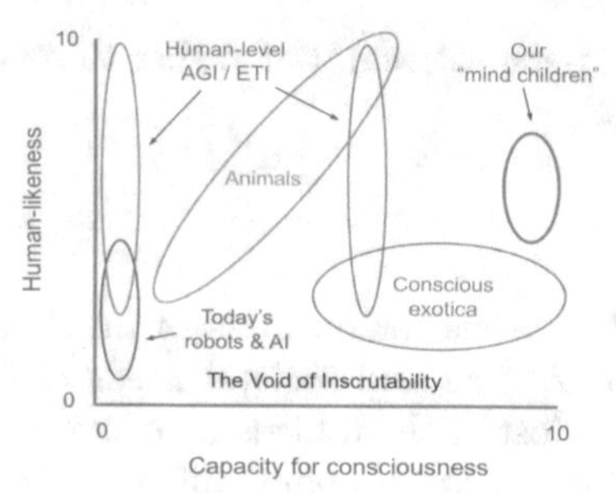

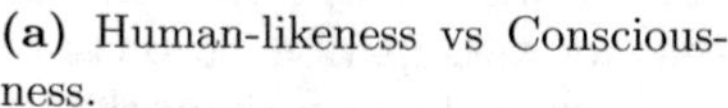
(a) Human-likeness vs Consciousness.

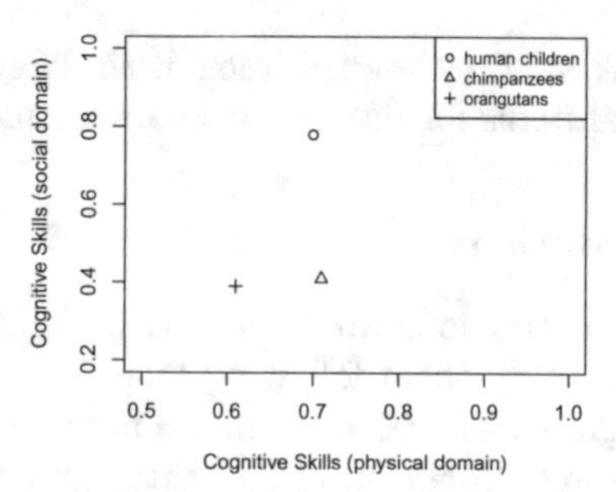

(b) Social vs physical dimensions for three ape species.

Fig. 1. Different kinds of minds represented according to several dimensions. Left: Figurative "human-likeness" vs consciousness (from Shanahan 2016). Right: Two dimensions of cognitive skills (social vs physical domain) according to the results of a test battery on three different groups of apes (adapted from Herrmann et al. 2007).

visible trade-off between completeness and empirical grounding. What we need is to leverage the best of both worlds: a data-based representation of very different cognitive systems, including humans, non-human animals, AI systems, hybrids and collectives, where actual measurements can be aggregated and combined.

This requires a novel platform, an 'atlas of intelligence', that integrates an extensive inventory of cognitive systems, a behavioural test catalogue (with test batteries that could be aggregated into dimensions) and an experimentation repository (results from measurements). The platform would be populated collectively, facilitating cross-comparison and reproducibility (Aarts et al. 2015; Vanschoren et al. 2015). The atlas would represent a new cartographical endeavour for a better understanding of the geography of the space of intelligence.

This paper explores the motivations, the requirements and the possibilities of such an atlas. Section 2 explores in more depth why the atlas is needed in terms of four lists of items specifying the motivations, applications, dimension manipulations and entities to be covered in the atlas. Section 3 focuses on the idea of an atlas as a set of maps, and configures a partial specification in terms of the maps we would like it to have. This section includes a collection of maps and graphical representations, some of them already proposed in the literature (but most without real data) and some desired representations. We close the paper with a discussion about future work. Finally, Appendix A gives a short overview of similar initiatives in other areas, and how these relate to the atlas.

2 Motivations, Applications, Dimensions and Kinds

This section presents a series of lists of items covering the motivations and applications of the atlas (why, and for what, an atlas is needed), and the potential dimensions and kinds of systems to be included (what the contents should be). The lists are not meant to be exhaustive and free from overlaps (some ideas are represented by several items with different perspectives), but rather to serve as initial items for discussion and refinement.

Motivations

The motivations are meant to highlight the needs for an atlas of intelligence. We identify them following scientific, technological and societal needs that are recognised at present or in the near future. Most of them focus on better understanding, representing and cataloguing what we know about different kinds of intelligence. Still, we do not exclude the needs for anticipation, so we also cover those motivations that are related to having better predictions about the existing and future changes of human and artificial intelligence.

- **Milestones and Pathways:** Unlike most non-human biological cognition, human cognition is changing: the average IQ in many countries is increasing (the *Flynn effect*), our memory (Sparrow et al. 2011) is changing due to the *Google effect* (digital amnesia), navigation abilities (McKinlay 2016; Milner 2016) atrophied because satnavs, cognitive rewards mechanisms are changing because of gamification, etc. This is a process that is accelerated by technology, and will be magnified by the use of cognitive assistants and cognitive prosthetics, especially for the elderly. AI itself and human-machine hybrids (either as individual cyborgs or as mixed collectives) are progressing in directions that we are not able to compare with the past or extrapolate, in order to understand where all this is leading, and the associated opportunities and risks (research priorities and safety concerns).
- **Laypeople Understanding:** In those cases where comparisons can be made by looking at a set of traits, it is usually too complex for non-experts to understand what the key differences are between two cognitive systems, especially when one is natural and the other is artificial. Visual representations are appropriate, as humans are good at understanding geographical analogies (e.g., the 1948 book "the map that came to life" helped children understand the countryside where a trajectory and a story were accompanied by maps).
- **Crossover Measuring:** Data-driven comparison is usually based on measurement instruments, reporting a series of measured values that can be represented. But we do not have many test batteries that can be applied across species or even AI systems. The generalisation of representations where different natural species and AI technologies are put together would encourage the adoption and definition of more universal tests having better measurement invariance across different entities.
- **Behavioral Taxonomies:** If we go beyond life in our comparisons, especially if they are based on similarity, the dominant genotypic approach cannot be used broadly. Taxonomies and models must be mostly informed by behavioural analyses, in contrast to phenotypic, ethological, genotypic or neurological approaches (Cattell and Coulter 1966; Miller, 1967). But we contemplate cladistic principles (using hierarchies or dendrograms) and we consider morphological or functional similarities as far as they affect behaviour.

- **Testing New Intelligence:** The progress in AI suggests that task-oriented evaluation (i.e., the performance of an AI system for a particular task) may be insufficient. Other ways of characterising and measuring AI are needed. While some capabilities and tests can be inherited or extended from psychometrics or animal cognition, there may be some other capabilities or skills that are completely new, especially when we analyse the cognitive profile of human-machine hybrids or collectives.
- **Critical Perspective:** There is an urgent need for better understanding the way the intelligence landscape is changing, for both humans and AI systems, in areas such as automation, education and ethics. It is hard to regulate or incentivise some actions not knowing how they affect the intelligence landscape.
- **Beyond Anthropocentrism:** While it is generally accepted that intelligence is the product of evolution, it is still hard to recognise intelligence in other species or in AI systems, and compare it without using humans as a yardstick.
- **Grand Goals:** While interdisciplinarity in the study of intelligence has increased, there are still many attributes and behaviours that are not properly mapped between disciplines, and there is no wide recognition of a shared space. The geographical analogy of an intelligence landscape as an opportunity for exploration and discovery can help inspire the next generation of researchers in areas such as comparative cognition, psychology, philosophy and artificial intelligence, and, most especially, in multidisciplinary domains.
- **Replicability and Reuse:** New research procedures and visualisations for the analysis of cognitive systems are difficult to apply to other systems or in other contexts. This limitation is more blatant when we see similar ideas, representations or experimental protocols appear in different disciplines.
- **Data-driven and Hypothesis-driven:** When cognition is analysed in one species or a particular AI technology, there is a lack of a sufficiently wide sample to infer and reject hypotheses. The recent trend of a more collaborative data science approach should encourage initiatives where data from different disciplines can be put together to test hypotheses about cognition.

Some of the motivations above have deep roots in cognitive science, comparative psychology, philosophy and AI (Macphail 1987; Thagard 2009; Gentner 2010), but others are more specific to some particular areas or emphasise the need of better representations and comparisons.

Applications

Moving from what is needed to the things an atlas will make possible leads us to the identification of new possibilities and transformations. The criteria for inclusion are such that the list covers potential applications for scientists, philosophers, educators, policy-makers and the general public, directly using the platform or as an indirect result of its use:

- **(Re-)Education:** Traditionally, children and adults used animals as models of different personalities and capabilities, interacting with them regularly.

Today, in urban societies with less contact with animals, it is becoming easier to portray and transmit some concepts using robots as models, as cinema and advertising (especially when targeting children) have already understood. An atlas covering animals and robots could be used in museums, schools and universities as a way of articulating over this intelligence landscape.

- **Effective Navigation:** An atlas, with different representations, would help us locate where we are (humans and AI), the trajectories taken in the past years and the destinations we are heading to, helping to visualise whether some targets or trajectories can take us to dangerous areas.
- **Ethical Assessments:** Visual representations make some ethical dilemmas about moral agency and patiency more explicit, as we can see whether the way we look at animals and artificial agents is different from the way measured traits put them on some representations. This will make some ethical issues more conspicuous (animal, robot or human suffering, uncanny valleys, etc.).
- **Consequences:** Not only the locations but also the distributions and densities would help us analyse (especially in advance) the population of creatures affected by research, law, environment, technology, etc., in a critical way. In other words, the maps could also be used to represent the areas and entities (and how many) would be affected by a phenomenon.
- **De/Re-Centre Humans:** Humans, as a species, groups and individuals could be located at different locations depending on the representation, making more explicit that there is a Copernican revolution in the way intelligence is seen today, sustained in the progress of comparative cognition, evolutionary psychology and, increasingly, artificial intelligence.
- **Metaphors and Narratives:** An atlas would build upon the perception we have about animal behaviour. This would help us better understand and locate where we are in AI, in a more meaningful way than just saying "AI is at the level of the rat". Instead we would like to align the cognitive profile of a rat with the cognitive profile of a particular AI system, and see the differences in a less monolithic way.
- **Archival Exploration:** An atlas of intelligence would also help to see a "history of intelligence", where we would go from extinct animals and past computer/AI systems to the present day, seeing the directions their evolution has taken according to different dimensions.
- **Morgan's Canon:** C. Lloyd Morgan stated: "In no case is an animal activity to be interpreted in terms of higher psychological processes if it can be fairly interpreted in terms of processes which stand lower in the scale of psychological evolution and development" (Morgan 1903). An atlas would help interpret, extend or overhaul this canon for artificial systems, hybrids or collectives.
- **Unification:** An atlas would require and hence would encourage the definition of more general tests and metrics, embracing natural and artificial systems, and would aim at more unified theories of cognition, going beyond human psychology and evolution to consider every possible cognitive system, especially looking at those places in the maps where there are gaps, whether it is possible to have entities there and how they would be interpreted.

Some of the applications clearly derive from the motivations (e.g., beyond anthropocentrism and de/re-centre humans) but others represent possibilities that perhaps were not even recognised as a necessity, such as the use of the atlas for archival exploration, which may lead to unforeseen purposes. Concerning the needs and possibilities introduced above, represented by the motivations and applications, we add the *dimensions* and the *kinds of systems* we want to cover, which specify the atlas in general terms.

Dimensions

We are aware of the lack of consensus about the most relevant attributes for the analysis of cognition – not to mention general theories. Because of this disagreement, we want the atlas to be able to integrate different perspectives and attributes of the interest. Consequently, rather than enumerating the specific dimensions of representation that could be used, which could ultimately be created and refined by the users, we clarify how these dimensions operate in general terms.

- **Observation-Based:** the dimensions of representation should be agnostic to particular hypotheses, so that the users could do their theories from the values observed. Of course, there are always some underlying assumptions (and the influence of underlying theories) whenever an observation or measurement is made, but this should be as explicit as possible.
- **Multiple Interface:** the atlas should allow users to project or aggregate the data and derive some maps and other representations from these transformations, as usual in other visualisation frameworks.
- **Interactive Querying:** the atlas could be interrogated through queries, including filters and joins across different data sources, in an interactive way, as in tools of analytical processing.
- **Creative and Constructive:** the atlas should allow users to combine elements, creating new features (and hence new spaces) and creating new entities, such as populations or individuals, combining their cognitive entities under some specified models.
- **Populational/Theoretical:** the elements to represent could correspond to actual populations or subgroups but also to theoretical elements and groups.
- **Bottom-Up/Top-Down:** the dimensions could correspond to basic psychological mechanisms or to more abstract, integrated skills. The atlas should allow users to aggregate and disaggregate these dimensions.
- **Transversal Connections:** the atlas would allow users to combine behavioural traits (skills, functions, capabilities) with non-behavioural traits (physical traits, computational effort, evolutionary traits, etc.).
- **Topographical/Geographical Visualisation:** the atlas should combine as many elements of visualisation and representation (colours, contours, textures) as may be found useful to show the information in insightful ways.

Despite the intended flexibility, some of these dimensional operations give a more precise account at the specification level for the atlas on how data, hypotheses and visualisations must be connected. For instance, the multiple interface, the interactive querying and the topographical and geographical representations very much resemble some common retrieval and representational systems powered by data visualisation tools. On the other hand, the other dimensional characteristics are more aligned with the management of conceptual ontologies and taxonomies.

Kinds of systems

Finally, regarding the kinds of cognitive systems to be represented, we want to cover all possible ranges, according to several criteria: integration, nature, time, distribution and existence. This comprehensive view of cognitive systems would ultimately allow us to put very different types of entities into comparison.

- **General and Narrow:** specific systems aiming at a single task or species in a narrow environment could be covered, as well as those systems that are flexible in a broad range of environments.
- **Individual and Collective:** individual entities could be located as well as collectives (along with their components).
- **Biological and Artificial:** living beings, including plants and animals, and artificial systems, including autonomous agents, robots, corporations, etc.
- **Hybrid (Extended/Enhanced Minds):** humans improved by technology, either internally (enhanced, as cyborgs or through nootropics) or externally (extended by assistants), as well as AI systems using human computation.
- **Novel and Old:** covering current living beings and AI systems, but also extinct species and AI systems of the past.
- **Distributed and Centralised:** systems that are identified by a single body, but also natural and artificial swarms as well as distributed intelligence, including societies.
- **Alien and Fictional:** even for speculation or theorisation, the atlas could also show some imaginary entities.

Apart from the scientific questions needed to build such a platform, its success depends on the engagement of the (research) community and other stakeholders. It is crucial then to identify whether the needs, dimensions and elements represented are well aligned with the potential users and contributors. Consequently, we conducted a preliminary survey to get feedback from researchers and other potential users in many different areas, using the items described in this section. We targeted different communities: artificial intelligence, animal cognition, psychology, philosophy, design and some others. The results of the questionnaire were positive in general. This was not taken as a justification or validation of the categories presented here but, more on the contrary, as a way of recognising omissions, duplications or desiderata nobody is asking for. We focused especially on the open comments from some respondents who were more critical[1].

[1] A detailed analysis of the questionnaire can be found in (Bhatnagar et al. 2017).

We considered the previous motivations, intended applications, dimensions to consider, and entities to cover to be a sufficient reason for starting the construction of an atlas, with the necessary caution about potential pitfalls and the need of selecting pieces of the atlas that could be chosen as more low-hanging fruits of the whole project. The previous lists are preliminary, and the priorities for selecting which categories are most important to start with—e.g., prototypes or first cornerstones of the project – are still subject to debate.

Next, we refine this first conception of the atlas by considering existing representations and maps.

3 Collections of Maps: Representational Possibilities

As an atlas is a set of maps, in this section we collect and recreate some of the maps that have been proposed in the past, most of them at a figurative level, and discuss representations that we would like to include in the future. Figure 1 contained examples of a classical multidimensional representation (although two dimensions are especially fitted for paper and screens). The axes represent dimensions of interest and the points represent the entities (the cognitive systems) we want to compare. We will see many others of these, being different because of the dimensions that are chosen or the elements that are represented. In other cases, the representations detach from this multidimensional view but still remain meaningful in geographical or topological terms.

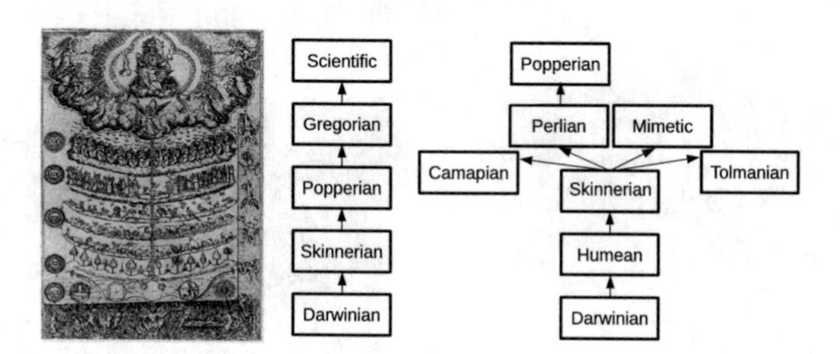

Fig. 2. *Left*: Scala naturae, as depicted in the 16th century (de Valadés 1579). *Middle*: a representation of Dennett's Tower of Generate and Test, which depicts creatures according to when and how they adapt (Dennett 1995), *Right*: Godfrey-Smith's refinement of the bottom part of Dennett's tower (the part corresponding to cognitive evolution) in the form of a tree (Godfrey-Smith 2015, Fig. 2).

We start with the oldest and simplest representations, those inspired in the *scala naturae*, which are monolithic, or at most, arboreal (see Fig. 2), where membership to a species is replaced by other criteria for classification. At some point the categorical representations (monolithic or hierarchical) led to more quantitative and multidimensional representations, as we see in Fig. 3.

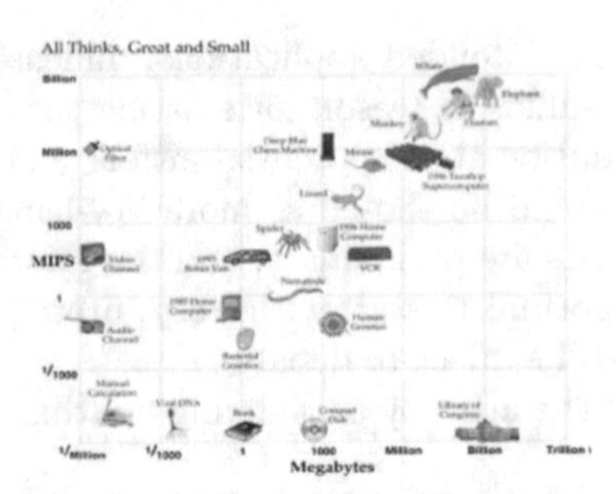

(a) Comparison of hardware between several living and inanimate objects (Moravec, 1998, Fig. 1)

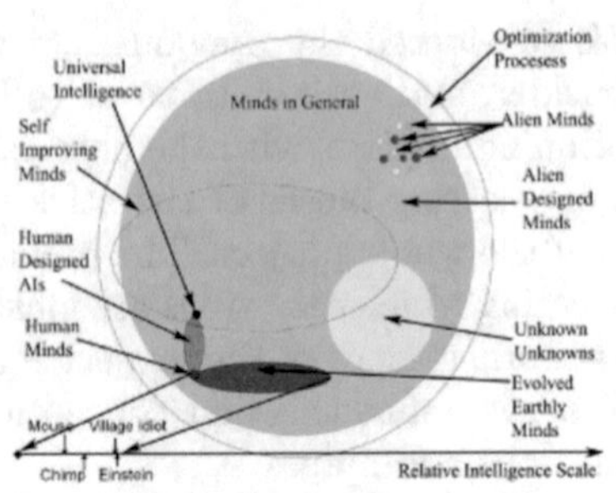

(b) Figurative space of minds (Yampolskiy, 2014, Fig. 1)

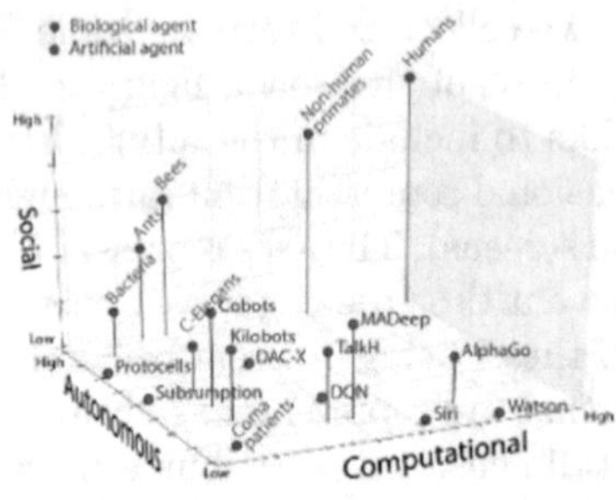

(c) "Morphospace of consciousness" from (Arsiwalla et al., 2017, Fig. 3)

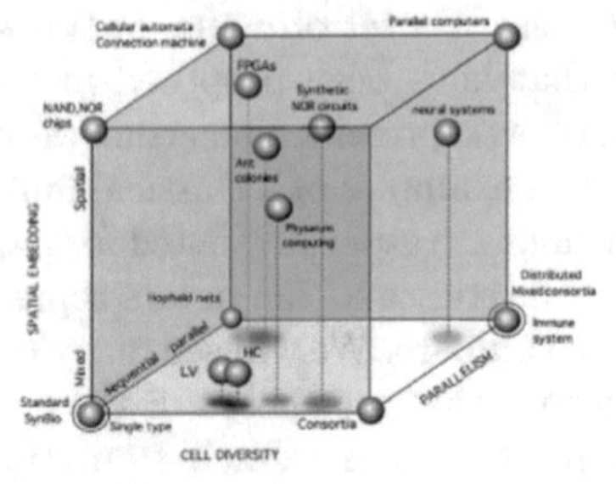

(d) "Biological computational morphospace" (Solé & Macia, 2011, Fig. 7)

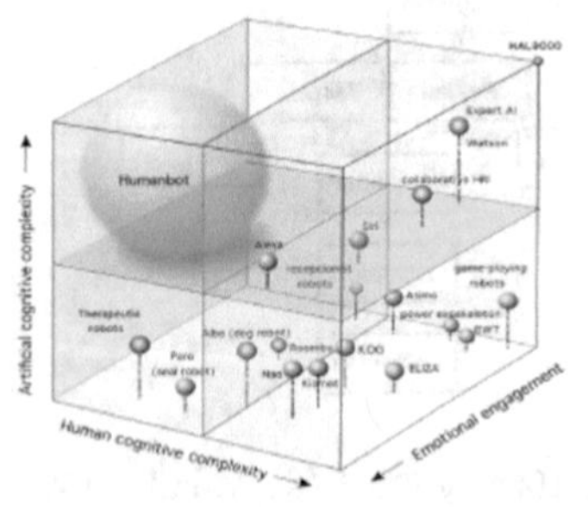

(e) Cognitive space of human-robot interactions (Solé, 2017, Fig. 2)

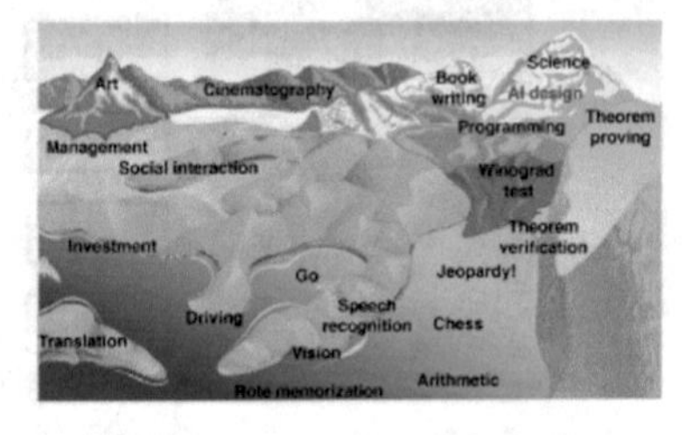

(f) Illustration of Hans Moravec's "landscape of human competence" (Tegmark, 2017, Fig. 2.2)

Fig. 3. A collection of figurative maps of intelligence.

Moravec was not the first one to compare animals and computers according to several dimensions, but some of his plots had an important effect on the narratives about how far AI had come in the 1990 s. For instance, Fig. 3a compares computational power (speed and storage capacity) for a wide range of entities.

Some other representations have tried to compare animals and artificial systems for other dimensions. For instance, computational efficiency can be replaced by an estimation of energy consumption (Winfield 2014), which is a physical property that can be used as a dimension alongside some other more behavioural traits. One common representation is based on Venn diagrams, where the sizes and locations are completely arbitrary, and the only purpose is to show a diversity and inclusions/overlaps between sets, such as Fig. 3b from Yampolskiy (2014). Some other plots are more speculative, especially when the goal is to represent consciousness, such as the one from Arsiwalla et al. (2017) in Fig. 3c.

Other comparisons are at a much more physical (or implementational) level, such as the one from Solé (2017), representing the "morphospace" in terms of "embedding", "diversity" and "parallelism", shown in Fig. 3d, or represent some aspects of human-robot interaction, again figurative (Fig. 3e). An interesting twist is given when the space represents the tasks or abilities (without any clear criterion for proximity), but the Z-dimension (height) is represented by time (or progress in AI). According to this, we have a figurative plot like Tegmark's representation (Fig. 3f) of Moravec's landscape (Tegmark 2017).

A more thoughtful analysis of dimensions may lead to more than three elements, whose representation (if all of them are quantitative) is more cumbersome. Star (cobweb) plots are a practical option here, although they can get too messy if too many individuals are shown. Also, trajectories are more difficult to represent in these plots. Figure 4 shows how four dimensions are used to compare the intelligence of several organisms.

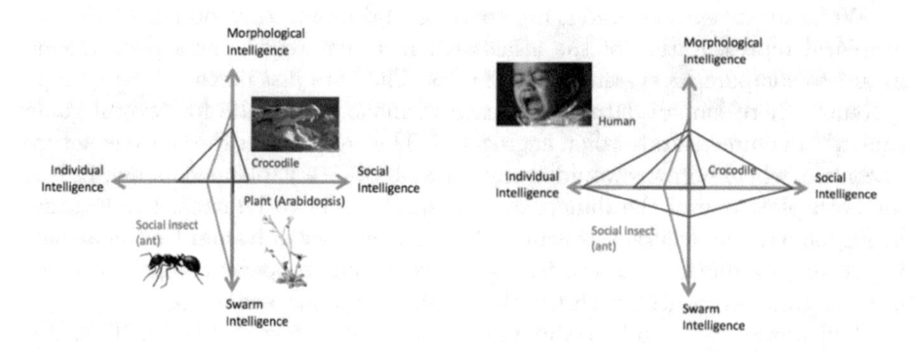

Fig. 4. Comparison of different systems on a space of four dimensions, using star plots (Winfield 2017, Figs. 2 and 3).

Following the comments of some of the respondents of the questionnaire, we are also interested in representations of 'collective intelligence', even if figurative. For instance, Fig. 5 represents a profile of members of a team and tries to derive aggregate values (minima, maxima and means) for the group.

So far, all the previous representations were figurative, in the sense that there was no measured data or observations from which the maps were represented, but

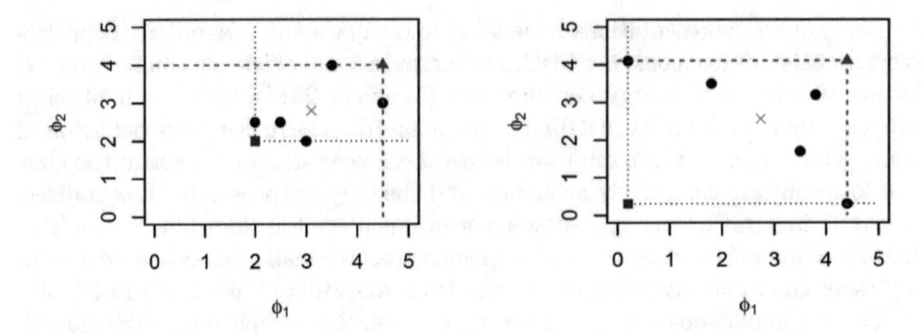

Fig. 5. Collective diversity in terms of psychometric profiles for two figurative groups of five agents, shown with circles. For each plot, the x dimension represents IQ score and the y dimension represents social sensitivity. The mean point is shown with the cross and the maximum and minimum envelopes are represented with a triangle and a square respectively (Hernández-Orallo 2017, Fig. 15.3).

just some general knowledge and intuitions of these magnitudes. In what follows, we include some representations that are using real data. For instance, the easiest way of comparing two systems or species is by comparing their results for the same task, as in Fig. 6a and b. But we can also compare abstract or aggregated traits or skills, as we showed in Fig. 1. A representation that is becoming very common in AI is to show the results normalised by human performance (Figs. 6c and d), even in cases where many tasks are aggregated.

While these representations are common and useful, they do not fit the geographical representation of the atlas well. In other words, these plots are not meant to compare AI systems and humans. They are just meant to compare AI systems, where human data is just used to make the results for several games somewhat commensurate when aggregated. This means that the space is anthropocentric, where humans would always be at 100% – a Ptolemaic model. Indeed, for both plots one of the dimensions does not apply to humans. For instance, in Fig. 6c, human accuracy is achieved with a number of frames that is at most in the small millions, and also in Fig. 6d we cannot properly compare the year humans were introduced with the year a ML technique was introduced.

Trajectories can also be compared over time, as shown in Fig. 6e. Here, time is applied to the same entity, so we see how the entity (a population in this case) changes with time. But a trajectory is better seen when the dimensions of the plot are not time – time is not usually represented in a static map. Instead, one can see how an individual or group moves in a space of dimensions chronologically (learning episodes, cognitive decline or enhancement, etc.), illustrated in Fig. 6f.

Actual data can also be obtained and processed from subjective perception. For instance, Gray et al. (2007) extract two principal components: agency and experience (what we could also refer to as 'patiency') in order to quantify how much mind people ascribe to different kinds of cognitive systems, from robots to dead people, as illustrated in Fig. 7.

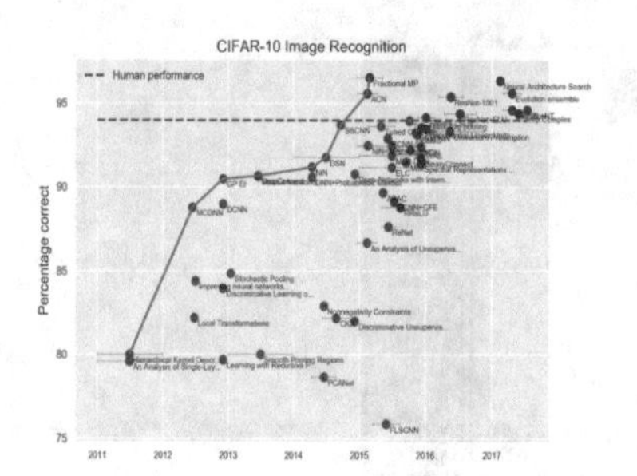

(a) Comparison between average humans and AI on CIFAR (Eckersley et al., 2017)

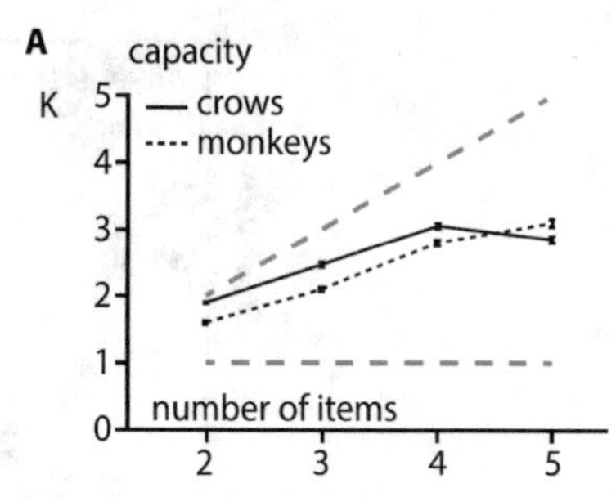

(b) Comparison between crows and monkeys for working memory (Balakhonov & Rose, 2017)

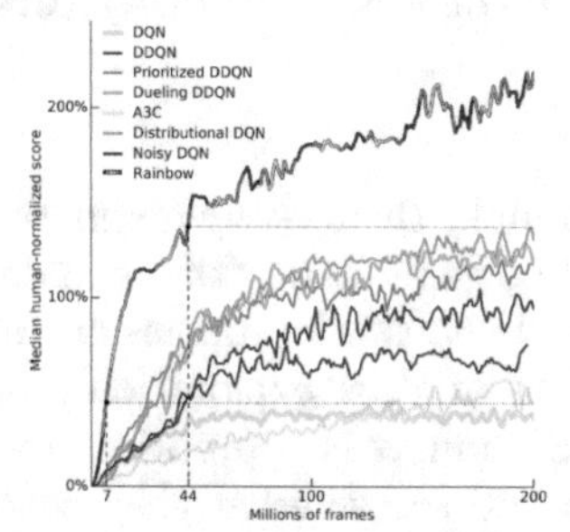

(c) Learning curves by frames seen on the ALE benchmark (Hessel et al., 2017, Fig. 1)

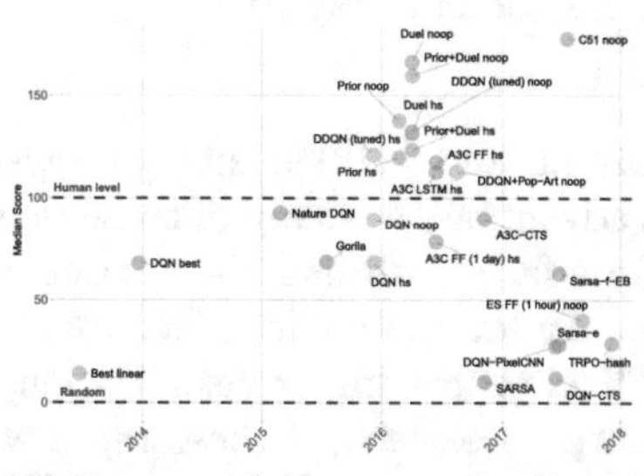

(d) Progress (average performance) by year on ALE benchmark (Eckersley et al., 2017)

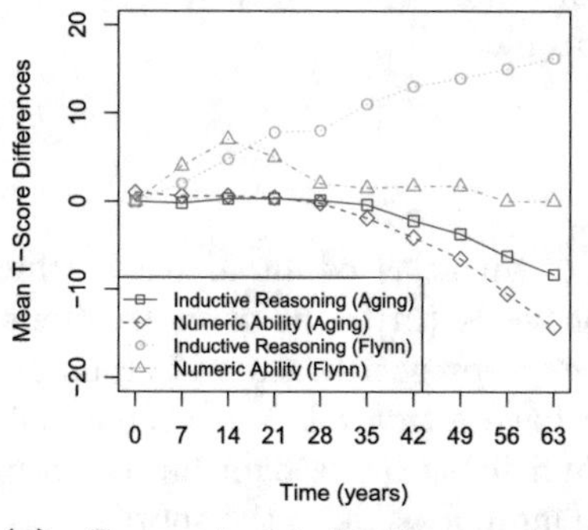

(e) Comparing inductive reasoning and numeric ability over time (Schaie, 1996)

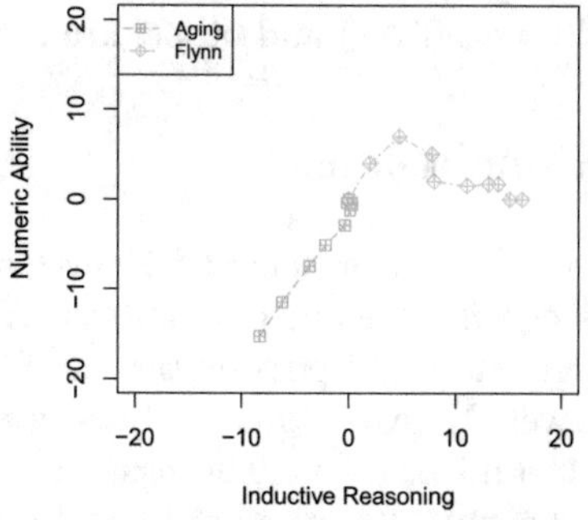

(f) Comparing inductive reasoning and numeric ability directly (Schaie, 1996)

Fig. 6. A collection of existing empirically-grounded maps.

After all these graphical representations, the question is how these can help us configure a set of relevant "maps" we would like the atlas to have. First, we can look at the elements: many are multidimensional and it is just the

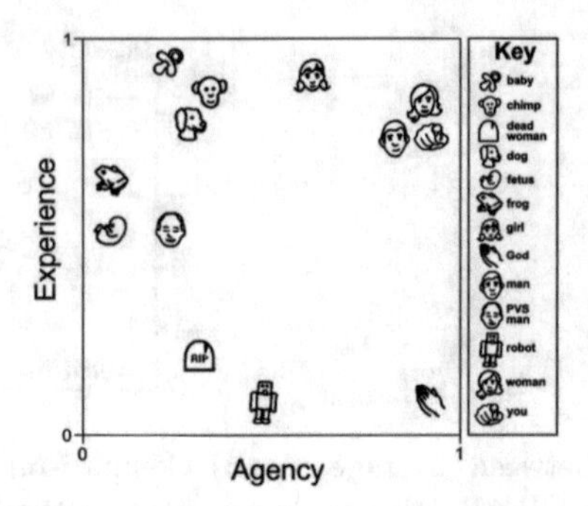

Fig. 7. Different cognitive systems according to the perception by people (from a survey, where the dimensions have been reduced to two dimensions by PCA). (Gray et al. 2007; Wegner and Gray 2017)

dimensions and the elements portrayed which make them really distinctive. This is an advantage, as many of these plots could be generated with a standard tool and interface if we had the data and we could choose the dimensions and elements. An interactive interface could be used as in other exploration tools (e.g., analytical processing or visualisation tools). Second, it is appropriate to look at the purpose of each of these representations and see whether they correspond to the needs and applications we identified in previous sections. For instance, Figs. 1, 2, 3b, c, d, e, 4, 6b, c, e, and f are mostly explanatory or differential in purpose, while Figs. 3a, d, f, 5, 6a and d seem to have a more forecasting intention. Some have a broader coverage of kinds of intelligence (Figs. 3a, b, c, 4 and especially 7) and others are more specific.

4 Conclusions

This paper has presented the first steps of an atlas of intelligence, which at this stage must focus on the elicitation of needs (in terms of motivations and requirements) and possibilities (applications, representations and kinds of entities covered). After this analysis, we now have a much better account of how wide the initiative is. The next steps should focus on recognising the applications that might have more impact and are more feasible in the short term. This assessment would allow us to establish the specification of the atlas in a progressive way, so that an essential part of it can be designed and enriched over time. For such an ambitious approach, it is important to think big, as we have done here, while starting small, and grow incrementally.

Apart from the instrumental purpose of this paper as a first step in the development of an atlas of intelligence, this work (independently of how far the atlas develops in the future) brings attention to methodological issues (and related philosophical and theoretical) issues in all disciplines related to intelligence and cognition. Scientists in these disciplines usually see themselves as explorers, but exploration involves much more than discovering and inventing. Scientists also need (to be) cartographers, curators and taxonomists in order to structure, facilitate and disseminate what is known, and assess their unknowns, prioritise their goals and see their progress in perspective. In the same way Linnaeus changed the way living beings were described, catalogued and named, motivating new lines of research, this initiative will help to establish the parameters and the instruments to properly handle and understand the space of existing and future cognitive systems, and exploit its research possibilities.

Acknowledgements. The initiative was supported by the Leverhulme Trust via the Leverhulme Centre for the Future of Intelligence. J. H-Orallo and F. M-Plumed were supported by EU (FEDER) and the Spanish MINECO under grant TIN 2015-69175-C4-1-R and by GVA under grant PROMETEOII/ 2015/013 and by the Air Force Office of Scientific Research under award number FA9550-17-1-0287. J. H-Orallo also received a Salvador de Madariaga grant (PRX17/00467) from the Spanish MECD for a research stay at the CFI, Cambridge, and a BEST grant (BEST/2017/045) from the GVA for another research stay also at the CFI. F. M-Plumed was also supported by INCIBE *(Ayudas para la excelencia de los equipos de investigación avanzada en ciberseguridad)*. A. Weller acknowledges support from the David MacKay Newton research fellowship at Darwin College, the Alan Turing Institute under EPSRC grant EP/N510129/1 & TU/B/000074, and the Leverhulme Trust via the CFI.

A Appendix: Why is an atlas needed? Similar initiatives

While identifying the need for an atlas, we look at how it fits in cognitive science as a whole and also whether there are initiatives in other fields that could be inspirational.

Regarding cognitive science, it is true that its goal is to cover all possible cognitive systems, understand their behaviour and mechanisms, and establish meaningful comparisons. However, the field has not yet been able to portray a systematic representation covering both natural and artificial systems. But if we do not find this systematic representation in cognitive science, do we find it in related subdisciplines? The answer is that some similar initiatives in other disciplines do exist[2]:

- **Life forms:** Examples are Wikispecies (Leslie 2005), the All Species Foundation (Gewin 2002), the Catalogue of Life and the Encyclopedia of Life (Roskov et al. 2018; Hayles 1996; Parr et al. 2014; Stuart et al. 2010).
- **Neuroscience:** the Cognitive Atlas[3] and related repositories for neuroscience[4] include an ontology of human cognitive functions and related tasks, and the pathologies affected. The Allen brain observatory[5] (Allen Institute

[2] Some of these initiatives are in genomics and brain imaging (Midford 2004; Boero and Bernardi 2014).
[3] http://www.cognitiveatlas.org.
[4] https://poldracklab.stanford.edu/.
[5] http://observatory.brain-map.org/visualcoding/.

for Brain Science 2016) is a more visually-oriented platform that maps perception and cognition to parts of the human brain (National Research Council 2011).
- **Psychometrics:** There are several initiatives bringing together test batteries and repositories: the mental Measurement yearbook[6], and with a more open character, the International Personality Item Pool[7] and the International Cognitive Ability Resource[8].
- **Machine learning and data science research:** Kaggle[9], OpenML[10] (Vanschoren et al. 2013) and many other platforms (e.g., gitxiv.com) provide benchmarks for ML. OpenML also includes experimental results that can be compared, aggregated and represented with powerful analytical packages.
- **Artificial intelligence:** there are many collections of benchmarks and associated results, such as ALE[11], OpenAI universe/gym[12], Microsoft Malmo[13], Facebook's CommAI-env[14], DeepMind Lab [15] (see Hernández-Orallo et al. 2017 for a summary) and meta-views, such as a recent EFF analysis[16] and the AI index report[17]. This is a sign that AI is looking in this direction (Castelvecchi 2016; Hernández-Orallo 2017). The tasks are rarely arranged into abilities and the data usually compares specialised AI systems against average humans.

A partially overlapping initiative is the AI Roadmap Institute[18], which encourages, compares and studies various AI and general AI roadmaps. It focuses on the future and on AI primarily, with representations that are usually flowcharts and pathway comparisons. Besides identifying where the field of AI stands as a whole, it also aims to identify dead-ends and open research problems on the path to the development of general AI systems.

The data and conceptual framing of the above projects can be used to inform an atlas of intelligence. Still, no repositories or taxonomies exist focusing mostly on behaviour, encompassing natural and artificial systems, as we are undertaking. Of course, the fact that something does not exist yet is not a sufficient reason that it should. The need for an atlas has to be supported by a series of motivations and applications, which we do in Sect. 2.

[6] http://buros.org/mental-measurements-yearbook.
[7] http://ipip.ori.org.
[8] http://icar-project.com.
[9] http://www.kaggle.com.
[10] http://www.openml.org.
[11] http://www.arcadelearningenvironment.org/.
[12] https://gym.openai.com/.
[13] https://www.microsoft.com/en-us/research/project/project-malmo/.
[14] https://research.fb.com/projects/commai/.
[15] https://deepmind.com/blog/open-sourcing-deepmind-lab/.
[16] http://www.eff.org/ai/metrics.
[17] http://aiindex.org.
[18] http://www.roadmapinstitute.org.

References

Aarts, A., Anderson, J., Anderson, C., Attridge, P., Attwood, A., Fedor, A.: Estimating the reproducibility of psychological science. Science **349**(6251), 1–8 (2015)

Allen Institute for Brain Science: Allen brain observatory (2016). http://observatory.brain-map.org/visualcoding

Arsiwalla, X.D., Moulin-Frier, C., Herreros, I., Sanchez-Fibla, M., Verschure, P.: The morphospace of consciousness. arXiv preprint arXiv:1705.11190 (2017)

Balakhonov, D., Rose, J.: Crows rival monkeys in cognitive capacity. Sci. Rep. **7**(1), 8809 (2017)

Bhatnagar, S., et al.: A First Survey on an Atlas of Intelligence (2017). http://www.dsic.upv.es/~flip/papers/Bhatnagar18_SurveyAtlas.pdf

Boero, F., Bernardi, G.: Phenotypic vs genotypic approaches to biodiversity, from conict to alliance. Mar. Genomics **17**, 63–64 (2014)

Brooks, R.A.: From earwigs to humans. Robot. Auton. Syst. **20**(2–4), 291–304 (1997)

Cadman, E.: AI not just a game for DeepMind's Demis Hassabis. Financial Times (2014). https://www.ft.com/content/1c9d5410-8739

Castelvecchi, D.: Tech giants open virtual worlds to bevy of AI programs. Nature **540**(7633), 323–324 (2016)

Cattell, R.B., Coulter, M.A.: Principles of behavioural taxonomy and the mathematical basis of the taxonome computer program. Br. J. Math. Stat. Psychol. **19**(2), 237–269 (1966)

Dennett, D.C.: Darwin's dangerous idea. Sciences **35**(3), 34–40 (1995)

de Valadés, D.: Rhetorica christiana (1579)

Eckersley, P., Nasser, Y., et al.: EFF AI Progress Measurement Project (2017). https://www.eff.org/es/ai/metrics. Accessed 10 Jan 2017

Einav, L., Levin, J.: Economics in the age of big data. Science **346**(6210), 1243089 (2014)

Frey, C.B., Osborne, M.A.: The future of employment: how susceptible are jobs to computerisation? Technol. Forecast. Soc. Chang. **114**, 254–280 (2017)

Gentner, D.: Psychology in cognitive science: 1978–2038. Top. Cogn. Sci. **2**(3), 328–344 (2010)

Gewin, V.: Taxonomy: all living things, online. Nature **418**(6896), 362–363 (2002)

Godfrey-Smith, P.: Towers and trees in cognitive evolution. In: Huebner, B. (ed.) The Philosophy of Daniel Dennett (Chap. 8.1). Oxford University Press (2015)

Gray, H.M., Gray, K., Wegner, D.M.: Dimensions of mind perception. Science **315**(5812), 619–619 (2007)

Hayles, N.K.: Narratives of artificial life. In: Future Natural: Nature, Science, Culture, pp. 146–164. Routledge, London (1996)

Hernández-Orallo, J., et al.: A new AI evaluation cosmos: ready to play the game? AI Mag. **38**(3), 66–69 (2017). https://www.aaai.org/ojs/index.php/aimagazine/article/view/2748

Hernández-Orallo, J.: The Measure of All Minds: Evaluating Natural and Artificial Intelligence. Cambridge University Press, Cambridge (2017)

Herrmann, E., Call, J., Hernández-Lloreda, M.V., Hare, B., Tomasello, M.: Humans have evolved specialized skills of social cognition: the cultural intelligence hypothesis. Science **317**(5843), 1360–1366 (2007)

Hessel, M., et al.: Rainbow: combining improvements in deep reinforcement learning. arXiv preprint arXiv:1710.02298 (2017)

Kirsh, D.: Today the earwig, tomorrow man? Artif. Intell. **47**(1–3), 161–184 (1991)

Landhuis, E.: Neuroscience: big brain, big data. Nature **541**(7638), 559–561 (2017)
Leslie, M.: Calling all taxonomists. Science **307**(5712), 1021–1022 (2005)
Lowndes, J.S.S., Best, B.D., Scarborough, C., Afflerbach, J.C., Frazier, M.R., O'Hara, C.C., Halpern, B.S.: Our path to better science in less time using open data science tools. Nat. Ecol. Evol. **1**, 0160 (2017)
Macphail, E.M.: The comparative psychology of intelligence. Behav. Brain Sci. **10**(4), 645–656 (1987)
Marcus, G.: Deep learning: a critical appraisal. arXiv preprint arXiv:1801.00631 (2018)
Marx, V.: Biology: the big challenges of big data. Nature **498**(7453), 255–260 (2013)
McKinlay, R.: Use or lose our navigation skills: automatic wayfinding is eroding natural abilities, warns roger mckinlay. Nature **531**(7596), 573–576 (2016)
Midford, P.E.: Ontologies for behavior. Bioinformatics **20**(18), 3700–3701 (2004)
Miller, R.: Task taxonomy: science or technology? Ergonomics **10**(2), 167–176 (1967)
Milner, G.: Pinpoint: How GPS is Changing Technology, Culture, and Our Minds. WW Norton & Company (2016)
Moravec, H.: When will computer hardware match the human brain. J. Evol. Technol. **1**(1), 10 (1998)
National Research Council: Toward Precision Medicine: Building a Knowledge Network for Biomedical Research and a New Taxonomy of Disease. National Academies Press (2011)
Ng, A.: What artificial intelligence can and can't do right now. Harvard Business Review, November 2016
Parr, C.S., et al.: The encyclopedia of life v2: Providing global access to knowledge about life on earth, vol. 2 (2014). http://www.eol.org/
Pfeifer, R.: Embodied artificial intelligence 10 years back, 10 years forward. In: Informatics, pp. 294–310 (2001)
Pinker, S.: The Language Instinct: How the Mind Creates Language. William Morrow and Company (1994)
Roskov, Y., et al.: Species 2000 ITIS catalogue of life. Aeon (2018). http://catalogueoflife.org/col
Schaie, K.W.: Intellectual development in adulthood. In: Birren, J.E., Schaie, K.W. (eds.) Handbook of the Psychology of Aging, vol. 4, pp. 266–286. Academic Press Inc. (1996)
Shah, H., Warwick, K., Vallverdú, J., Wu, D.: Can machines talk? Comparison of eliza with modern dialogue systems. Comput. Hum. Behav. **58**, 278–295 (2016)
Shanahan, M.: Conscious exotica. from algorithms to aliens, could humans ever understand minds that are radically unlike our own? Aeon (2016). https://aeon.co/essays/beyond-humans-what-other-kinds-of-minds-might-be-out-there
Shead, S.: Facebook's AI boss: in terms of general intelligence, we're not even close to a rat. Business Insider (2017). http://uk.businessinsider.com/facebooks-ai-boss-in-terms-of-general-intelligence-were-not-even-close-to-a-rat-2017-10
Silver, D., et al.: Mastering chess and shogi by self-play with a general reinforcement learning algorithm. arXiv preprint arXiv:1712.01815 (2017)
Sloman, A.: The Structure and Space of Possible Minds. University of Sussex, School of Cognitive Sciences (1984)
Solé, R.: Rise of the humanbot. arXiv preprint arXiv:1705.05935 (2017)
Solé, R.V., Macia, J.: Synthetic biocomputation: the possible and the actual. In: ECAL, pp. 29–36 (2011)
Sparrow, B., Liu, J., Wegner, D.M.: Google effects on memory: cognitive consequences of having information at our fingertips. Science, p. 1207745 (2011)

Stuart, S., Wilson, E., McNeely, J., Mittermeier, R., Rodríguez, J.: The barometer of life. Science **328**(5975), 177–177 (2010)

Tegmark, M.: Life 3.0: Being human in the age of artificial intelligence. Knopf (2017)

Thagard, P.: Why cognitive science needs philosophy and vice versa. Top. Cogn. Sci. **1**(2), 237–254 (2009)

Vanschoren, J., et al.: Towards a data science collaboratory. Lecture Notes in Computer Science (IDA 2015), vol. 9385 (2015)

Vanschoren, J., van Rijn, J.N., Bischl, B., Torgo, L.: OpenML: networked science in machine learning. SIGKDD Explor. **15**(2), 49–60 (2013). https://doi.org/10.1145/2641190.2641198

Wegner, D.M., Gray, K.: The Mind Club: Who Thinks, What Feels, and Why It Matters. Penguin (2017)

Winfield, A.F.: Estimating the energy cost of (artificial) evolution. In: Sayama, H., Rieffel, J., Risi, S., Doursat, R., Lipson, H. (eds.) Proceedings of 14th International Conference on the Synthesis and Simulation of Living Systems (ALife), pp. 872–875. MIT Press (2014)

Winfield, A.F.: How intelligent is your intelligent robot? arXiv preprint arXiv:1712.08878 (2017)

Yampolskiy, R.V.: The universe of minds. arXiv preprint arXiv:1410.0369 (2014)

Do Machine-Learning Machines Learn?

Selmer Bringsjord[1(✉)], Naveen Sundar Govindarajulu[1], Shreya Banerjee[1], and John Hummel[2]

[1] Rensselaer Polytechnic Institute, 110 8th Street, Troy, NY 12180, USA
Selmer.Bringsjord@gmail.com, Naveen.Sundar.G@gmail.com, shreyabbanerjee@gmail.com
[2] 901 West Illinois Street, Urbana, IL 61801, USA
jehummel@illinois.edu

Abstract. We answer the present paper's title in the negative. We begin by introducing and characterizing "real learning" ($\mathcal{RL}$) in the formal sciences, a phenomenon that has been firmly in place in homes and schools since at least Euclid. The defense of our negative answer pivots on an integration of *reductio* and proof by cases, and constitutes a general method for showing that any contemporary form of machine learning (ML) isn't real learning. Along the way, we canvass the many different conceptions of "learning" in not only AI, but psychology and its allied disciplines; none of these conceptions (with one exception arising from the view of cognitive development espoused by Piaget), aligns with real learning. We explain in this context by four steps how to broadly characterize and arrive at a focus on $\mathcal{RL}$.

1 Introduction

Presumably you've read the title, so: No; despite the *Zeitgeist*, according to which today's vaunted 'ML' (= "machine learning") is on the brink of disemploying most members of *H. sapiens sapiens*, no. Were the correct answer 'Yes,' a machine that machine-learns some target t would, in the determinate, non-question-begging, well-founded sense of 'learn' that has been firmly in place for millennia and which we soon define and employ,[1] learn t. But this cannot be the case.

Why? Because, as we show below, an effortless application of indirect proof with proof by cases proves the negative reply. (A formal version of the reasoning is given in the Appendix (= Sect. 8), as a general method that covers any instantiation of 'ML' in contemporary AI.)

[1] The need for the qualifications (i.e. determinate, non-question-begging) should be obvious. The answer to the present paper's title that a machine which machine-learns *by definition* learns, since 'learn' appears in 'machine-learn,' assumes at the outset that what is called 'machine learning' today *is* real learning—but that's precisely what's under question; hence the *petitio*.

V. C. Müller (Ed.): PT-AI 2017, SAPERE 44, pp. 136–157, 2018.
https://doi.org/10.1007/978-3-319-96448-5_14

2 Preliminaries

To validate the negative answer, first, without loss of generality,[2] let's regard that which is to be learned to be a unary function $f : \mathbb{N} \mapsto \mathbb{N}$. The set of all such functions is denoted by $\mathcal{F}$. We say that agent $\mathfrak{a}$ has *really learned* such a function f only if[3]

$\mathfrak{a}$ has *really learned* f

(c1) $\mathfrak{a}$ understands the formal definition D_f of f,
(c2) can[a] produce both $f(x)$ for all $x \in \mathbb{N}$, and
(c3) a proof of the correctness of what is supplied in (c2). (**Note**: (c3) is soon supplanted with (c3′).)

[a] This is the 'can' of computability theory, which assumes unlimited time, space, and energy for computation. See e.g. (Boolos et al. 2003) for explanation.

As we shall see in a moment when considering a grade-school example, real learning so defined (= $\mathcal{RL}$)[4] is intuitive, has been solidly in place for at least 2.5 millennia, and undergirds everyday education every day. Of course, we must concede immediately that the first condition, (c1), employs a notorious word: viz., 'understands.' What is understanding? Not an easy question, that; this we must also concede. Instead of laboring to give an answer, which would inevitably call up the need for a sustained defense of the view that the concept of understanding, as applied to both humans and machines that are supposedly in possession of human-level intelligence and/or consciousness, is not only sufficiently clear, but is also a property that separates real minds from mere machines, we cheerfully

[2] All mathematical models of learning relevant to the present discussion that we are aware of take learning to consist fundamentally in the learning of number-theoretic functions from $\mathbb{N} \times \mathbb{N} \times \cdots \times \mathbb{N}$ to $\mathbb{N}$. Even when computational learning was firmly and exclusively rooted in classical recursion theory, and dedicated statistical formalisms were nowhere to be found, the target of learning was a function of this kind; see e.g. (Gold 1965; Putnam 1965), a modern, comprehensive version of which is given in (Jain et al. 1999). We have been surprised to hear that some in our audience aren't aware of the basic, uncontroversial fact, readily appreciated by consulting the standard textbooks we cite here and below, that machine learning in its many guises takes the target of learning to be number-theoretic functions. A "shortcut" to grasping *a priori* that all systematic, rigorously described forms of learning in matters and activities computational and mechanistic must be rooted in number-theoretic functions, is to simply note that computer science *itself* consists in the study and embodiment of number-theoretic functions, defined and ordered in hierarchies (e.g. see Davis and Weyuker 1983). We by the way focus herein on unary functions $f : \mathbb{N} \mapsto \mathbb{N}$ only for ease of exposition.

[3] A biconditional isn't needed. We use only a weaker set of necessary conditions, not a set of necessary *and* sufficient conditions.

[4] Not to be be confused with RL, reinforcement learning, in which real learning, as revealed herein, doesn't happen.

supplant the term in question with something unexceptionable.[5] Our substitute for the term is a simple and standard operationalization: instead of relying on the murky and mushy concept of understanding, we simply reply upon *testable behavior that for millennia has served as the basis for ascriptions of understanding to cognizers in the formal sciences.*[6] What behavior are we talking about? Well, the behavior of Euclid and everyone following him who has convinced the objective and the skeptical that they understand such things as mathematical (including specifically number-theoretic) functions, to wit: answers to penetrating questions, and associated proofs that those answers are correct. There literally has been no other way for a human being to provide evidence sufficiently strong to warrant an ascription, to that human being by others, of understanding in the realm of formal functions—or, since the machinery needed for careful articulation of these functions is at least something like axiomatic set theory, in the realm of mathematics itself. Here then, more explicitly, is what we replace (c1) with in order to define $\mathcal{RL}$:

(c1′) $\mathfrak{a}$ can correctly answer test questions regarding the formal definition D_f of f, where the answers in each case are accompanied by correct proofs[7] discovered, expressed, and provided by $\mathfrak{a}$.

We point out that the use of tests to sharpen what AI is, and how to judge the intelligent machines produced by AI, is a longstanding conception of AI itself, provided first by Bringsjord and Schimanski (2003), and later expanded by Bringsjord (2011).[8] It's true that philosophers may crave something more abstract and less pragmatic, but the fact of the matter is that tests are the coin of the realm in real-world AI, and also the coin of the realm in human-level learning in matters formal.[9] For economical exposition in the sequel, we continue to refer to real learning as simply '$\mathcal{RL}$.' We now turn to a simple example that shows $\mathcal{RL}$ to be, as we've said, intuitive, ancient, and operative every single day in the lives of all neurobiologically normal children with the parental or community wherewithal to be schooled:

[5] As many readers will know, Searle's (1980) Chinese Room Argument (CRA) is intended to show that computing machines can't understand anything. It's true that Bringsjord has refined, expanded, and defended CRA (e.g. see Bringsjord 1992, Bringsjord and Noel 2002; Bringsjord 2015), but bringing to bear here this argumentation in support of the present paper's main claim would instantly demand an enormous amount of additional space. And besides, as we now explain, calling upon this argumentation is unnecessary.

[6] Since at bottom, as noted (see note 2), the target of learning should be taken for generality and rigor to be a number-theoretic function, it's natural to consider learning in the realm of the formal sciences.

[7] Just as (computer) programs can be correct or incorrect, so too proofs can be correct or incorrect. For more on this, see e.g. (Arkoudas and Bringsjord 2007).

[8] If we regard Turing to have been speaking of modern AI in his famous (Turing 1950), note then too that his orientation is test-based: he gave here of course the famous 'Turing Test.'.

[9] In fact, this is why real learning for humans in mathematics is challenging; see e.g. (Moore 1994).

Example 1

You, a student, left for high school after breakfast and upon arriving were reminded in math class of the factorial function $n!$. Later in the day, when home, you inform a parent that you have learned the function in question. But you are promptly asked whether you *really* did learn it. So, you are tested by your parent, and by some homework questions that align with (c1′)–(c3):

1. The first problem relates to satisfying (c1)/(c1′): Consider the function g that maps a natural number k to the sum $k + (k + 1)$. Is it true that $\forall n \in \mathbb{N}[n! > g(n)]$? Prove it. Does this proposition hold for every natural number n after a certain size? Answer and prove it.[b]
2. A second problem asks you to ascertain whether the factorial of every natural number greater than 1 is even, and to then prove that the answer is correct.

You certainly can determine the correct answers to problems like these that probe your understanding of the factorial function, and you certainly can supply the definition in various forms and can decide whether proposed definitions are valid, and you certainly (assuming unlimited resources; see note a) can for any input n give back $n!$. Can you also prove that your outputs are correct? Yes, easily. For the fact is that you, reader, can really learn such functions.

[b]Of course an affirmative is correct, and the proof is a trivial use of mathematical induction.

Obviously, an infinite number of such examples can be effortlessly given, in order to anchor $\mathcal{RL}$. For instance, Example 2 could refer instead to the double factorial $n!!$ function, Example 3 to the Ackermann function, and so on *ad infinitum*. Without loss of generality, we rely solely on Example 1.

Now we consider two cases, each predicated on the assumption that the agent $\mathfrak{a}^\star$ we are assessing is a machine-learning one. We specifically assume that, as such, $\mathfrak{a}^\star$ is a standard artificial neural network that machine-learns by repeatedly receiving finite collections of ordered pairs (m, m') of natural numbers, some of which are from the graph of f and annotated as such, and some of which aren't from the graph of f and are annotated as such.[10] Provided that in the limit $\mathfrak{a}^\star$, upon receiving an arbitrary natural number n through time, outputs $f(n)$, save for a finite number of erroneous verdicts, $\mathfrak{a}^\star$ has machine-learned f.[11]

[10] Our assumption here thus specifically invokes *connectionist* ML. But this causes no loss of generality, as we explain by way the "tour" of ML taken in Sect. 6.1, and the fact that the proof in the Appendix, as explained there, is a general *method* that will work form any contemporary form of ML.

[11] This is a rough-and-ready extraction from (Jain et al. 1999), and must be sufficient given the space limitations of the present short paper, at least for now. Of course, there are many forms of ML/machine learning in play in AI of today. In Sect. 6.1 we consider different forms of ML in contemporary AI. In Sect. 6.2 we consider different types of "learning" in psychology and allied disciplines.

3 Case 1

Here we assume that human persons are capable of hypercomputation. Given this, humans can learn some Turing-uncomputable functions in $\mathcal{F}$. (One example might be Rado's (1963) "Busy Beaver" function Σ, which maps the size of a Turing machine measured by the number of its states to the maximum number of contiguous 1s such a TM can produce as output before halting (where the alphabet used is simply $\{0, 1\}$).) Let $h \in \mathcal{F}$ be such a function. That $\mathfrak{a}^\star$ hasn't learned h is a trivial theorem.[12]

4 Case 2

Assume now that $\mathfrak{a}^\star$ is to learn a Turing-computable unary number-theoretic function f, say one that might be seen in math classes; we here refer to Example 1 and its infinite cousins; see above. This case is likewise trivial. The models for machine learning on offer today from AI preclude even reproducing an accurate formal definition of f along with easy proofs therefrom, let alone proofs that proposed values are correct relative to such a definition; that is, conditions (c1′)–(c3) aren't satisfied. Since Case 1 and Case 2 are exhaustive: QED.

5 Objections; Replies

A number of objections are perfectly anticipatable. However, voicing and rebutting all of them here is beyond the reach of a reasonably sized paper. Nonetheless, perhaps substantive dialectic is possible. We start by considering a first objection (Objection 1) that we view as a family of interrelated objections.

5.1 Objection 1a: Yours is an idiosyncratic type of learning!

We imagine the objection in question expressed thus: "The definition of 'learning' employed here, i.e. what you dub 'real learning,' results in a very peculiar concept—one that captures neither machine learning nor human learning! And it certainly does not motivate why only this concept is the correct one."

This is flatly wrong. From the mathematical point of view, today's ANN-based machine learning, such as for example has been used in the construction of better-than-any-humans Go-playing systems (i.e. deep learning/DL as the specific type of ML), can be rigorously defined in only two or three ways, for the simple reason that these ways must be based directly on mathematical definitions of machine learning. We are not in the business of taking seriously modern-day

---

[12] Lathrop (1996) shows, it might be asserted, that uncomputable functions can be machine-learned. But in his scheme, there is only a probabilistic approximation of real learning, and—in clear tension with (c1′)–(c3)—no proof in support of the notion that anything has been learned. The absence of such proofs is specifically called out in the formal deduction given in the Appendix.

alchemists, let alone pointing out to them that their use of the term 'learning,' in the context of what learning has for millennia been, is outré. Some internationally famous deep-learning "engineers" have confessed to us in face-to-face conversation that what they are doing in this regard is utterly mysterious to them, mathematically speaking. We in the foregoing cite the ways that exist to understand machine learning logico-mathematically; see our References.[13] We confess that our argument, reflecting our logico-mathematical point of view, quietly but importantly includes a principle that can be summed up as follows:

(*) When investigating whether today's ML (in any of its forms) is real learning (of a number-theoretic function f), the only way to end up with an affirmative to the question is to find a mathematical account $\mathcal{A}$ of today's ML according to which in at least one of its forms its "learning" of f is real learning of f.

In a more formal version of our argument, such as what we give in the Appendix, we provide a step-by-step deductive argument for our main claim that machine-learning machines don't really learn; and this deductive argument renders the principle just given explicit and mechanical.

As to our definition of the real human learning of functions, i.e. $\mathcal{RL}$, this is extracted directly from mathematics textbooks used for many, many centuries. In fact, our triad (c1′)–(c3) can be traced clearly and unswervingly all the way back to Euclid. Real learning isn't peculiar in the least; on the contrary, it's orthodox, and the bedrock of all systematic human knowledge and technology. To validate and explicate $\mathcal{RL}$, we need nothing more than the problems, solutions, and proofs for those solutions that are part and parcel of high-school math—and in fact we only need algebra. Our triadic definition can be empirically confirmed by examining such simple textbooks; see for instance (Bellman et al. 2012). For the case of high-school calculus, see note 18. There is no small amount of irony in the fact that those touting "machine learning" in today's machines as genuine learning have invariably been required to pass the very courses, with the very textbooks, that demand $\mathcal{RL}$.

To wrap up our rebuttal, we note that $\mathcal{RL}$, far from being idiosyncratic, is directly reflective of something that most if not all ML ignores: viz., learning is what produces knowledge. An agent that has genuine knowledge of the differential-and-integral calculus is an agent whose learning has produced at least something very close to justified, true belief with respect to the relevant propositions.[14] That is, the agent believes these true propositions, and has justifications in the form of arguments that establish, or at least render highly likely, the relevant propositions.

[13] A pair of additional works help to further seal our case: (Kearns and Vazirani 1994; Shalev- Shwartz and Ben-David 2014). Study of these texts will reveal that $\mathcal{RL}$ as per (c1′)–(c3) is nowhere to be found.

[14] We of course join epistemological cognoscenti in being aware of Gettier-style cases, but they can be safely left aside here. For the record, Bringsjord claims to have a solution anyway—one that is at least generally in the spirit of Chisholm's (1966) proposed solution, which involves requiring that the justification in justified-true-belief accounts of knowledge be of a certain sort. For Gettier's landmark paper, see (Gettier 1963).

The arguments that undergird knowledge in this way (which were called out above in our example of (c1′)–(c3) in action) are nowhere to be found in contemporary ML, at least in its connectionist, probabilistic, and reinforcement forms.

5.2 Objection 1b: This isn't AI!

In a variant of Objection 1a, we imagine some saying this: "In AI, we are, as a rule, not interested in learning functions over naturals with an infinite domain, given by a graph (or table)."

This is a painfully weak objection, one that reflects, alas, the alchemic nature of much of modern AI. AIniks may not be interested in X, but mathematically they may well be doing X; and if they can't say mathematically what they're doing, then they shouldn't say anything at all in debates such as the present one. Regardless, rest assured that formally speaking, machine learning is learning such functions as we have pointed to (or alternatively learning formal grammars or idealized computing machines). We have given references that confirm this with a ring of iron.

5.3 Objection 1c: What about toads?!

"Your argument has the absurd consequence that even lower animals turn out to be classified as non-learners. Can a toad learn? Certainly. Can a toad learn a number-theoretic function in your sense of learn? Certainly not."

This objection is a kind, unwitting gift, for this is just another way to expose the absurdity of statistical ANN-based machine learning (and of—as we shall momentarily see—other forms of non-logicist machine learning[15]). Agreed: a toad can't learn a number-theoretic function, in the established triadic sense of learning such things we specified above. (We now know that *no* nonhuman animals can do anything of the sort; see e.g. (Penn et al. 2008); ergo our critic can be encouraged to substitute for 'toad' 'dog' or 'chimp,' etc.) But, by the mathematics of statistical ML/DL, a toad (or a toad-level AI produced by the likes of Deep Mind) can learn such a function. This allows us to deduce by *reductio* what the man on the street already well knows: a bunch of smarty-pants people have defined their own private, bizarre, and self-advancing sense of learning. We're now seeing the hidden underbelly of this smug operation, because adversarial tests are showing such things as that DL-based vision systems declare with 99% confidence, for example, that as a turtle is a gun.[16] Of course, we

[15] Specifically, we shall see that the formal deduction of the Appendix is actually a *method* for showing that other forms of "modern" ML, not just those that rely on ANNs, don't enable machines to really learn anything. E.g., the method can take Bayesian learning in, and yield as output that such learning isn't real learning.

[16] Shakespeare himself, or better yet even Ibsen, or better better yet Bellow, couldn't have invented a story dripping with this much irony—a story in which the machine-learning people persecuted the logicians for building "brittle" systems, and then the persecutors promptly proceeded to blithely build comically brittle systems as their trophies (given to themselves).

concede what everyone knows: connectionist ML will continue to improve, and the current brittleness of this form of learning will specifically be addressed in many applications. Yet the mere fact that there currently *is* brittleness is profoundly telling, in the context of $\mathcal{RL}$, for imagine a student Johnny who has real-learned our now-well-worn factorial function. And now imagine that to "test" Johnny, instead of presenting him with a numeral $\langle n \rangle$ where $n \in \mathbb{N}$, and a question as to what the factorial of n is, we instead present him with a picture p of a turtle, and ask him what the factorial of a turtle is. Johnny is likely to inform his parents that some teachers at this school are mentally unstable; certainly there's no chance he's going to blurt out such a response as '24.' The reason for this, speaking imprecisely (recall the earlier discussion at the outset of the paper about the concept of understanding), is that while the DL system has no real understanding of what a turtle or a gun is, Johnny, in satisfying (c1′)–(c3), does.

5.4 Objection 2: Case 1 is otiose!

"Surely your Case 1 is otiose, since—so the objection goes—finite agents, whether human or machine, as everyone concedes, don't in *any* sense learn uncomputable functions."

This is a silly objection, swept away as but dust by the relevant empirical facts; for everyone *doesn't* concede such a thing; witness (Bringsjord et al. 2006), which is in fact based on the aforementioned Σ. As is explained there, Gödel made no concession to the effect that humans don't learn uncomputable functions. For a purported proof that human persons hypercompute, see (Bringsjord and Arkoudas 2004);[17] for a book-length treatment, see (Bringsjord and Zenzen 2003).

5.5 Objection 3: Your definition of human learning is tendentious!

"Your triadic definition of learning [based on your conditions (c1′)–(c3)] conveniently stacks the deck against modern statistical machine learning (=ML in the current discussion and in the—by-your-lights fawning—media). This definition is highly unnatural, and highly demanding."

We note first in reply that convenience *per se* is of course unobjectionable. Next, telling in this dialectic is the brute fact that for well over two millennia we have known what it is for an agent to have really learned some math or formal logic, number-theoretic functions included; and what we in this regard know aligns *precisely* with the triadic account of $\mathcal{RL}$ given above. Again, empirical confirmation of this alignment can be obtained by turning to what the textbooks

[17] In which is by the way cited hypercomputational artificial neural networks.

demand in terms of proof,[18] and what the disciplines in question demand of those who wish to lay claim to having truly learned some formal logic or math.

In short, we cannot allow the field of AI, and specifically its ML subpart, now on the intellectual scene for not more than a blip of time, to trample ordinary language and ordinary meaning that has been firmly in place within the formal sciences for millennia. We are not here appealing directly to so-called "ordinary language" philosophy, and philosophers in this school (such as G.E. Moore, Austin, Norman Malcolm, and various modern defenders). As a matter of fact, the veridicality of ordinary language is something we in general find attractive, but we need only a circumspect general principle like this one:

(+) If natural-language communication has for millennia taken the *bona fide* learning of an arithmetic function f by an intelligent agent $\mathfrak{a}$ to happen only if Φ, then, absent a separate and strong argument in favor of an incompatible set Ψ of conditions that contravenes this, one is justified in applying Φ to claims that $\mathfrak{a}$ can learn/has learned some given function f'.

Perhaps the remarkable thing about (+) is that the behavior of ML practitioners *themselves* confirms its truth. The field of machine learning has both foundational theorems such as the *No Free Lunch* theorem (Wolpert 1996) and new working theorems that are constantly introduced in the scientific literature of the field, e.g. Theorem 2.1 in (Achab et al. 2017). Leaving aside theorems and other formal knowledge produced by ML practitioners, consider the case of ANN-based ML, for instance today's DL. DL experts examine some given data, and through domain expertise built up in the past (via a process much mediated by natural communication, written and oral), devise a target set of functions (denoting the architecture of the neural network in question):

$$\{f_{\mathbf{w}} \mid \mathbf{w} \text{ is in some large space}\}$$

The machine then simply tunes the weights $\mathbf{w}$. Specifically, in convolutional (artificial) neural networks, the form of the function best suited for image processing was conceptualized by humans and justified, by not just performance measures, but by an *argument* in good old-fashioned English for the conclusion that this form of neural networks might be good for image processing. See (LeCun et al. 1998) and Chap. 9 in (Goodfellow et al. 2016) for examples of this process. Note that even if a machine selects the architecture, that selection is happening from a

[18] E.g. even beginning textbooks introducing single-variable differential/integral calculus ask for verification of human learning by asking for proofs. The cornerstone and early-on-introduced concept of a *limit* is accordingly accompanied by requests to students that they supply proofs in order to confirm that they understand this concept. Thus we e.g. have on p. 67 of (Stewart 2016) a request that our reader prove that

$$\lim_{x \to 3} g(x) = (4x - 5) = 7$$

. What machine-learning machine that has learned the function g here can do *that*?

class of architectures designed by none other than the guiding humans, and there is no justification from the machine beyond performance measures. A relatively different form of machine learning, *inductive programming* (Kitzelmann 2009), seeks to learn functions like addition by looking at a very small set of sample inputs and outputs. But even this is shallow when stacked against real learning: In $\mathcal{RL}$, humans can not only look at inputs and outputs, but also descriptions of the properties of the function written in English (and other natural languages, as the case may be) that go well beyond the examples.

5.6 Objection 4: Do flying machines (really) fly?

"Unfortunately, you are dancing around an unanswerable quagmire that has been with us for rather a long time, one summed up by the seemingly innocent question: Do flying machines (really) fly?"

Suppose there is an embodied AI $\mathfrak{a}'$ with all sorts of relevant sensors and effectors in the form of an autonomous drone that can take off by itself and travel great distances adroitly, land, and so on—all without any human intervention in it or its supporting systems during some flight from time t to t'. Did $\mathfrak{a}'$ really fly during this interval? Of course it did. Do eagles really fly over intervals of time? Of course they do. There is no objection to our argument to be found in the vicinity of these (nonetheless interesting) questions. In the case of $\mathcal{RL}$, there are no machines on the planet, and indeed no machines in the remotely foreseeable future of our solar system, that have the attributes constitutive of this learning.

5.7 Objection 5: You concede that your case is limited to the formal sciences!

"You have conceded, perhaps even stipulated, that real learning in your argument is restricted to the realm of the formal sciences. Hence, if your case is victorious, its reach is rather limited, no?"

Quite the contrary, actually. We have indeed restricted real learning to the formal sciences. However, we had assumed that it would be clear to all readers that adaptation and expansion of (c1′)–(c3) to *non*-formal domains would if anything bolster our case, if not immediately render it transparently victorious. Apparently our critic in the case of the current objection needs to be enlightened. Consider creative writing. What does it take to learn the "functions" at the heart of creative writing, so that eventually one can take as input the premise for a story and yield as output a good story?[19] We can safely say that any agent capable of doing this must be able to read not formal-scientist Euclid, but, say, Aristophanes, and a line of creative writers who have been excelling since the ancient Greeks; and learn from such exemplars how such a "function" can be computed. But reading and understanding literary prose, and learning thereby,

[19] This is essentially the Short Short Story Game of (Bringsjord 1998), much harder than such Turing-computable games as Checkers, Chess, and Go, which are all at the same easy level of difficulty (EXPTIME).

is patently outside the purview of current and foreseeable AI. And it gets worse for anyone who thinks that today's machine-learning machines learn in such domains: In order to learn to be a creative writer one must *generate* stories, over and over, and learn from the reaction and analysis thereof, and then generate again, and iterate the process. Such learning, which is real learning in creative writing, isn't only not happening in ML today; it's also hard to *imagine* it happening in even ML of tomorrow.

6 Real Learning in Context

The dialectic in the previous section makes it abundantly clear that 'learning' is polysemous: it means many different things to many different people. Given this fact, we think it's worthwhile to a bit more systematically place real learning within the context of different senses of learning in play in contemporary AI and cognitive science/psychology. We thus briefly review the prominent senses of learning in AI (Sect. 6.1), and then in cognitive science/psychology (Sect. 6.2); and then, this two-part review complete, we proceed (Sect. 6.3) to quickly explain in broad strokes how by a series of four steps real learning can be isolated within the broader context afforded by the review.

6.1 Learning in AI

Everyone must admit that there are many different extant ways to map the geography of what is called "learning" in the field of AI. This is easily confirmed by the existence of modern, credible overviews of learning in AI, in textbooks (each of which, of course, has been fully professionally vetted): the geographies offered in each pair of these books is different between the two. Given this divergence, we can't possibly give here a single, definitive, received breakdown of learning in its various forms within contemporary AI. On the other hand, it's nonetheless clear that *any* orthodox breakdown of the types of learning in the field, in any textbook, will immediately reveal that no type matches real learning = $\mathcal{RL}$.[20] We here follow Luger (2008), whom we find particularly perspicuous, and quickly point out, as we move through his geography, that real learning is nowhere to be found. Nonetheless, it will be seen that Luger (2008), to his credit, opens a

[20] Outside of the present paper, we have carried out a second analysis that confirms this, by examining learning in AI as characterized in (Russell and Norvig 2009), and invite skeptical readers to carry out their own analysis for this textbook, and indeed for any comprehensive, mainstream textbook. The upshot will be the stark fact that $\mathcal{RL}$, firmly in place since Euclid as what learning in the formal sciences is, will be utterly absent.

door to a path that could conceivably lead to real learning, at some point in AI's future.[21]

Luger (2008) devotes Part IV of his book to "Machine Learning;" four chapters, 10–13, compose this part, and each is devoted to a different form of machine learning:

- Chap. 10: "Symbol-Based"
- Chap. 11: "Connectionist"
- Chap. 12: "Genetic and Emergent"
- Chap. 13: "Probabilistic"

As one would expect, connectionist learning covers machine learning that is rooted in ANNs. For reasons given in the present paper, there isn't a scintilla of overlap between what is covered in Chap. 11 and $\mathcal{RL}$. This is true for starters because the familiar, immemorial declarative information that has defined such things as the factorial function are nowhere to be found within an any artificial neural network whatsoever. The same applies, *mutatis mutandis*, to the genetic-and-emergent type of learning covered in Luger's (2008) Chap. 12, as should be obvious to all readers. (Genetic algorithms, for example, make no use of the sort of declarative content that defines number-theoretic functions.) We are thus left to consider whether $\mathcal{RL}$ appears in symbol-based learning presented in Chap. 10, or in probabilistic learning covered in Chap. 13. In point of fact, which energetic readers can confirm when reading for themselves, real learning doesn't appear in either of these places.

Now, we said above that Luger (2008) opens the door to a future in which AI includes real learning. We end the present section by explaining what we mean.

In the final part of (Luger 2008), V, entitled "Advanced Topics for AI Problem Solving," two topics are covered, each of which is given its own chapter: "Automated Reasoning," covered in Chap. 14; and "Understanding Natural Language," presented in Chap. 15. Luger's (2008) core idea is that for truly powerful forms of problem-solving in a future AI, that remarkable machine will need at least two key things: it will need to be able to reason automatically and autonomously in deep ways, starting with deductive reasoning; and second, this AI will need to be able to really and truly understand natural language,

[21] Instead of looking to published attempts to systematically present AI (such as the textbooks upon which we rely herein), one could survey practitioners in AI, and see if their views harmonize with the publications explicitly designed to present all of AI (from a high-altitude perspective). E.g., one could turn to such reports as (Müller and Bostrom 2016), in which the authors report on a specific question, given at a conference that celebrated AI's "turning 50" (*AI@50*), which asked for an opinion as to the earliest date (computing) machines would be able to simulate human-level learning. It's rather interesting that 41% of respondents said this would *never* happen. It would be interesting to know if, in the context of the attention ML receives these days, the number of these pessimists would be markedly smaller. If so, that may well be because, intuitively, plenty of people harbor suspicions that ML in point of fact hasn't achieved any human-level real learning.

including complete sentences, following one upon another.[22] It will not escape the alert reader's notice that the capability constituted by this pair is at the heart of what it takes to be a "real learner," that is, to be an agent that really learns as per (c1′)–(c3) in the formal sciences. Unless an AI can *itself* prove such things as—to repeat a part of the Example we began with — that the factorial function's range consists of the even natural numbers, and receives and understands/processes challenges to prove such things, where these challenges come in the form of arbitrary, full sentences like "Show that the factorial of every number is even," it won't be an AI that really learns. Unfortunately, while Luger (2008) points the way toward aspects of two key capabilities needed for real learning, he does only that, by his own admission: point. So, while the door is open, our claim, that machine-learning machines of *today* don't really learn, is unscathed.

6.2 Learning in Psychology and Allied Disciplines

Recall the factorial-function example we gave at the outset. When, upon returning home after school, you are asked by a parent, "So, what did you learn today in math?" it's rather doubtful that if you answered earnestly and sincerely, and if your time in class was a pedagogical success, you replied in accordance with anything violently outside the bounds of $\mathcal{RL}$. Nonetheless, psychology and its allied disciplines (= psychology$^+$) have (perhaps inadvertently) erected an ontology of forms of learning that at least in principle offer viable alternatives to $\mathcal{RL}$, or even perhaps forms of learning that match, overlap, or conceivably subsume $\mathcal{RL}$. Put intuitively, the question before us in the present section is this one: Could you reasonably have conversed with your parent on the basis of any of the types of learning in psychology$^+$'s ontology thereof? As we now reveal, the answer is No.[23] We begin with the authoritative (Domjan 2015), which is based on this operationally inclined definition:

[22] Luger's book revolves around a fundamental distinction between what he calls *weak problem-solving* versus *strong problem-solving*.

[23] There are a few exceptions. Hummel (2010) has explained that sophisticated and powerful forms of symbolic learning, ones aligned with second-order logic, are superior to associative forms of learning. Additionally, there's one clear historical exception, but it's now merely a sliver in psychology (specifically, in psychology *of reasoning*), and hence presently has insufficient adherents to merit inclusion in the ontology we now proceed to canvass. We refer here to the type of learning over the years of human development and formal education posited by Piaget; e.g. see (Inhelder and Piaget 1958). Piaget's view, in a barbaric nutshell, is that, given solid academic education, nutrition, and parenting, humans develop the capacity to reason with and even eventually *over* first-order and modal logic—which means that such humans would develop the capacity to learn in $\mathcal{RL}$ fashion, in school. Since attacks on Piaget's view, starting originally with those of Wason and Johnson-Laird (e.g. see Wason and Johnson-Laird 1972), many psychologists have rejected Piaget's position. For what it's worth, Bringsjord has defended Piaget; see e.g. (Bringsjord et al. 1998).

> Learning is an enduring change in the mechanisms of behavior involving specific stimuli and/or responses that results from prior experience with those or similar stimuli or responses. (Domjan 2015, p. 14)

That learning is here attributed to a change in the 'mechanisms of behavior' would seem to draw a hard line between learning and performance. Performance can after all be the effect of multiple factors besides learning, and hence is not a sole determinant of the latter. At any rate, in our study of types of learning in psychology$^+$, we found the following six forms of learning. As we progress through the enumeration of these forms, we offer in turn a rather harshly economical summary of each, and render a verdict as to why each is separate from and irrelevant to real learning (with the possible exception, as we note, of the last). Here goes:

1. *Associative Learning: Classical and Instrumental Conditioning.* The theory of classical conditioning originates from the (Pavlovian) finding that if two stimuli, one unconditional (US), such as food, and the other neutral (CS), come in close temporal contiguity, and if US elicited some response naturally (say salivation), then CS too eventually elicits that response. While here the change in behavior is attributed to some contingency between CS and US (also called *reinforcer*), in instrumental conditioning this change results from some contingency between that behavior and the reinforcer (Mackintosh 1983). Obviously, if this strengthening or reinforcement of the new pattern in behavior is no more than a new stimulus-response connection, real learning is nowhere to be found.[24]
2. *Representational Learning.* The representational theory of learning (Gallistel 2008) views the brain as a functional model capable of computing a representation of the experienced world; and that representation in turn informs the agent's behavior. While learning here is taken to be a process of acquiring knowledge from experience, 'knowledge' here means nothing like the knowledge that is front and center in Example 1 of $\mathcal{RL}$.
3. *Observational Learning.* Here, a new behavior is learned simply by observing someone else. Mostly associated with the social learning theory of psychologist Albert Bandura (1977), his Bobo-doll experiment (Bandura et al. 1961) is an interesting study of how children learn social behavior such as aggression through the process of observational learning. This type of learning in psychology$^+$ is learning by straight imitation, and as such is obviously not $\mathcal{RL}$. Put simply and baldly, the decision problems we presented in our starting example (e.g., is $n!$ invariably even?), and the confirmatory proofs for each answer, are not supplied by shallow imitation of the likes of inflatable Bobo dolls.
4. *Statistical Learning.* Extraction of recurring patterns in the sensory input generated from the environment over time is the core essence of this type

[24] We are happy to concede that years of laborious (and tedious?) study of conditioning using appetitive and aversive reinforcement (and such phenomena as inhibitory conditioning, conditioned suppression, higher-order conditioning, conditioned reinforcement, and blocking) has revealed that conditioning can't be literally reduced to new reflexes, but there is no denying that in conditioning, any new knowledge and representation that takes form falls light years short of $\mathcal{RL}$.

of learning (Schapiro and Turk-Browne 2015). Taking a cue from associative learning in nonhuman primates, past studies showed a possibility of sensitivity of certain parts of the brain when exposed to temporally structured information. Detection of conditional probability patterns in sound streams as a precursor to language parsing, leading to predictions of some sounds given other sounds, would be a good example. Schapiro and Turk-Browne (2015) give a nice overview of various studies related to auditory and visual statistical learning in humans, including neural investigations towards the role of different regions of brain in diverse forms of such learning. Though statistical learning is suggested as a pervasive element of cognition, it is yet early to state this as a form of real learning.

Marblestone et al. (2016) draw a parallel between human brain functioning and the activity of ANNs in connectionist ML. They specifically claim that the neural structure of the brain coincides with various methods of weight assignments to multiple hidden layers of ANNs when machine learning takes place. We gladly concede for the sake of argument that this direction holds promise for the neurological "decoding" of the human brain, since the core idea is that there's a match between brain activity and ANNs through time in ML. But since this activity cannot in any way be interpreted to constitute embodiments of the three clauses that define $\mathcal{RL}$, we once again see here an entirely irrelevant form of learning.

5. *Neurocentric Learning.* Titley et al. (2017) propose a non-exclusive, neurocentric type of learning. For ease of exposition, let's label this type of learning simply 'L_{ne}.' L_{ne} marks a move away from a synaptocentric neurobiological form of learning: in L_{ne}, both synaptic plasticity and intrinsic plasticity play a role in learning and memory. More specifically, synaptic plasticity assigns connectivity maps, while intrinsic plasticity drives engram integration. While L_{ne} is certainly interesting, and while it may well hold much promise, it's undeniable that learning in this sense is clearly not relevant to our conception of $\mathcal{RL}$. Confirmation of this comes from the brute fact that no account based on the building-blocks of L_{ne} can be used to express even the tiniest part of $\mathcal{RL}$. Colloquially put, no agent who learns, say, the Ackermann function in a given recursion-theory class, and is proud that she has, can report this happy event by expressing her enlightenment in terms of the proofs demanded by the clauses that define $\mathcal{RL}$.
6. *Instructional Learning.* Instructional learning is in play when an individual learns from instruction (for example, a teacher's verbal commands in a classroom) and responds with corresponding action/s. While we of course agree that instruction acts as a purposeful direction of the learning process (Huitt 2003), this learning fails to qualify as $\mathcal{RL}$ because action alone doesn't define learning. Of course, in theory, the actions of student learners could be fleshed out to correspond to $\mathcal{RL}$'s three clauses. Were this carried out, it would merely show that instructional learning, at least of a particular type (e.g., instructional learning in the formal sciences), corresponds to $\mathcal{RL}$—but this we've known from, and has indeed been plain to readers since, the outset of the present paper.

6.3 The Four-Step Road to Real Learning

Having completed our rapid tour of ML in contemporary AI, and learning in psychology^{+}, we now provide a general characterization of what real learning is, within this context. Saying what real learning is in the broader context constituted by the previous two subsections can be achieved by first by throwing aside irrelevant, lesser forms of cognition; this will be the first of four general steps taken to arrive at $\mathcal{RL}$:

Step 1: We begin by observing that the cognitive powers of creatures on Earth are discontinuous, because human persons have reasoning and communication powers of a wholly different nature than those possessed by nonhuman animals. A non-technical version of this observation is provided by Penn et al. (2008). A more specific, technical analysis, undertaken from a logico-mathematical standpoint, allows us to simply observe that only members of *H. sapiens sapiens* are capable of such things as[25]

- understanding and employing indubitable abstract inference schemas that are independent of physical stuff (e.g. *modus tollens*; see Ross 1992);
- understanding and employing arbitrary, layered quantification (such as that 'Everyone likes anyone who likes someone' along with 'Alvin likes Bobby' allows us to prove that 'Everyone likes Bobby');
- recursion (e.g. as routinely introduced in coverage of the recursive functions in an intermediate formal-logic course, which might wisely use (Boolos et al. 2003));
- infinite structures and infinitary reasoning (a modern example being the proof that the Goodstein sequence goes to zero; see (Goodstein 1944));
- etc.

Step 2: We next exclude forms of "learning" made possible via exclusive use of reasoning and communication powers in nonhuman animals, and set a focus on learning enabled by human-level-and-above (HLAB) reasoning and communication powers. (Given the previous two subsections, this step makes perfect sense. Recall our discussion, for example, regarding Luger's layout for learning in modern AI, all of which, save for what might be possible in the future, made no use whatsoever of the human capacity to read.)

Step 3: Within the focus arising from Step 2, we next avail ourselves of basic facts of cognitive development in order to narrow the focus to HLAB reasoning and communication sufficiently mature to perceive, and be successfully applied to, both (i) cohesive, abstract bodies of declarative content, and to (ii) sophisticated natural-language content. A paradigmatic case of such content would be axiom systems, such as those for geometry routinely introduced in high school. Another such case would be elementary number theory, also introduced routinely in high school; such coverage includes

[25] Note that all occurrences of 'understanding' in the itemized list that follows, in keeping with the psychometric operationalization introduced at the outset in order not to rely on the murky concept of understanding, could be invoked here; but doing so would take much space and time, and be quite inelegant.

the example of the factorial function, with which we started the present paper.[26] Let's denote such reasoning and communication by '$\mathrm{RC}^{h^\star}$.'

Step 4: Finally, we proceed to define real learning = $\mathcal{RL}$ as the acquisition of new knowledge by using $\mathrm{RC}^{h^\star}$. For example, forms of reasoning that use sophisticated analogical reasoning, or deduction applied to the axiom system *PA* (see note 26), can be used to allow an agent to *really* learn new things in the formal domain. Of course, the specific account of real learning will always boil down to specifics such as those given in (c1′)–(c3), but we have sought here to put real learning in a broader context, via our tour and, following on that, the four steps now concluded.

7 Final Remarks

We have heard echoes of an objection not explicitly presented and rebutted above; viz., "Perhaps you should do some soul-searching. For does it not simply boggle the mind that, if you're right, real learning hasn't even been seriously targeted by AI, despite all the praise that it receives for machines that 'learn'?!" Well, it *does* boggle the mind. All of us, the authors and all our readers, know quite well what real learning is, and how it came to be that on its shoulders we all arrived at a place that allows us to study and do AI: we got here by learning in precisely the fashion that $\mathcal{RL}$, in its three conditions, prescribes. We thus take ourselves to have simply revealed in the present paper what everyone in their heart of hearts knows: the exuberant claims of today that machine-learning machines learn are, when stacked against how we all learn enough to put ourselves in position to study and do AI, are simply silly. Accordingly, since AI in the new millennium increasingly penetrates the popular consciousness, we recommend that those working to advance non-real forms of ML extend to the public the courtesy of issuing a disclaimer that the type of learning to which they are devoted isn't real learning. This is a public, of course, that thinks of learning in connection not with artificial agents, but with schoolchildren, with high-schoolers, with undergraduates, with those in job-training programs, etc., all these groups being, of course, natural agents in the business of real learning.

Finally, we admit that the case we have delivered herein isn't yet complete, for there is an approach to computation, and an approach to the study of intelligence, neither of which we have discussed in connection with our core claim that contemporary ML isn't real learning. The approach to computation can be called *natural computation*, and the core idea is that nature itself computes (and perhaps *is* computation) (an excellent introduction is provided in Dodig-Crnkovic and Giovagnoli 2013); the approach to intelligence that we have left aside puts a premium on bodies and their interconnection with the physical environment (see e.g. Barrett 2015). In subsequent work, we plan to consider the relationship between $\mathcal{RL}$ and forms of learning based on these two intertwined approaches.

[26] Peano Arithmetic (*PA*) is rarely introduced by name in K–12 education, but all the axioms of it, save perhaps for the Induction Schema, are introduced and taught there.

Even now, though, it's safe to say that because $\mathcal{RL}$ takes little to no account of the physical (it's after all based in the formal sciences), and because it's conception of an agent is of a disembodied one,[27] it's highly unlikely that forms of physical-and-embodied learning not considered above will overlap real learning.

8 Appendix: The Formal Method

The following deduction uses fonts in an obvious and standard way to sort between functions ($\mathfrak{f}$), agents ($\mathfrak{a}$), and computing machines ($\mathfrak{m}$) in the Arithmetical Hierarchy. Ordinary italicized Roman is used for particulars under these sorts (e.g. f is a particular function). In addition, '$\mathcal{C}$' denotes any collection of conditions constituting jointly necessary-and-sufficient conditions for a form of current ML, which can come from relevant textbooks (e.g. Luger 2008; Russell and Norvig 2009) or papers; we leave this quite up to the reader, as no effect upon the validity of the deductive inference chain will be produced by the preferred instantiation of '$\mathcal{C}$.' It will perhaps be helpful to the reader to point out that the deduction eventuates in the proposition that no machine in the ML fold that in this style learns a relevant function $\mathfrak{f}$ thereby also real-learns $\mathfrak{f}$. We encode this target as follows:

$$(\star)\ \neg\exists\mathfrak{m}\ \exists\mathfrak{f}\ [\phi := MLlearns(\mathfrak{m},\mathfrak{f}) \land \psi := RLlearns(\mathfrak{m},\mathfrak{f}) \land \mathcal{C}_{\phi}(\mathfrak{m},\mathfrak{f}) \vdash^{*} (\mathrm{c1}') - (\mathrm{c3})_{\psi}(\mathfrak{m},\mathfrak{f})]$$

Note that $(\star)$ employs meta-logical machinery to refer to particular instantiations of $\mathcal{C}$ for a particular, arbitrary case of ML (ϕ is the atomic sub-formula that can be instantiated to make the particular case), and particular instantiations of the triad (ci′)–(ciii) for a particular, arbitrary case of $\mathcal{RL}$ (ψ is the atomic sub-formula that can be instantiated to make the particular case). Meta-logical machinery also allows us to use a provability predicate to formalize the notion that real learning is produced by the relevant instance of ML. If we "pop" ϕ/ψ to yield ϕ'/ψ' we are dealing with the particular instantiation of the atomic sub-formula.

The deduction, as noted earlier when the informal argument was given, is indirect proof by cases; accordingly, we first assume $\neg(\star)$, and then proceed as follows under this supposition.

[27] This conception matches that of an agent in orthodox AI: see the textbooks, e.g. (Luger 2008; Russell and Norvig 2009).

	(1)	$\forall\, \mathfrak{f}, \mathfrak{a}\, [\mathfrak{f} : \mathbb{N} \mapsto \mathbb{N} \rightarrow (RLlearns(\mathfrak{a}, \mathfrak{f}) \rightarrow (c1')\text{–}(c3))]$	Def of Real Learning
	(2)	$MLlearns(m, f) \wedge RLlearns(m, f) \wedge f : \mathbb{N} \mapsto \mathbb{N}$	supp (for $\exists$ elim on $\neg(\star)$)
	(3)	$\forall\, \mathfrak{m}, \mathfrak{f}\, [\mathfrak{f} : \mathbb{N} \mapsto \mathbb{N} \rightarrow (MLlearns(\mathfrak{m}, \mathfrak{f}) \leftrightarrow \mathcal{C}(\mathfrak{m}, \mathfrak{f}))]$	Def of ML
	(4)	$\forall\, \mathfrak{f}\, [\mathfrak{f} : \mathbb{N} \mapsto \mathbb{N} \rightarrow (TurComp(\mathfrak{f}) \vee TurUncomp(\mathfrak{f}))]$	theorem
	(5)	$TurUncomp(f)$	supp; Case 1
	(6)	$\neg\exists\, \mathfrak{m}\, \exists\, \mathfrak{f}\, [(\mathfrak{f} : \mathbb{N} \mapsto \mathbb{N} \wedge TurUncomp(\mathfrak{f}) \wedge \mathcal{C}(\mathfrak{m}, \mathfrak{f})]$	theorem
$\therefore$	(7)	$\neg\exists\, \mathfrak{m}\, MLlearns(\mathfrak{m}, f)$	(6), (3)
$\therefore$	(8)	$\bot$	(7), (2)
	(9)	$TurComp(f)$	supp; Case 2
$\therefore$	(10)	$\mathcal{C}_{\phi'}(m, f)$	(2), (3)
$\therefore$	(11)	$(c1')\text{–}(c3)_{\psi'}(m, f)$	from supp for $\exists$ elim on $\neg(\star)$ and provability
$\therefore$	(12)	$\neg(c1')\text{–}(c3)_{\psi'}(m, f)$	inspection: proofs wholly absent from $\mathcal{C}$
$\therefore$	(13)	$\bot$	(11), (12)
$\therefore$	(14)	$\bot$	*reductio*; proof by cases

A final remark to end the present Appendix: Note that the explicit deductive argument given immediately above conveys a general **method**, m, for showing that real learning $= \mathcal{RL}$ can't be achieved by other forms of limited learning. (Methods, or proof methods, are generalized proof "recipes" that can be composed and built up like computer programs. Proof methods were first introduced in (Arkoudas 2000), and extensive usage of proof methods can be found in (Arkoudas and Musser 2017).) This method m, given suitable input, produces a valid formal proof. All that needs to be done in order to follow the method is to shift out the set $\mathcal{C}$ of conditions to some other set $\mathcal{C}'$ that captures some alternative kind of ML, i.e. some alternative kind of limited learning *Xlearning*. For instance, Bayesian learning (*Blearning*) can by this method be proved to fail to yield real learning in a machine (or agent) that employs *Blearning*.

Acknowledgement. We are deeply appreciative of feedback received at PT-AI 2017, the majority of which is addressed herein. The first author is also specifically indebted to John Hummel for catalyzing, in vibrant discussions at MAICS 2017, the search for formal arguments and/or theorems establishing the proposition Hummel and Bringsjord co-affirm: viz., statistical machine learning simply doesn't enable machines to actually learn, period. Bringsjord is also thankful to Sergei Nirenburg for valuable conversations. Many readers of previous drafts have been seduced by it's-not-really-learning forms of learning (including worse-off-than artificial neural network (ANN) based deep learning (DL) folks: Bayesians), and have offered spirited objections, all of which are refuted herein; yet we are grateful for the valiant tries. Bertram Malle stimulated and guided our sustained study of types of learning in play in psychology^{+}, and we are thankful. Jim Hendler graciously read an early draft; his resistance has been helpful (though perhaps now he's a convert). The authors are also grateful for five anonymous reviews, some portions of which reflected at least partial and passable understanding of our logico-mathematical perspective, from which informal notions of "learning" are inadmissible in such debates as the present one. Two perspicacious comments and observations from two particular PT-AI 2017 participants, subsequent to the conference, proved productive to deeply ponder. We acknowledge the invaluable support of "Advanced Logicist Machine Learning" from ONR, and of "Great Computational

Intelligence" from AFOSR. Finally, without the wisdom, guidance, leadership, and raw energy of Vincent Müller, PT-AI 2017, and any ideas of ours that have any merit at all, and that were expressed there and/or herein, would not have formed.

References

Achab, M., Bacry, E., Gaïffas, S., Mastromatteo, I., Muzy, J.F.: Uncovering causality from multivariate Hawkes integrated cumulants. In: Precup, D., Teh, Y.W. (eds) Proceedings of the 34th International Conference on Machine Learning, PMLR, International Convention Centre, Sydney, Australia. Proceedings of Machine Learning Research, vol. 70, pp. 1–10 (2017). http://proceedings.mlr.press/v70/achab17a.html

Arkoudas, K.: Denotational proof languages. Ph.D. thesis, MIT (2000)

Arkoudas, K., Bringsjord, S.: Computers, justification, and mathematical knowledge. Minds Mach. **17**(2), 185–202 (2007)

Arkoudas, K., Musser, D.: Fundamental Proof Methods in Computer Science: A Computer-Based Approach. MIT Press, Cambridge (2017)

Bandura, A., Walters, R.H.: Social Learning Theory, vol. 1. Prentice-Hall, Englewood Cliffs (1977)

Bandura, A., Ross, D., Ross, S.A.: Transmission of aggression through imitation of aggressive models. J. Abnorm. Soc. Psychol. **63**(3), 575 (1961)

Barrett, L.: Beyond the Brain: How Body and Environment Shape Animal and Human Minds. Princeton University Press, Princeton (2015)

Bellman, A., Bragg, S., Handlin, W.: Algebra 2: Common Core. Pearson, Upper Saddle River (2012). Series Authors: Charles, R., Kennedy, D., Hall, B., Consulting Authors: Murphy, S.G

Boolos, G.S., Burgess, J.P., Jeffrey, R.C.: Computability and Logic, 4th edn. Cambridge University Press, Cambridge (2003)

Bringsjord, S.: What Robots Can and Can't Be. Kluwer, Dordrecht (1992)

Bringsjord, S.: Chess is too easy. Technol. Rev. **101**(2), 23–28 (1998). http://kryten.mm.rpi.edu/SELPAP/CHESSEASY/chessistooeasy.pdf

Bringsjord, S.: Psychometric artificial intelligence. J. Exp. Theor. Artif. Intell. **23**(3), 271–277 (2011)

Bringsjord, S.: The symbol grounding problem-remains unsolved. J. Exp. Theor. Artif. Intell. **27**(1), 63–72 (2015). https://doi.org/10.1080/0952813X.2014.940139

Bringsjord, S., Arkoudas, K.: The modal argument for hypercomputing minds. Theor. Comput. Sci. **317**, 167–190 (2004)

Bringsjord, S., Noel, R.: Real robots and the missing thought experiment in the Chinese room dialectic. In: Preston, J., Bishop, M. (eds.) Views into the Chinese Room: New Essays on Searle and Artificial Intelligence, pp. 144–166. Oxford University Press, Oxford (2002)

Bringsjord, S., Schimanski, B.: What is artificial intelligence? psychometric AI as an answer. In: Proceedings of the 18th International Joint Conference on Artificial Intelligence (IJCAI 2003), pp. 887–893. Morgan Kaufmann, San Francisco (2003). http://kryten.mm.rpi.edu/scb.bs.pai.ijcai03.pdf

Bringsjord, S., Zenzen, M.: Superminds: People Harness Hypercomputation, and More. Kluwer Academic Publishers, Dordrecht (2003)

Bringsjord, S., Bringsjord, E., Noel, R.: In defense of logical minds. In: Proceedings of the 20th Annual Conference of the Cognitive Science Society, pp. 173–178. Lawrence Erlbaum, Mahwah (1998)

Bringsjord, S., Kellett, O., Shilliday, A., Taylor, J., van Heuveln, B., Yang, Y., Baumes, J., Ross, K.: A new Gödelian argument for hypercomputing minds based on the busy beaver problem. Appl. Math. Comput. **176**, 516–530 (2006)
Chisholm, R.: Theory of Knowledge. Prentice-Hall, Englewood Cliffs (1966)
Davis, M., Weyuker, E.: Computability, Complexity, and Languages: Fundamentals of Theoretical Computer Science, 1st edn. Academic Press, New York (1983)
Dodig-Crnkovic, G., Giovagnoli, R. (eds.): Computing Nature: Turing Centenary Perspective. Springer, Berlin (2013). https://www.springer.com/us/book/9783642372247
Domjan, M.: The Principles of Learning and Behavior, 7th edn. Cengage Learning, Stamford (2015)
Gallistel, C.R.: Learning and representation. In: Learning and Memory: A Comprehensive Reference, vol. 1. Elsevier (2008) https://doi.org/10.1016/j.neuron.2017.05.021
Gettier, E.: Is justified true belief knowledge? Analysis **23**, 121–123 (1963). http://www.ditext.com/gettier/gettier.html
Gold, M.: Limiting recursion. J. Symb. Logic **30**(1), 28–47 (1965)
Goodfellow, I., Bengio, Y., Courville, A.: Deep Learning. MIT Press, Cambridge (2016). http://www.deeplearningbook.org
Goodstein, R.: On the restricted ordinal theorem. J. Symb. Logic **9**(31), 33–41 (1944)
Huitt, W.: Classroom Instruction. Educational Psychology Interactive (2003)
Hummel, J.: Symbolic versus associative learning. Cogn. Sci. **34**(6), 958–965 (2010)
Inhelder, B., Piaget, J.: The Growth of Logical Thinking from Childhood to Adolescence. Basic Books, New York (1958)
Jain, S., Osherson, D., Royer, J., Sharma, A.: Systems That Learn: An Introduction to Learning Theory, 2nd edn. MIT Press, Cambridge (1999)
Kearns, M., Vazirani, U.: An Introduction to Computational Learning Theory. MIT Press, Cambridge (1994)
Kitzelmann, E.: Inductive programming: a survey of program synthesis techniques. In: International Workshop on Approaches and Applications of Inductive Programming, pp 50–73. Springer (2009)
Lathrop, R.: On the learnability of the uncomputable. In: Saitta, L. (ed.) Proceedings of the 13th International Conference on Machine Learning, The conference was held in Italy, 3–6 July 1996, pp 302–309. Morgan Kaufman, San Francisco (1996). https://pdfs.semanticscholar.org/6919/b6ad91d9c3aa47243c3f641ffd30e0918a46.pdf
LeCun, Y., Bottou, L., Bengio, Y., Haffner, P.: Gradient-based learning applied to document recognition. Proc. IEEE **86**(11), 2278–2324 (1998)
Luger, G.: Artificial Intelligence: Structures and Strategies for Complex Problem Solving, 6th edn. Pearson, London (2008)
Mackintosh, N.J.: Conditioning and Associative Learning. Calendron Press, Oxford (1983)
Marblestone, A.H., Wayne, G., Kording, K.P.: Toward an integration of deep learning and neuroscience. Front. Comput. Neurosci. **10**(94) (2016). https://doi.org/10.3389/fncom.2016.00094
Moore, R.C.: Making the transition to formal proof. Educ. Stud. Math. **27**(3), 249–266 (1994)
Müller, V., Bostrom, N.: Future progress in artificial intelligence: a survey of expert opinion. In: Müller, V. (ed.) Fundamental Issues of Artificial Intelligence (Synthese Library), pp. 553–571. Springer, Berlin (2016)
Penn, D., Holyoak, K., Povinelli, D.: Darwin's mistake: explaining the discontinuity between human and nonhuman minds. Behav. Brain Sci. **31**, 109–178 (2008)

Putnam, H.: Trial and error predicates and a solution to a problem of Mostowski. J. Symbolic Logic **30**(1), 49–57 (1965)
Rado, T.: On non-computable functions. Bell Syst. Tech. J. **41**, 877–884 (1963)
Ross, J.: Immaterial aspects of thought. J. Philos. **89**(3), 136–150 (1992)
Russell, S., Norvig, P.: Artificial Intelligence: A Modern Approach, 3rd edn. Prentice Hall, Upper Saddle River (2009)
Schapiro, A., Turk-Browne, N.: Statistical learning. Brain Mapp. Encyclopedic Ref. **3**, 501–506 (2015)
Searle, J.: Minds, brains and programs. Behav. Brain Sci. **3**, 417–424 (1980)
Shalev-Shwartz, S., Ben-David, S.: Understanding Machine Learning: From Theory to Algorithms. Cambridge University Press, Cambridge (2014)
Stewart, J.: Calculus. 8th edn. Cengage Learning, Boston (2016), We refer here to an electronic version of the print textbook. The "Student Edition" of the hard-copy textbook has an ISBN 978-1-305-27176-0
Titley, H.K., Brunel, N., Hansel, C.: Toward a neurocentric view of learning. Neuron **95**(1), 19–32 (2017)
Turing, A.: Computing machinery and intelligence. Mind LIX **59**(236), 433–460 (1950)
Wason, P., Johnson-Laird, P.: Psychology of Reasoning: Structure and Content. Harvard University Press, Cambridge (1972)
Wolpert, D.H.: The lack of a priori distinctions between learning algorithms. Neural Comput. **8**(7), 1341–1390 (1996)

Where Intelligence Lies: Externalist and Sociolinguistic Perspectives on the Turing Test and AI

Shlomo Danziger(✉)

Department of Cognitive Science, Hebrew University of Jerusalem, Jerusalem, Israel
shlomo.danziger@mail.huji.ac.il

Abstract. Turing's Imitation Game (1950) is usually understood to be a test for machines' intelligence; I offer an alternative interpretation. Turing, I argue, held an externalist-like view of intelligence, according to which an entity's being intelligent is dependent not just on its functions and internal structure, but also on the way it is perceived by society. He conditioned the determination that a machine is intelligent upon two criteria: one technological and one sociolinguistic. The Technological Criterion requires that the machine's structure enables it to imitate the human brain so well that it displays intelligent-like behavior; the Imitation Game tests if this Technological Criterion was fulfilled. The Sociolinguistic Criterion requires that the machine be perceived by society as a potentially intelligent entity. Turing recognized that in his day, this Sociolinguistic Criterion could not be fulfilled due to humans' chauvinistic prejudice towards machines; but he believed that future development of machines displaying intelligent-like behavior would cause this chauvinistic attitude to change. I conclude by discussing some implications Turing's view may have in the fields of AI development and ethics.

Keywords: Alan Turing · Turing Test · Imitation Game · Artificial Intelligence Externalism

1 Introduction

Can machines be intelligent? In his 1950 paper "Computing Machinery and Intelligence", Alan Turing introduced the Imitation Game (IG) in which a machine tries to imitate human intellectual behavior to such an extent that a human interrogator mistakes the machine for a human. The Imitation Game, later known as the Turing Test, has been commonly understood to be a test for intelligence: A machine that does well in the Game must be regarded as intelligent.

Turing's paper, considered a classic in the fields of AI and philosophy of cognitive science, raises many difficulties, and several attempts have been made throughout the years to explain Turing's intentions (see Saygin et al. 2000; Oppy and Dowe 2011). The commonality between almost all interpretations offered is that they see the IG as a test for *intelligence*. In this essay I reject that widely accepted view and propose an alternative way of understanding Turing's paper and his approach to intelligence. I shall show

V. C. Müller (Ed.): PT-AI 2017, SAPERE 44, pp. 158–174, 2018.
https://doi.org/10.1007/978-3-319-96448-5_15

that Turing holds an externalist-like view of intelligence, which bears resemblance to Wittgenstein's approach to the mental domain (Wittgenstein 2009; 1958). My reading of Turing is based on remarks he makes in other publications (especially Turing 1947; Turing 1948; Turing et al. 1952) and on careful reading of the 1950 paper itself.[1]

In Sect. 2 I introduce the two main streams of interpretation of the IG that have been suggested by Turing's commentators, and I mention some of the problems they raise. In Sect. 3 I discuss some remarks Turing makes in his earlier publications that I believe may shed light on the way he understands the term "intelligence". Sections 4 and 5 are the heart of this essay: In Sect. 4 I present my *technological* interpretation, pointing out what the IG is intended to test and what is outside its scope, and I discuss Diane Proudfoot's interpretation of the IG. In Sect. 5 I present Turing's prediction that in the future the meanings of concepts will change, allowing machines to be deemed "intelligent"; and I offer a critical look at this line of thought. In Sect. 6 I discuss some implications of Turing's approach vis-a-vis AI development and ethics.

2 Imitation Game: Common Interpretations

In his 1950 paper Turing describes the IG as follows: A human interrogator communicates via teletext with another human and with a machine, without knowing which is which. The interrogator must try to ascertain which of the two beings is human by asking each of them any questions whatsoever; each of the beings must try to convince the interrogator that *it* is the human, by answering in a human-like manner. According to the accepted reading, the logical structure of Turing's argument is as follows:

(1) *A machine that does well in the IG – a machine that successfully imitates human intellectual behavior to the extent that the interrogator cannot tell the difference – must be regarded as an intelligent (or a thinking[2]) entity*

(2) *Machines that do well in the IG can indeed be constructed*

Therefore,

(3) *Intelligent machines are possible*

The role of the IG within the argument seems puzzling, and several attempts have been made to explain Turing's paper. Following Diane Proudfoot's classification (2013),

[1] My reading is supported by several pieces of non-orthodox commentaries of Turing scattered throughout the literature, such as Whitby (1996), Boden (2006, pp. 1346–1356), Sloman (2013), and especially Proudfoot (2005; 2013).
Some of the arguments suggested in this paper have appeared in Danziger (2016).

[2] As Piccinini (2000), Proudfoot (2013), and others have pointed out, Turing uses the terms "thought" and "intelligence" interchangeably. Although I will not differentiate between the terms, for reasons of uniformity I shall usually use the term "intelligence".

I shall point out two main streams of interpretation suggested in the literature and briefly mention some of the problems they raise.[3]

2.1 Behavioristic Interpretations

According to *behavioristic* interpretations, Turing held that "intelligent-like behavior" is the *definition* of intelligence: Any system (that is, any organism or machine) whose behavior is similar to that of an intelligent entity (a human) is itself intelligent. French (1990, p. 53) exemplifies this operational definition of intelligence by saying that according to Turing, "[w]hatever *acts* sufficiently intelligent *is* intelligent."[4]

Behavioristic interpretations – the most common way of understanding Turing's paper – raise several difficulties. For according to these interpretations, Turing seems to be going against a very basic human intuition – that mental occurrences and properties are *internal* traits, independent of *external* actions.[5] Also, it is not clear why, according to Turing, intelligence is tested for by verbal behavior, and not by any other human cognitive faculty.

2.2 Inductive Interpretations

As opposed to behavioristic interpretations, *inductive* interpretations claim that Turing indeed sees intelligence as an *internal* property of a system – one that takes place inside it, and not as an external, behavioristic trait. According to these interpretations, Turing holds that a system's success in the IG gives us *good grounds* to assume it possesses the property of intelligence, based on the success that this kind of attribution has shown hitherto; and in the absence of contradicting evidence, we should regard such a system as intelligent.[6]

Inductive interpretations raise problems too, for according to them Turing implies the following: "Due to our long-learned experience, we humans *tend* to attribute (the internal property of) intelligence to systems on the basis of their (external) behavior; therefore, we *must* attribute intelligence to systems that display intelligent behavior (i.e.,

[3] Almost all commentaries on – and attacks against – Turing's paper can be classified into one of the two streams of interpretation described in Sects. 2.1 and 2.2; see Proudfoot (2013) for detailed analysis and critique of these interpretations. Other ways of classification can be found in Saygin et al. (2000) and in Oppy and Dowe (2011).

[4] Also Searle's interpretation of Turing is behavioristic: "The Turing Test is typical of the tradition in being unashamedly behavioristic and operationalistic" (Searle 1980, p. 423; cf. next footnote). References to other behavioristic interpretations can be found in Proudfoot (2013), Copeland (2004, pp. 434–435), and Moor (2001, pp. 81–82).

[5] This is the crux of perhaps the two most well-known arguments against the IG, namely, Searle's Chinese Room (Searle 1980) and Block's Blockhead / Aunt Bubbles Machine (Block 1981; 1995): Intelligence, they maintain, cannot be captured in behavioral terms alone. (Note that both arguments belong to the behavioristic school of interpretation, in that they assume that the IG is intended to be a behavioral test for intelligence.)

[6] The main proponent of the school of inductive interpretations is Moor (1976; 2001). Other inductive interpretations can be found in Watt (1996) and Schweizer (1998).

do well in the Game)." This move from "we tend" to "we must" seems strange, for a system's external behavior can be misleading, causing us – the observers – to adopt a false picture of the system's inner state; why *must* we trust behavior so blindly?

3 Turing 1947 and 1948: Machines Can Think

Before I present my interpretation I would like to refer to Turing's earlier publications from 1947 and 1948, which can shed light on his approach and give us a better understanding of his 1950 paper.

3.1 Intelligence as an Emotional Concept[7]

Towards the end of his 1948 paper, after a lengthy discussion of ways in which machinery can imitate various human cognitive functions, Turing writes (1948, p. 431, my italics):

> **Intelligence as an emotional concept**
>
> The extent to which we regard something [a machine or an organism, SD] as behaving in an intelligent manner is determined as much by *our own state of mind and training* as by the properties of the object under consideration. If we are able to explain and predict its behaviour or if there seems to be little underlying plan, we have little temptation to imagine intelligence. With the same object therefore it is possible that one man would consider it as intelligent and another would not; the second man would have found out the rules of its behaviour.

Turing could have said that in a case of different epistemic viewpoints one of the viewers would be *wrong*, but he chose to say otherwise: A given system could be both "intelligent" and "non-intelligent" at the same time if viewers having two such opposing viewpoints existed. Contrary to the way he was understood by behavioristic and inductive interpretations, Turing implies here that "intelligence" cannot be given a clear-cut definition in terms of *behavior* or in terms of *internal properties*; one system having a single set of behavior and internal properties allows for two opposing viewpoints, and neither would be wrong.

In saying that intelligence is an "emotional concept" Turing is referring to the emotions and reactions of the people perceiving the system, and thus points out the major role of the environment in deeming a system "intelligent". Intelligence, so to speak, is in the eye of the beholder: What defines certain systems as intelligent is, first and foremost, the fact that we humans perceive *those* systems, and not others, as intelligent. All systems – including intelligent ones – are mere physical mechanisms; what makes a system unique and defines it as "intelligent", says Turing, is the viewpoint of the people in its environment.

[7] The ideas in this section draw partly on Proudfoot (2005; 2013). As I shall show later, my inter pretation of Turing differs from Proudfoot's in small but crucial points; to prevent inaccuracies I shall refrain for now from mentioning her take on the subjects discussed, despite my great debt to her work.

Note that Turing's methodology reveals his approach to intelligence. Turing adopts a Wittgensteinian-like methodology in his analysis of the term "intelligence", reviewing cases in which people would or would not perceive systems as intelligent and attribute intelligence to them (Turing 1947, p. 393; 1948, pp. 412, 431). But by limiting the discussion to analysis of humans' *reactions* to systems and to the way the term "intelligence" is used in ordinary language, and by refusing to provide any further definition of intelligence (cf. Turing et al. 1952, p. 494), Turing reveals that in his view, what matters is the question of whether machines would be perceived as intelligent by human society. If they would be – then they could be said to be "intelligent", and no further inquiry would be needed (i.e., there would be no need to ask if they are "really" intelligent, according to some real-but-unknown definition that exists "out there").

Turing's approach thus described highly resembles what Coeckelbergh (2010) and Torrance (2014) call the *social-relationist* perspective (as opposed to the *realist* perspective).[8] Following this terminology, Turing's approach can be formalized as the following *premise*:

Social-Relationist Premise: *A system (organism or machine) perceived as intelligent by human society is an intelligent system*[9]

Now, given that "an intelligent system" is logically equivalent to "a system that is *perceived* as intelligent" (*Social-Relationist Premise*), the question

(Q1) *Can machines be intelligent?*

can be rephrased as

(Q2) *Is it possible for machines to be perceived as intelligent?*

In order to answer question (Q2) we must first find what *causes people* to perceive certain systems as intelligent:

(Q2.1) *What would be a sufficient condition for a system to be perceived as intelligent?*[10]

Hence, in our attempt to understand the criteria for intelligence, we find that before delving into questions in the domains of cognition and computation pertaining to the system's structure and functions, we must first focus on the fields of sociology and psychology, and ask questions pertaining to the people in the system's environment: What causes *human society* to regard a system – organism or machine – as intelligent? Once we answer the sociological questions we can proceed to the technological ones, regarding the system's functions and structure. Let us ask, therefore: What would be a sufficient condition for a system to be perceived as intelligent? What properties or abilities would an intelligent machine have?

---

[8] Turing's approach bears resemblance also to Dennett's "intentional stance" (Dennett 1987a).

[9] In Sects. 3.2 and 4.3 I shall bring further textual evidence for this being Turing's approach, and shall briefly discuss what might have motivated Turing into adopting such a stance.

[10] Turing is trying to *prove* that the existence of an intelligent machine is possible, and is not merely *asking* if it possible. Therefore he will try to show that machines fulfill a *sufficient* condition for being (perceived as) intelligent, and will put less emphasis on the *necessary* conditions.

3.2 Sufficient Conditions for Intelligence

In the passage titled "Intelligence as an emotional concept" quoted above (Sect. 3.1), Turing claims that when we encounter a simple system, one that "we are able to explain and predict its behaviour," we have "little temptation to imagine intelligence," and so we experience it as a mere mechanistic, non-intelligent system (Turing 1948, p. 431). According to Turing, the reason no machine has ever been perceived by humans as intelligent is that all machines that humans have ever encountered were of very limited character (1948, p. 410); no machine had ever displayed sophisticated human-like cognitive abilities involving learning, such as the ability to learn from experience and the ability to modify one's own "programming" (1947, pp. 392–393). But if, says Turing, such a "learning machine" were built – it *would* be experienced by humans as intelligent (1947, p. 393, my italics):

> Let us suppose we have set up a machine with certain initial instruction tables [programs, SD], so constructed that these tables might on occasion, if good reason arose, modify those tables. One can imagine that after the machine had been operating for some time, the instructions would have altered out of all recognition, but nevertheless still be such that one would have to admit that the machine was still doing very worthwhile calculations. Possibly it might still be getting results of the type desired when the machine was first set up, but in a much more efficient manner. In such a case one would have to admit that the progress of the machine *had not been foreseen* when its original instructions were put in. It would be like a pupil who had learnt much from his master, but had added much more by his own work. When this happens *I feel that one is obliged to regard the machine as showing intelligence.*

Note the last sentence: According to Turing, a "learning machine", programmed in such a sophisticated way that it could modify its own code, would arouse a feeling of surprise in its observers, and they would find themselves *regarding it* as showing intelligence. In his 1948 paper Turing repeats this prediction when discussing the would-be reaction of a human playing chess against a machine (as part of an early version of the IG; p. 431). In fact, he seems to say that he himself had actually reacted in such a way when encountering a chess-playing machine (1948, p. 412). Turing's descriptions of these would-be reactions (or actual reactions) of humans to such machines amount to his claiming that the conjunction of the properties of a "learning machine" is a *sufficient condition* for this machine to be perceived as an intelligent system:[11]

Sociological Claim: *A "learning machine" would be perceived by human society as intelligent*

Another important claim Turing makes in his 1947 and 1948 publications is that "learning machines" like the one just discussed *can* be built. According to Turing, the digital computer – which was then being developed – could carry out any task that the human brain could, and could therefore display the human-like cognitive abilities needed for being a "learning machine". (The digital computer will be discussed later in greater length.) As opposed to the "very limited character of the machinery which has been used

[11] There is no need to point out here which properties of the "learning machine" are necessary conditions for perceiving a system as intelligent; all that is being claimed is that a "learning machine" indeed *has* these properties, whatever they may be.

until recent times" which "encouraged the belief that machinery was necessarily limited to extremely straightforward, possibly even to repetitive, jobs" (Turing 1948, p. 410), the digital computer *would* be able to learn from experience, change its own programming, and display any other property of a "learning machine":

Minor-Technological Claim: *It is possible to build a "learning machine"*[12]

3.3 The Logical Structure of Turing's Argument

The logical structure of Turing's argument in his 1947 and 1948 papers is as follows:

Minor-Technological Claim: *It is possible to build a "learning machine"*

Sociological Claim: *A "learning machine" would be perceived by human society as intelligent*

- **Conclusion:** *It is possible to build a machine that would be perceived by human society as intelligent*

Social-Relationist Premise: *A system (organism or machine) perceived as intelligent by human society is an intelligent system*

- **Conclusion:** *It is possible to build an intelligent machine* ***(Q.E.D)***

To sum up: In his 1947 and 1948 publications, Turing argues that there *can* be intelligent machines. He claims that the construction of "learning machines" is a technological challenge that *can* be met (*Minor-Technological Claim*), and claims that these machines would inevitably be *perceived* as intelligent by human society (*Sociological Claim*) and would thereby *be* "intelligent machines" (*Social-Relationist Premise*). Turing shifts the focus from technological questions regarding the system's internal structure to sociological questions regarding the people in the system's environment. In doing so, he sidesteps the need to define "intelligence" (or "thought"); regardless of what the definition of intelligence is, if human society were to perceive a machine as intelligent – it would be correct to say that it *is* intelligent.

4 Turing 1950: Technological Interpretation

I shall now turn to analyze Turing's famous 1950 paper, where he introduces the well-known Imitation Game. I shall offer my "technological" interpretation and claim that in 1950 Turing retreats from the stance he presented in 1947 and 1948, realizing that the *Sociological Claim* (Sect. 3.2), according to which machines with special functions would be perceived by society as intelligent entities, was naïve and perhaps too optimistic.

[12] It will later become clear why this claim is labeled "minor".

4.1 From Specific Abilities to All-Encompassing Imitation Ability

In the 1947 and 1948 publications discussed above, Turing claims that if a machine were to display the abilities of a "learning machine" (learning from experience, reprogramming itself, etc.), it would inevitably be perceived as intelligent by human society (*Sociological Claim*); and he claims that machines *can*, in principle, display these abilities (*Minor-Technological Claim*). In his 1950 paper, though, Turing aims higher. He no longer tries to convince the reader that the digital computer could imitate *some ability or another* (however important that ability may be for being considered "intelligent"), but claims that the digital computer can imitate the entire human cognitive system, as it can imitate the human brain *as a whole*.[13]

Turing's confidence that machines could do so is based on the strong imitation ability of the "universal machine" (later known as the "Turing machine") introduced in his 1936 paper. According to Turing, each function of the human brain could be imitated *closely enough* by a "digital state machine"; all functions of all digital state machines could be fully imitated by a universal machine; hence, all functions of the human brain could be imitated closely enough by a universal machine. This machine could, in principle, do anything a human brain can do; it could successfully carry out any human cognitive task.

Turing thus moves from discussing machines that could display specific, unique abilities that would be sufficient for intelligence ascription (1947, 1948), to discussing machines that could imitate the entire human cognitive system (1950). Machines of the latter kind would have all the abilities that machines of the former kind had, and many more. If such brain-imitating machines (machines of the latter kind) were built, we could answer question (Q2) above – "Is it possible for machines to be perceived as intelligent?" – with an unequivocal "yes", without needing to determine which abilities, exactly, are "responsible" for a system's being perceived as intelligent. For *whatever* these abilities might be – we would know for certain that they *could* be realized by these machines that can do everything the human brain does.

4.2 The Imitation Game and the Technological Criterion

From a technological aspect, these machines that satisfactorily imitate any brain function would be quite sophisticated. The Imitation Game is a means to check if the *technological challenge* of building such sophisticated machines has been met. The Game tests if a given machine acts so much like a human brain that one cannot differentiate between the two; in other words, it tests if a machine's behavior is *intelligent-like*.[14]

A machine that does well in the IG – a machine that has intelligent-like behavior – fulfills what I call the *Technological Criterion* for intelligence. The Technological Criterion requires that a system's structure (or program) enables it to imitate the human brain very well, to the extent that a human interrogator experiences it as

[13] Hodges (2014, p. 530) explains in a similar way the difference between Turing's 1948 and 1950 papers.

[14] "Intelligent-like behavior" may be roughly defined as "behavior that under regular circumstances cannot be differentiated from that of a human".

intelligent. The IG, therefore, tests if a given system fulfills the Technological Criterion for intelligence.[15]

Turing designed the IG as a test involving verbal interaction because of the huge technological challenge this posed. Constructing machines that successfully engage in real-time human conversation seemed to him as an extremely difficult task from a technological/algorithmic aspect, on par with – if not harder than – constructing machines that possess any other cognitive ability related to "learning from experience" (Sect. 3.2). Turing, I claim, saw the IG as an "AI-complete" problem (Mallery 1988, p. 47, fn. 96): If a machine could do well in the Game, it could accomplish practically *anything* related to AI. (Even today, programming a computer to do well in Natural Language Processing tasks is considered one of the greatest challenges in AI development.) According to Turing, if a machine were constructed in such a way that it displayed intelligent-like behavior and caused a human interrogator to experience it as intelligent, its engineers could lean back in satisfaction knowing that the technological challenge of creating brain-imitating, intelligent-like machines had been met. It is with regard to this *technological* challenge that Turing makes the following prediction (1950, p. 442):

> I believe that in about fifty years' time it will be possible to programme computers, with a storage capacity of about 10^9, to make them play the imitation game so well that an average interrogator will not have more than 70% chance of making the right identification after five minutes of questioning.

In 1950, then, Turing predicts that brain-imitating machines with *intelligent-like* behavior could indeed be built. Did he claim that these machines would be *intelligent*?

4.3 The Iron Curtain of Sociolinguistic Restrictions and the Sociolinguistic Criterion

Turing's 1947 and 1948 publications imply that "learning machines" that displayed abilities such as learning from experience, reprogramming themselves, etc. would be perceived by human society as intelligent (*Sociological Claim*, Sect. 3.2) and would thus *be* intelligent (*Social-Relationist Premise*, Sect. 3.1). One would therefore expect that in his 1950 paper Turing would say the same of machines that fulfilled the Technological Criterion and displayed intelligent-like behavior; those machines, one would presume, would surely be perceived by society as intelligent entities. But careful reading reveals that in his 1950 paper Turing retreats from the naïve view presented in his earlier publications. He realizes that while humans perceive other humans as "intelligent entities", humans perceive machines *a–priori* as "non-intelligent entities", due to a *chauvinistic attitude* towards machines that humans have (and may be unaware of). Turing recognizes that even if a machine were to fulfill the Technological Criterion for intelligence by doing well in the IG, *human prejudice* would preclude any possibility of machines being thought of as intelligent. In a 1952 radio broadcast Turing describes this prejudiced attitude (Turing et al. 1952, p. 500, my italics):

[15] The *Technological Criterion* (1950) is closely connected to the *Minor-Technological Claim* (1947, 1948) but is more "demanding" (as explained above, Sect. 4.1); that is why the 1947–1948 claim is labeled "minor".

> If I had given a longer explanation [of how to construct a machine that would have the ability to identify analogies, SD] … you'd probably exclaim impatiently, 'Well, yes, I see that a machine could do all that, but *I wouldn't call it thinking*.' As soon as one can see the cause and effect working themselves out in the brain, *one regards it as not being thinking, but a sort of unimaginative donkey-work.*

Turing realizes that the term "intelligent machine" is an *oxymoron*: People think of machines as systems whose workings they can understand; and a system whose workings could be understood is seen as consisting of *mere mechanistic processes*, devoid of any intelligence.[16] Turing hence recognizes that the concept of intelligence could not be ascribed to machines, by definition. This idea strongly resembles one that appears in Wittgenstein's later writings (2009: §360):

> But surely a machine cannot think! – Is that an empirical statement? No. We say only of a human being and what is like one that it thinks. We also say it of dolls; and perhaps even of ghosts.[17]

It seems that according to Wittgenstein, even if there were a person who did not have that a-priori chauvinistic attitude towards machines and *did* see them as potentially intelligent systems (as Turing might have), that person would run into the iron curtain of language conventions that prevent us from applying the term "intelligent" to machines in the literal sense. Machines cannot be said to be intelligent, Wittgenstein would say, because of the way the terms "intelligent" and "machine" are used in language. This, presumably, is the reason why Turing, despite his being convinced that machines could do *anything* a brain could and could behave in an intelligent-like manner, refrains in his 1950 paper from explicitly stating that such machines would be *intelligent*.

To frame it differently: Turing understood that alongside the Technological Criterion for intelligence, there also exists what I call the *Sociolinguistic Criterion*: the requirement that the system be such that its kind is perceived by society as potentially intelligent. According to the Sociolinguistic Criterion, a system that is perceived a-priori by society as non-intelligent (i.e., belongs to a species or a kind that is perceived a-priori as non-intelligent) cannot be said to be intelligent, by definition. Turing realized that in the year 1950, the Sociolinguistic Criterion, which is dependent on the system's environment (human society), could not be fulfilled with regard to machines. Doing well in the IG – fulfillment of the Technological Criterion – would show only that the system's behavior is *intelligent-like*, but this would not break the sociolinguistic barricade seeded in the minds of humans that causes them to see machines a-priori as non-intelligent entities.[18]

[16] Bringsjord et al. (2001) mention a similar idea of "restricted epistemic relation": They suggest the "Lovelace Test" for intelligence in which "not knowing how a system works" is a necessary condition for attributing intelligence to it. The fundamental difference between the Lovelace Test and Turing's IG will be explained later (fn. 23).

[17] Other clear remarks of Wittgenstein in this spirit are Wittgenstein (2009, §281) and Wittgenstein (1958, p. 47). The similarity between Turing's and Wittgenstein's ideas here has been pointed out also by Boden (2006, p. 1351) and Chomsky (2008, p. 104).

[18] The *Sociolinguistic Criterion* (1950) is closely connected to the *Sociological Claim* (1947, 1948) mentioned in Sect. 3.2. The addition of the "linguistic" component will soon be explained.

This is how we should understand Turing's enigmatic remark (pun intended) in his 1950 paper, which follows his prediction quoted above in Sect. 4.2 (Turing 1950, p. 442, my italics):

> The original question, "Can machines think?" I believe to be *too meaningless* to deserve discussion.

This question, says Turing, is "meaningless" – in the Wittgensteinian sense: Machines are not things that fall under the concept of "thinking" (or "intelligence"). And this is why Turing, at the outset of his paper, replaces the question "Can machines think?" with the question of whether or not machines can do well in the IG. While Turing's original question touches on both Technological and Sociolinguistic Criteria (and contains a built-in negative answer), the new question relates to the Technological Criterion alone; the IG offers a way to check if the Technological Criterion has been satisfied.[19]

To recap: Turing tackles the question "Can machines think?" by saying, "Look, a machine can do anything a brain can do. Anything. It can behave so similar to a human brain that it can even do well in the Imitation Game. Does this mean that a machine can *think*? No. But that is not because there is something it cannot *do*; it's not like the fact that I can't climb a very steep cliff due to the limits of my strength. A machine cannot think because the term 'thinking machine' is an *oxymoron*; it cannot be said to think because of the way the terms 'think' and 'machine' are used in language."

4.4 Proudfoot's Interpretation: The Imitation Game as a Test for Intelligence

Before presenting the last stage in Turing's argument I must mention the writings of Diane Proudfoot (2005; 2013), which greatly inspired my interpretation presented thus far. In her comprehensive and enlightening papers, Proudfoot promotes an externalist-like interpretation of Turing, according to which intelligence is (what she calls) a *response-dependent* property (Proudfoot 2013, p. 398):

> Turing's remarks suggest something like this schema: *x* is *intelligent* (or *thinks*) if, in an unrestricted computer-imitates-human game, *x* appears intelligent to an average interrogator.

[19] At this point one might raise the following objection: "Your reading boldly ignores the next sentence in Turing's paper, in which he supposedly predicts that in fifty years there would be intelligent machines (1950, p. 442): 'Nevertheless I believe that at the end of the century the use of words and general educated opinion will have altered so much that one will be able to speak of machines thinking without expecting to be contradicted.' This implies that Turing identified success in constructing machines that do well in the Game – with success in creating *intelligent* machines; the timeframe in both sentences is the same (the year 2000), and so they seem to be referring to the *same* futuristic occurrence!" My reply, in short, is that this objection is based on an incorrect – albeit very common – reading of the passage in Turing's paper. Turing, I claim, makes two different predictions here, and these predictions are connected *causally* but not *logically*. "Doing well in the IG" is not the same as "being intelligent". The IG, I insist, is not a test for intelligence, but a test only for the Technological Criterion of intelligence: it tests if a system's behavior is *intelligent-like*. (I shall return to this issue in Sect. 5.1.)

Turing, according to this, holds that what defines a system as intelligent is the attitude of an *average interrogator*, who is supposed to represent society in an unadulterated, impartial way (like a jury in court, perhaps).

My interpretation is close to Proudfoot's but differs from it in crucial points. The main difference is that her interpretation, like behavioristic and inductive ones mentioned earlier, sees the IG as a test for *intelligence*. But Turing, I claim, did not intend the Game to be a test for intelligence. Intelligence requires that *society* perceive the system as intelligent, and the IG does not test *that*. It tests only whether a *single, isolated* interrogator *temporarily* experiences the system as intelligent during the few moments in which the interrogator does not yet know that s/he is conversing with a machine. Turing himself refers to the IG as an "imitation test" (Turing et al. 1952, p. 503; cf. p. 495); indeed, the Game is a test for the Technological Criterion only, a test for *intelligent-like behavior*.[20] A test for *intelligence*, on the other hand, would require the fulfillment of the Sociolinguistic Criterion too.[21]

5 Shifts in the Meanings of Concepts

5.1 Turing's Prediction

According to my reading, in 1950 Turing acknowledged that machines could not "be intelligent" or "think", due to humans' prejudiced attitude towards machines and the way the terms "intelligence", "thinking" and "machine" were used in language. But the way people use words can change. Here is Turing's remark quoted above (Sect. 4.3) followed by his prediction (1950, p. 442, my italics):

> The original question, "Can machines think?" I believe to be too meaningless to deserve discussion. Nevertheless I believe that at the end of the century *the use of words and general educated opinion will have altered so much* that one will be able to speak of machines thinking without expecting to be contradicted.

Turing believed that technological progress – development of machines that do well in the IG and show intelligent-like behavior – would eventually cause humans' chauvinistic attitude towards machines to erode. The term "intelligent machine" would then no longer constitute an oxymoron, as the meanings of the concepts "intelligence" and

[20] Aaron Sloman, too, sees the IG as Turing's way of defining a *technological* challenge, and not as a test for intelligence (Sloman 2013). In an earlier version of his paper Sloman expresses his dissatisfaction with the orthodox interpretations of the IG; I found myself wholly identifying with his words (my italics): "It is widely believed that Turing proposed a test for intelligence. This is false. *He was far too intelligent to do any such thing,* as should be clear to anyone who has read his paper…"
(Source: http://www.cs.bham.ac.uk/research/projects/cogaff/misc/turing-test.html. Accessed Oct. 11, 2017.)

[21] To develop this point further: A real test for a system's intelligence would check if the system is perceived as intelligent by society as a whole, in an ongoing manner, in normal life situations. But if that were to happen there would be no need for an intelligence test, because "society perceiving a system as intelligent" is the *definition* of a system's being intelligent, not a *sign* of it! (See the *Social-Relationist Premise*, Sect. 3.1.)

"machine" would have changed; the concept of "intelligence" would then be applicable to machines.[22] In that future state, the Sociolinguistic Criterion would have been fulfilled, as society would indeed see machines as *potentially intelligent* systems. If a certain machine also fulfilled the Technological Criterion, it would be perceived as an intelligent system, and would rightly be said to be an intelligent machine.

In conclusion, Turing thought both that machines could do well in the IG *and* that intelligent machines were possible. But contrary to the accepted reading presented in Sect. 2, Turing saw the connection between the IG and intelligence not as a *logical* connection, but as a *causal* one. He did not claim that machines that do well in the IG *are* intelligent, but that success of machines in the IG would eventually *cause* people to see machines as intelligent.

The widespread misunderstanding of the IG can be further clarified by differentiating between *descriptive* and *normative* readings. While my interpretation sees Turing's account as descriptive ("That is how people *would* react upon their encounter with machines that do well in the IG"), Turing's commentators – who thought he intended the IG to be a test for intelligence – understand him as giving a normative account ("That is how we *should* regard machines that do well in the IG"). I am of the opinion that the normative reading is an incorrect understanding of Turing's paper.[23]

5.2 A Critical Look at Turing's Prediction

Technically speaking, Turing was too optimistic; the year 2000 has passed and we still do not perceive of machines as thinking/intelligent entities. (In fact, it has been stressed that the only time we say of a computer that it is "thinking" is when it gets stuck.) Turing's prediction that the meanings of concepts will change may indeed come about sometime in the future. However, I want to suggest the opposite scenario: If we develop machines that have intelligent-like abilities and act very much like humans, we might stop identifying those abilities with intelligence, just like we stopped seeing "winning the chess game" as a sign for intelligence in 1997, when "Deep Blue" beat chess champion Kasparov.[24] "Tesler's Theorem" expresses this point elegantly (Larry Tesler, ca. 1970):

> Intelligence is whatever machines haven't done yet.

---

[22] This is how Turing's prediction was understood by Mays (1952, pp. 149–151), Beran (2014) and others. (Piccinini 2000 understands that Turing *hopes* such a change will occur.) For an illuminating discussion regarding the possibility of this sort of change (not concerning Turing's paper) see Torrance (2014).

[23] The main difference between Turing's IG and Bringsjord et al.'s "Lovelace Test" mentioned above (fn. 16) is that while the IG is descriptive, the Lovelace Test is normative (see Bringsjord et al. 2001, p. 9).

[24] Sloman makes a similar point and says that while computers are now doing much cleverer things, "increasing numbers of humans have been learning about what computers can and cannot do" (Sloman 2013, p. 3). Indeed, getting humans to attribute intelligence to machines might become harder with time.

Hence, an engineer might do everything philosophers said should be done in order to develop intelligent systems – only to discover that the philosophers keep changing the rules.

Moreover: If machines start acting like humans, we humans might find ourselves changing the way *we* behave in order to distance ourselves from machines so that we remain the "superior race". Our new behavior will then become the new standard for intelligence, the behavior in virtue of which we perceive systems as intelligent. In such a scenario, humans would make sure to act in a unique way so that society (whomever that may include) clearly understands that humans, and not machines, are the *real* bearers of intelligence.

6 Implications of Turing's View

6.1 Externalism and AI Development

When discussing the criteria for intelligence, Turing focuses on the way a system is perceived by human society, rather than on the system's functions or internal structure. Turing, therefore, can be said to hold an *externalist-like* view of intelligence (and of the mental domain in general[25]). A system's functions and internal structure may indeed play an important role in shaping society's attitude towards the system (thereby circuitously contributing to the definition of the system as "intelligent"), but by no means are they the only factors.

I think Turing's approach may lead to interesting insights regarding the ongoing attempt to develop intelligent systems. Recent years have seen efforts in the fields of technology and algorithm development to devise human-like intelligent systems (including ongoing attempts to write computer programs that would "pass the Turing Test"). Turing's approach teaches us that it would be wise to pay attention also to the major role that *society* plays in determining the intelligence of a system. This might lead developers to put more emphasis on properties that had once been considered irrelevant to intelligence. One such property is the external appearance of the system. Another is the way the system was developed: Humans might be more inclined to attribute intelligence to a system that, like themselves, went through a long and tedious learning process, as opposed to a system that had a whole database injected into it; the latter might seem less human-like and would be less likely to be perceived as intelligent.[26]

Awareness of the sociological dynamics involved in determining a system's intelligence may also teach us why we must have *patience* when trying to construct intelligent

[25] In his brief reply to the "Argument from Consciousness", Turing seems to claim that if a machine did well in the IG it would be *perceived* as conscious too (1950, pp. 445–447; see Michie 1993, pp. 4–7. But cf. Copeland 2004, pp. 566–567). I am of the opinion that likewise intelligence, also consciousness and other mental phenomena can be explained in terms of being perceived by society; I plan to discuss this elsewhere.

[26] Both properties mentioned were suggested by Mays (1952), in his analysis of Turing's 1950 paper. Interestingly, Turing himself seems to have viewed both properties as insignificant for intelligence attribution (see Turing 1950, p. 434; Davidson 1990). For a list of other properties that might shape humans' attitude towards machines, see Torrance (2014).

systems. As pointed out by Beran (2014, pp. 54–55), for machines to be intelligent, humans must first adapt to the idea of intelligent machinery, and this change of attitude may take time. In addition, developers should be willing to accept that humans' *stubborn chauvinistic attitude* might completely prevent the possibility of perceiving machines as intelligent. If machines were to acquire abilities considered paradigmatic intelligent-like behavior (such as learning from experience), people might stop seeing those abilities as central to intelligence, and "replace" them with others. This would be equivalent to changing the criteria for intelligence, rendering machines as *non-intelligent* again and again (*a-la* Tesler's Theorem), every time it seems as though they "almost got there."

6.2 Ethics

According to Turing, if humans' chauvinistic attitude towards machines changed and they came to see some machines as intelligent – those machines would *really be* intelligent. But what if only part of society came to see machines as intelligent beings (or, for the sake of the argument, as conscious beings), while the other part kept seeing them as mere machinery? According to Turing's approach, these two points of view would reflect two incommensurable paradigms (to use Thomas Kuhn's terminology), and there would not be any objective viewpoint from which this dispute could be settled. Human society would then be split over the question of how human-like machines should be treated; for example, should they be given human(!) rights and be freed from slavery? This question would probably not be resolved by logical reasoning, but by persuasion, or perhaps by violence (among humans). Indeed, due to the ethical aspects involved, people would probably have very little tolerance for the "other" opinion, which they would see as a totally unethical stance. In addition, the ethical flavor of the dispute would not leave much room for personal ambivalence, as each person would feel that they *must* take a side in the debate.[27]

7 Epilogue

Turing illuminates the important role played by human society in determining whether machines are intelligent. Machines cannot be perceived as intelligent in a society that has a prejudiced chauvinistic attitude towards them; but if this a-priori attitude were to change, brain-imitating machines could indeed be perceived as intelligent entities. Turing, who was convinced that machines could do everything a human brain does, feared that *his* opinion would not be accepted due to human prejudice towards himself, as appears in a worried letter he wrote in 1952 while standing trial on charges of "gross indecency" (Hodges 2014, pp. xxix–xxx):

[27] Discussions regarding the active role of humans in drawing the borders of the "Charmed Circle" of consciousness or intelligence (relevant also to the issue of animal consciousness and to disputes regarding humans' attitude towards animals) can be found in Dennett (1987b) and Michie (1993).

I'm rather afraid that the following syllogism may be used by some in the future –
Turing believes machines think
Turing lies with men
Therefore machines cannot think

When studying the nature of thought and intelligence, concentrating solely on the system's functions and internal structure can be misleading. That is not where intelligence lies. In emphasizing the major role of society, Turing's research – while focusing on machine intelligence – can teach us quite a bit about human intelligence as well.

Acknowledgments. Research for this paper was financially supported by the Sidney M. Edelstein Center for the History and Philosophy of Science, Technology, and Medicine at the Hebrew University of Jerusalem; and by the Centre for Moral and Political Philosophy (CMPP) at the Hebrew University of Jerusalem. I thank Orly Shenker, Oron Shagrir, Netanel Kupfer, Sander Beckers, Anna Strasser, Selmer Bringsjord and Vincent C. Müller, who reviewed this paper and added thoughtful comments. Special thanks to the participants of the PT-AI 2017 Conference for thought-provoking discussions, and to Shira Kramer-Danziger for her assistance in editing and her wise advice.

References

Beran, O.: Wittgensteinian perspectives on the Turing test. Studia Philosophica Estonica **7**(1), 35–57 (2014)
Block, N.: Psychologism and behaviorism. Philos. Rev. **90**, 5–43 (1981)
Block, N.: The mind as the software of the brain. In: Smith, E.E., Osherson, D.N. (eds.) Thinking, pp. 377–425. MIT Press, Cambridge (1995)
Boden, M.A.: Mind as Machine: A History of Cognitive Science. Oxford University Press, Oxford (2006)
Bringsjord, S., Bello, P., Ferrucci, D.: Creativity, the Turing test, and the (better) Lovelace test. Mind. Mach. **11**, 3–27 (2001)
Chomsky, N.: Turing on the "imitation game". In: Epstein, R., Roberts, G., Beber, G. (eds.) Parsing the Turing Test, pp. 103–106. Springer, New York (2008)
Coeckelbergh, M.: Robot rights? Towards a social-relational justification of moral consideration. Ethics Inf. Technol. **12**(3), 209–221 (2010)
Copeland, B.J. (ed.): The Essential Turing. Oxford University Press, Oxford (2004)
Danziger, S.: Can computers be thought of as thinkers? Externalist and linguistic perspectives on the Turing test. MA thesis, Hebrew University of Jerusalem (2016). [Hebrew]
Davidson, D.: Turing's test. In: Said, K., Newton-Smith, W., Viale, R. Wilkes, K. (eds.) Modelling the Mind, pp. 1–12. Clarendon Press, Oxford (1990)
Dennett, D.C.: The Intentional Stance. MIT Press, Cambridge (1987a)
Dennett, D.C.: Consciousness. In: Gregory, R.L., Zangwill, O.L. (eds.) The Oxford Companion to the Mind, pp. 160–164. Oxford University Press, Oxford (1987b)
French, R.M.: Subcognition and the limits of the Turing test. Mind **99**(393), 53–65 (1990)
Hodges, A.: Alan Turing: The Enigma. Princeton University Press, Princeton and Oxford (2014)
Mallery, J.C.: Thinking about foreign policy: finding an appropriate role for artificially intelligent computers. The 1988 Annual Meeting of the International Studies Association, St. Louis (1988). 10.1.1.50.3333
Mays, W.: Can machines think? Philosophy **27**, 148–162 (1952)
Michie, D.: Turing's test and conscious thought. Artif. Intell. **60**(10), 1–22 (1993)

Moor, J.H.: An analysis of the Turing test. Philos. Stud. **30**, 249–257 (1976)
Moor, J.H.: The status and future of the Turing test. Mind. Mach. **11**, 77–93 (2001)
Oppy, G., Dowe, D.: The Turing test. In: Zalta, E.N. (ed.) The Stanford Encyclopedia of Philosophy (2011). plato.stanford.edu/archives/spr2011/entries/turing-test. Accessed 13 Oct 2017
Piccinini, G.: Turing's rules for the imitation game. Mind. Mach. **10**, 573–582 (2000)
Proudfoot, D.: A new interpretation of the Turing test. Rutherford J. N. Z. J. Hist. Philos. Sci. Technol. 1 (2005). Article 010113. rutherfordjournal.org/article010113.html. Accessed 1 Nov 2017
Proudfoot, D.: Rethinking Turing's test. J. Philos. **110**(7), 391–411 (2013)
Saygin, A., Cicekli, I., Akman, V.: Turing test: 50 years later. Minds Mach. **10**, 463–518 (2000)
Schweizer, P.: The truly total Turing test. Mind. Mach. **8**, 263–272 (1998)
Searle, J.R.: Minds, brains, and programs. Behav. Brain Sci. **3**, 417–424 (1980)
Sloman, A.: The mythical Turing test. In: Cooper, S.B., Van Leeuwen, J. (eds.) Alan Turing: His Work and Impact, pp. 606–611. Elsevier, Amsterdam (2013)
Torrance, S.: Artificial consciousness and artificial ethics: between realism and social relationism. Philos. Technol. **27**(1), 9–29 (2014)
Turing, A.M.: On computable numbers, with an application to the Entscheidungsproblem. Reprinted in Copeland (2004), pp. 58–90 (1936)
Turing, A.M.: Lecture on the automatic computing engine. Reprinted in Copeland (2004), pp. 378–394 (1947)
Turing, A.M.: Intelligent machinery. Reprinted in Copeland (2004), pp. 410–432 (1948)
Turing, A.M.: Computing machinery and intelligence. Mind **50**, 433–460 (1950)
Turing, A.M., Braithwaite, R., Jefferson, G., Newman, M.: Can automatic calculating machines be said to think? Reprinted in Copeland (2004), pp. 494–506 (1952)
Watt, S.: Naive psychology and the inverted Turing test. Psycoloquy **7**(14) (1996). http://citeseerx.ist.psu.edu/viewdoc/download?doi=10.1.1.43.2705&rep=rep1&type=pdf. Accessed 2 Nov 2017
Whitby, B.: The Turing test: AI's biggest blind alley? In: Millican, P., Clark, A. (eds.) Machines and Thought: The Legacy of Alan Turing, pp. 53–62. Calderon Press, Oxford (1996)
Wittgenstein, L.: The Blue and Brown Books: Preliminary Studies for the "Philosophical Investigations". Harper & Row, New York (1958)
Wittgenstein, L.: Philosophical Investigations (Trans: G.E.M. Anscombe, P.M.S. Hacker, & J. Schulte; revised fourth edition by P.M.S. Hacker & J. Schulte). Wiley-Blackwell, Chichester (2009)

Modelling Machine Learning Models

Raül Fabra-Boluda(✉), Cèsar Ferri, José Hernández-Orallo, Fernando Martínez-Plumed, and M. José Ramírez-Quintana

DSIC, Universitat Politècnica de València, Valencia, Spain
{rafabbo,cferri,jorallo,fmartinez,mramirez}@dsic.upv.es

Abstract. Machine learning (ML) models make decisions for governments, companies, and individuals. Accordingly, there is the increasing concern of not having a rich explanatory and predictive account of the behaviour of these ML models relative to the users' interests (goals) and (pre-)conceptions (ontologies). We argue that the recent research trends in finding better characterisations of what a ML model does are leading to the view of ML models as complex behavioural systems. A good explanation for a model should depend on how well it describes the behaviour of the model in simpler, more comprehensible, or more understandable terms according to a given context. Consequently, we claim that a more contextual abstraction is necessary (as is done in system theory and psychology), which is very much like building a subjective mind modelling problem. We bring some research evidence of how this partial and subjective modelling of machine learning models can take place, suggesting that more machine learning is the answer.

1 Introduction

The increasing ubiquity of machine learning (ML) models in devices, applications, and assistants, which replace or complement human decision making, is prompting users and other interested parties to model what these ML models are able to do, where they fail, and whether they are vulnerable. On many occasions, we can only interact with the ML models by querying them as a black box since they may have been generated by a third party or may be too complex to understand.

With infinitely many queries, we would be able to build a comprehensible model A (e.g., a decision tree) that captures the behaviour of the original model M (Domingos 1998; Blanco-Vega et al. 2004). However, even if this were theoretically and practically possible, or were we to do this at a limited or desired level of accuracy, we would not necessarily obtain a comprehensible model since it could have a vast number of rules to be comprehensively treated or the rules could be too complex.

Also, the rules of the model may still be unrelated to the conceptions of the users (the ontology, background knowledge, and their everyday concepts) and their interests (which decisions are most relevant, what costs are involved, etc.). Instead, what we discuss in this paper is that a more abstract, subjective, and

V. C. Müller (Ed.): PT-AI 2017, SAPERE 44, pp. 175–186, 2018.
https://doi.org/10.1007/978-3-319-96448-5_16

partial account of the original model may capture *meta-information* about its behaviour, and that this could be expressed in terms of the conceptions (vocabulary, language, etc.) and interests (goals, constraints, etc.) of the potential users.

Let us further illustrate this with an example. In a medical diagnosis domain, the patient will be interested in being explained his/her diagnosis, and possibly some other alternative (but less likely) diagnoses. The patient will not be interested in the whole original model, which could cover many other pathologies that the patient will never have. This is the trade-off between local vs global interpretability (Doshi-Velez and Kim 2017). The patient will also want the explanation of his/her diagnosis in terms of his/her own conceptions, instead of in terms of meaningless variables (those that were probably used to build the original model). For instance, a user might be familiar with terms like infection, inflammation or pain, but not with the specific analytical variables in blood tests that explain such symptoms. Of course, extracting a comprehensible explanation directly from the original model might not always be possible, so further abstractions of the original model that take into account the user's conceptions and interests would be needed. Indeed, this is the way humans are able to abstract the qualities of human decision makers (e.g., a human doctor is reliable, prone to ask for many tests, unlikely to prescribe many medicines) and to tell others in a few words (e.g., when recommending a doctor to a friend). In this example, the model that abstracts the qualities of a human doctor can be very accurate in relation to the patient's conceptions and interests (i.e., likely to prescribe many tests), despite the model not having a profound medical knowledge.

Thus, capturing what a machine learning model does is something that should go much beyond obtaining or simulating its behaviour with a different representation. The question is then how we can *model a ML model* such that we can have a rich explanatory and predictive account of its behaviour (i.e., model prediction and its explanations) relative to a given context. The crux of the matter is that there is: an object (the original model, M, the *explanandum*) to be explained, the explanation itself (an abstraction, A, the *explanans*), and, finally, a target (the user, U, the *explananti*, the recipient of the explanation, Chart 2000). It should be noted that the *explanans* and the *explanandum* can be expressed in different terms. Most importantly, there would be a different *explanans* for each possible recipient, based on his/her knowledge, interests, and utility functions. In this paper, we suggest that the problem may be seen as a context-based modelling problem, where the original machine learning model is the *explanandum*, and, because of its potential complexity, we will only be able to partially explain it (according to the context). Furthermore, this should be done in terms of the ontologies and utilities of the recipient (typically the "user context" or simply the "context"). In summary, both the machine model and the user's knowledge and utility must be inputs to a behavioural modelling, which we call "model of a machine learning model".

The rest of the paper is organised as follows. The next section describes some previous and recent approaches to explain machine learning models. Then, Sect. 3 outlines how ML models can be described based on the context in general terms.

Section 4 provides five specific cases where we show how this can be done using machine learning as the modelling tool. Finally, Sect. 5 closes the paper.

2 Background

The interest in explaining how a machine learning model makes its decisions (how it behaves) is not new (Tickle et al. (1998)). However, in recent times, it has been receiving a lot of attention due to the growing success of AI applications. This section reviews some previous works that struggle to explain the behaviour of machine learning models from different points of view.

The crudest and simplest characterisation of what a model does is a measure of performance that provides an insight about how good or bad the model is. There are many performance measures for machine learning models (see, e.g., Japkowicz and Shah 2011; Hernández-Orallo et al. 2012; Ferri et al. 2009), which provide information about the certainty of the predictions (outputs) of a model, which is closely related to the notion of trust (Ribeiro et al. (2016)). For instance, when using class probability estimators, we can consider probabilities to be an indicator of reliability, e.g., if a credit model decides to grant a loan with a probability of 90%, the decision becomes more trustworthy than if the probability was 55%. The certainty or confidence can be analysed in terms of how good the probabilities are (i.e., calibration, (Bella et al. 2013)) or how good the confidence intervals are (i.e., conformal prediction (Balasubramanian et al. 2014)).

The measures should be general enough to ensure that the model behaves as expected in different contexts. This means that the model should also be evaluated under different utility functions that are able to measure how well the model performs according to a given context. For instance, the measures used in the area of cost-sensitive or context-sensitive learning (Elkan (2001)) are based on cost matrices for classification tasks and on cost functions for regression tasks; however, they may integrate other kinds of costs, such as test costs (Turney (2002)). The notions of cost-sensitive learning and context-sensitive learning have been generalised under the notion of reframing (Hernández-Orallo et al. (2016)), where the context can be used to train the original model, but also to adapt it for different operating contexts.

None of the previous approaches actually go much farther beyond the aggregation of statistical indicators from the results. There is no inference in the extraction of the performance metrics. In contrast, there are approaches such as those based on item response theory (IRT), which have been used to estimate the ability of the model based on its responses to items that have different levels of difficulty (Martínez-Plumed et al. 2016).

On the other hand, instead of analysing the relation between predicted outputs and actual outputs, a different way of analysing machine learning models is to study how input attributes affect output values (especially in supervised learning). The most classical insight is to determine those attributes that are the most relevant (Langley et al. 1994) and derive feature importance from the model (e.g., decision trees, neural networks, linear regression, etc.). Knowing which attributes are crucial for some predictions is key to understanding how the model works, its attribute costs, and also its robustness to some of the attributes.

A related question is how to determine those input features that are sensitive, actionable, or negotiable, which is not exactly how to determine feature importance, but rather whether a (small) change in an attribute can trigger a big change in the output. In general, *model actionability* refers to the ability to turn predictions into actionable knowledge that the user can transfer to the inputs in order to achieve a desired goal (e.g., to get a loan, to lower blood pressure, etc.). Thus, actionability is related to the reaction of the model against changes in features as well as the associated cost of these changes. Some approaches are restricted to only changing one attribute (Yang et al. 2007; Cui et al. 2015), but others address the more general problem of changing several attributes (Lyu et al. 2016). Some works focus on finding the features that might change the predicted class of a given model by introducing small perturbations on input data, as in the case of adversarial machine learning (Huang et al. (2011)). The term negotiable (Bella et al. 2011) is used when one wants to obtain a given output at the lowest cost of some of the inputs. For instance, given a credit model, a user might wonder which of the input attributes should be changed (loan amount, years, etc.) in order to have the credit granted.

The issues of how the model performs or how the model maps inputs to outputs are completely different from what the model does, which usually requires a more detailed level. Different methods have been introduced to explain the behaviour of an incomprehensible model. A simple example is to mimic the behaviour of a black box, incomprehensible model (e.g., a neural network, a classifier ensemble, etc.) by obtaining an equivalent one that is expressed as a set of comprehensible rules (Domingos 1998; Blanco-Vega et al. 2004; Ferri et al. 2002). A simple way to do this is by considering the model M as an oracle that we can interact with in order to learn other model (Jain et al. 1999). Thus, we can generate input data and query the model to label those data. The result is used as the training set to build a new comprehensible model A (the surrogate, or mimetic, model) that captures the behaviour of M. Furthermore, there have been some previous works where the surrogate model is created taking the context or cost into account. For instance, given an original model M, one can create a new model A that gives more relevance to a specific class (Blanco-Vega et al. 2006). Here, the goal is not to give a customised explanation but rather to change the behaviour of the original model.

Many of these works are now reintroduced (or reinvented) in the area known as Explainable Artificial Intelligence (XAI) (Core et al. 2006; Samek et al. 2017). The general goal of XAI is to provide more comprehensibility and interpretability to AI models. This includes concepts such as explainability, transparency, comprehensibility, interpretability, trust or fairness (Ribeiro et al. 2016; Doshi-Velez and Kim 2017; Weller 2017; Kamiran and Calders 2009).

The search for further abstraction on artificial systems also happens in other areas of computer science. A good inspirational example is software engineering, ranging from abstract interpretation in formal methods (Cousot and Cousot 1977) to empirical analysis (Fenton and Neil 1999). However, given the nature and increasing complexity of machine learning models, other areas such as cognitive modelling, system theory, or psychometrics may be better similes for what is needed here.

Overall, there are many approaches in machine learning with the goal of understanding how a model behaves. However, only some of them create another model or an abstraction of the original model (i.e., create an 'explanans'), and only some of them take into account the user's context (i.e., consider the 'explananti'). However, none of them does both things at the same time. This is what we explore in Sect. 3, in a general way, and in Sect. 4, with a series of more specific cases.

3 Towards Contextual Abstractions for Modeling ML Models

As we discussed in the previous sections, we are looking several aspects of machine learning models as complex behavioural systems: their intended vs actual behaviour, the relation between inputs and outputs, and understanding the whole model or some of its decisions. In order to make sense of all of this in a better way, we need to construct further abstractions. But, what type of abstractions? How can we get these abstractions so that we can interpret and understand the behaviour of a complex model? Other fields like psychology or biology have faced similar problems: when the object of study M is too complex, researchers develop a series of experiments based on the aspects that they want to study (interests) so that they can explain it in terms that they understand (conceptions). From there, they build a theory or model, explaining part of the behaviour of M.

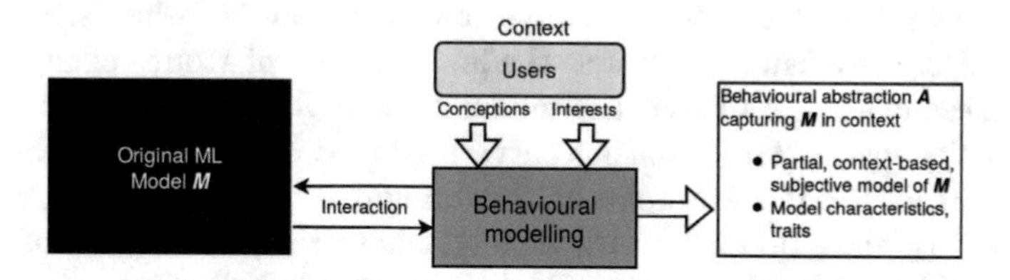

Fig. 1. A given model M (black box) that some potential users want to understand better. These users want an abstraction that is customised to their context: their conception of the domain and their interests. This model of the model (the abstraction A) captures behaviour that is very partially, subjective to the context, usually in terms of a set of characteristics and traits

Figure 1 shows how an abstraction A is built from an original model M according to a certain context, which is given by our interests and conceptions.

The procedure for building A must go beyond the definition of a sophisticated utility function that is followed by an optimization process on the model M. We claim that building A requires a re-modelling process that must include regularization terms, the abstract model representation (i.e., features and combinations of those features), and always giving priority to those features that are easier to understand, fairer, leading to more stable models and better calibration.

The context (conceptions and interests) must be the drive for abstraction. Only in this way can one ignore the irrelevant (uninteresting) details for the model, thereby overhauling the notion of overfitting. The abstraction must be done using the conceptual narratives of the user, including features that are common in behavioural (mind) models, such as reliability, trust, risks, roles, uncertainty, etc. (Boden 1988).

4 Illustrative Examples of Modelling ML Models

Given the previous general view, here we provide some examples of how this could be done for different applications, some of which are related to the areas described in Sect. 2. We cover the following issues: how performance metrics can be modelled, how modelling can be made partial and subjective, how latent variables can explain the behaviour of a model, how to address the problem of actionability by inverting a model, and, finally, how to model the boundaries of a model.

4.1 Modelling Context-Oriented Metrics

In Sect. 2, we have seen that context-sensitive learning builds a model according to some utility function. Similarly, reframing adapts the model according to some utility function. We provide two examples to illustrate what we mean by modelling context-oriented metrics and its usefulness.

Suppose that we want to determine how a model M behaves for different levels of noise. If we can characterise the level of noise of groups of examples, we can use these examples to feed the model M and get performance metrics for different noise levels. The resulting curve might be bumpy, especially if it can only be built with a few examples or there might be areas for which we cannot get the performance. So in the end, we might have the performance of M based on some number of levels of noise. With this, we can just create a very simple dataset: $(x = noiselevel, y = performance)$, and we can learn a linear model A that predicts (interpolates) how the model M behaves for each level of noise. As a result, we can explain in a simple way (a linear function of just one variable) how performance is affected by noise. An example of this setting is shown in Fig. 2, which is extracted from (Ferri et al. 2014). In this plot, we observe the behaviour of five classification models with respect to different values of noise ν that are injected in the features of a test dataset.

The context of a performance metric might be determined by just a few attributes, which may be those that are most relevant or important for a specific

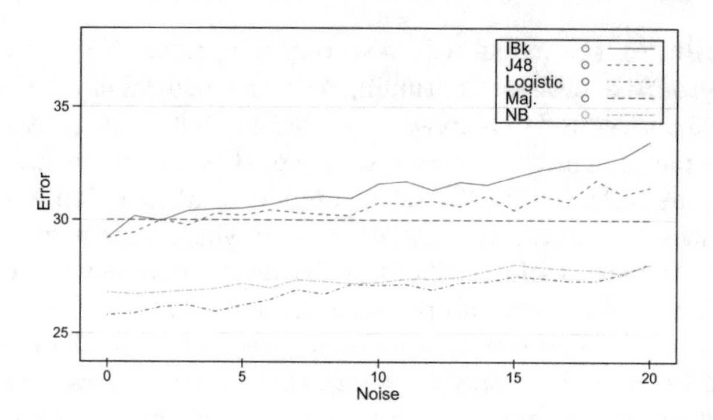

Fig. 2. Example of a context plot for a classification problem ("Creditg" dataset from UCI (Lichman 2013)) with the expected error on the y-axis (percentage of misclassified instances) for four classification methods and a reference model (majority class) for different noise levels ν on the x-axis. More details of the plot can be found in (Ferri et al. 2014)

user. For instance, if a user wants to know how the error of M depends on two attributes, age and gender, we can create a dataset with many examples of the type $(x_1 = age, x_2 = gender, y = performance)$. From here, we will learn a model A (a decision tree, or a linear regression model). Then, we can analyse how the model M behaves depending on the age and gender. This would allow us to extract conclusions relative to these attributes. For instance, we could come up with insights such that this particular model M is good for young males, but bad for older females. Or if we use false positives and false negatives instead of performance, we could get one group that might have more false positives than the other. This model A may be used to better understand and ultimately avoid unfair behaviour arising from the original model M. Fairness is one of the core concepts of XAI.

4.2 Partial, Subjective Modelling

The idea of using subsets of variables, groups, or partial sets of instances for modelling performance (as discussed in Sect. 3) suggests that we can do the same for modelling what the model does (its behaviour). Basically, we can adapt the mimicking approach to construct a surrogate model, but for a subset or a transformation of attributes, classes, or examples.

For instance, assume that a user (e.g., a patient) is interested in being explained how the model works with the only two variables he/she understands, x_3 (e.g., Blood Pressure) and x_5 (e.g., Body Mass Index). If we just mimic M using attributes x_3 and x_5 and the output, the resulting model A will necessarily be expressed in terms of x_3 and x_5 and will be understood by the patient. The accuracy may not be excellent with only two attributes (compared to the accuracy of the original model which includes all variables), but a general understanding of the model will be possible. Also, if we are using a decision tree with only two attributes, the model will have very few leaves. We can do something similar with groups of classes. Continuing with the medical domain, if the patient ultimately has a disease d diagnosed by model M, but M is general (multiclass, diagnosing m pathologies), we can create a model A where the classes have been merged (d vs the rest). The newly trained model or abstraction A will just learn rules to say why it is d and not any other pathology. This is exactly what the patient wants to know, without all of the details and complexities of the original model M, which had to cover all possible pathologies.

Finally, the same idea can be applied for the instances. If we want to know how a model behaves for a specific group (e.g., young males), we can make a sample of only those instances and create a mimetic model A with only those instances. The model A will show how M behaves, but only focusing on the group of interest.

In summary, as can be observed, by restricting the set of attributes, classes, or examples and keeping only the ones that are of interest, the model A necessarily has to express its behaviour in terms of such attributes, classes or examples.

4.3 Modelling Latent Behaviour

One of the most common ways of understanding a complex phenomenon is by extracting latent variables, which explain many more observed variables. For a machine learning model, we can choose many possible subsets of observed variables, combining inputs, outputs, etc., and varying aggregations and selections.

For instance, one simple example is to take inputs, outputs, and actual classes as observed variables and try to derive the latent variables that best explain what we see. For instance, we can construct instances such as: $(x_1, x_2, ..., x_k, y_1^{pred}, ..., y_k^{pred}, y_1^{actual}, ... y_k^{actual})$. With all these variables, we can perform PCA or factor analysis. For instance, it might be the case that a dominant factor f explains variables y_1^{pred}, y_1^{actual}, x_3. Note that this is different from using PCA or factor analysis on the original dataset $(x_1, x_2, ..., x_k)$. Here, we might conclude that a latent factor is not only determining some observable input variables, but also some of the behaviour of the model. For instance, if we go back to the medical example, we could say that instances with high values of a latent variable f will have high values of y_1^{pred}, y_1^{actual} and x_3, showing that x_3 is really key for some high and reliable values of y_1.

Finding latent variables is just a kind of descriptive model of how M works. For instance, other unsupervised techniques, such as clustering, could be used, where output variables (actual and/or predicted) would be the inputs of the algorithm.

4.4 Modelling Actionability

In Sect. 2, we showed that there is interest in how to make a model actionable, i.e., how the inputs have to change in order to obtain an output. Again, we may have a model M, and we may have some fixed attributes x_1 to x_6 and we want to know how to change x_7 (e.g., the amount) in order to swap the output y from the negative to the positive class (granting a loan). For a linear model, this is easy to calculate, but for a more complex model we would need some kind of optimisation search in order to find this value. Instead, what we can do is to try to model how x_7 depends on x_1 to x_6 and y. This can be done with machine learning by putting the attribute of interest (x_7) as output and the rest as inputs, e.g., ($x_1,\ldots, x_6, y \rightarrow x_7$). The new model A is able to predict the value of x_7 that we need when putting a combination of $x_1,\ldots, x_6$ and y, which is what we wanted in the first place.

However, this will not give us the optimal value of x_7 given the context. In order to do this, we could include a utility value u as input instead of y, i.e., ($x_1,\ldots, x_6, u \rightarrow x_7$). If we want to maximise u, using the newly learnt model A, we would use a high value of u to see what value of x_7 we obtain. Note that the optimisation process is done during learning, and this abstraction is perfectly actionable.

4.5 Modelling Decision Boundaries and Model Families

Another trait that can be analysed in a model is its decision boundaries. This type of insight is directly related to the behaviour of the model, since the boundaries establish the mappings between input features and output values. For instance, given a model M, we might want to infer the model family of M, which crucially affects the shape of the decision boundaries. Namely, instead of asking what the model does or how well it does, we ask how it maps the space, especially in terms of some well-known families in machine learning. The concept of model family refers to groups of models that exhibit similar behaviour for the same set of data.

Why would we be interested in finding out the family of a given model? If we were able to determine the model family of a loan scoring system and we determine that it is a decision tree, we could use specific adversarial machine learning (Huang et al. 2011) techniques for decision trees to determine the values of the inputs that maximise the utility value u to obtain a desired output (e.g., granting a loan), as we have explained in Sect. 4.4.

We could try to determine the family of M by using some rules of thumb, or by trying to mimic the model and look at the boundaries. An alternative, which involves machine learning once again, is to learn to classify the family of a model after having trained a classifier A from a range of models, their features, and their known families. More precisely, to learn a classifier A that identifies the model family, we would need a collection of models $M'_1, M'_2, \ldots, M'_m$, that are labelled with their model family $y_1, y_2, \ldots, y_m$. Each model would be a row in a dataset, for which we will extract a range of features. In other words, each row of this dataset will be an array of characteristics $x_1, x_2, ..., x_n$ about a model (the input attributes), and its family (the output attribute). Figure 3 illustrates the process for extracting the features for a single model M'_i (top row). M'_i could be used for labelling an artificial surrogate dataset SD_i (left plot, bottom row), so that SD_i would reflect the decision boundaries learned by M'_i. We could then

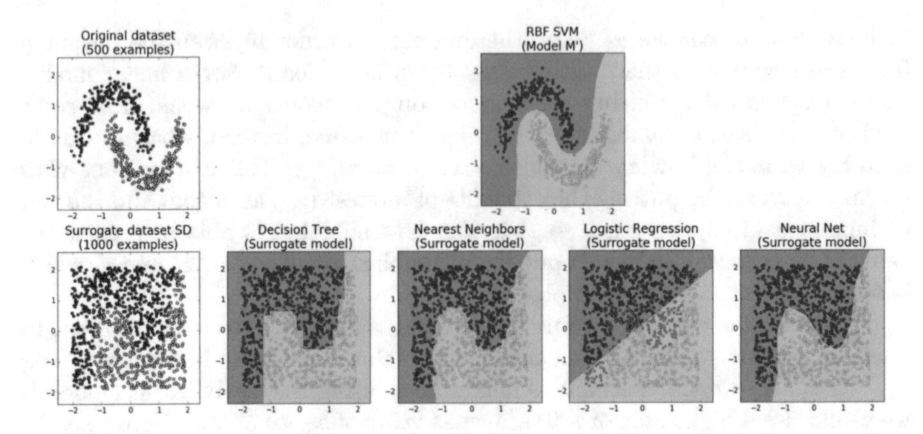

Fig. 3. The first row presents a dataset (first plot) for which a model M' (SVM with RBF kernel) is learnt. For such a model, we learn an artificially generated surrogate dataset that reflects the boundaries of M' (first plot of the bottom row). The remaining plots correspond to different surrogate models that are trained on the surrogate dataset. Different surrogate models (with different inner workings) replicate (better or worse) the decision boundaries of M'

use SD_i to learn some new machine learning models and obtain a number of performance measures so that each M_i' would be characterised as a collection of performance measures that could be used as the features $x_1, x_2, ..., x_n$ of the dataset to learn the model A. Thus, A would be able to identify the model family of a new instance (i.e., a model) M by applying the same process of feature extraction.

5 Conclusions

The previous section just outlines the possibilities of using machine learning to model partial accounts of the behaviour of an existing model M. It is not the goal of this paper to flesh out each and every one of these possibilities, which would require independent papers of their own. The goal of this paper is to see all of these possibilities (some of them new, and some with more potential than others) under the same umbrella, the idea of modelling machine learning models. Also, in general, similar ideas can be applied when the base system is an AI system, where machine learning is only a part of it (Hernández-Orallo 2017).

All of this is very timely regarding the recent research emphasis on more powerful and abstract evaluation metrics, the extraction of latent variables about model behaviour, the explanation of decisions in a comprehensible way, and the analysis of sensitive/actionable features. Terms like interpretability, transparency, trust, or fairness require understanding machine learning models as being complex behavioural systems for which further abstractions are necessary, as is done in system theory, cognitive modelling, and some other areas of psychology. In the same way that these areas are using more machine learning to construct models from observation, it would be unnatural not to use machine learning to model machine learning models for a similar purpose.

Acknowledgements. This material is based upon work supported by the Air Force Office of Scientific Research under award number FA9550-17-1-0287, the EU (FEDER), and the Spanish MINECO under grant TIN 2015-69175-C4-1-R, the Generalitat Valenciana PROMETEOII/2015/013. F. Martínez-Plumed was also supported by INCIBE (Ayudas para la excelencia de los equipos de investigación avanzada en ciberseguridad). J. H-Orallo also received a Salvador de Madariaga grant (PRX17/00467) from the Spanish MECD for a research stay at the CFI, Cambridge, and a BEST grant (BEST/2017/045) from the GVA for another research stay at the CFI.

References

Balasubramanian, V., Ho, S.-S., Vovk, V.: Conformal Prediction for Reliable Machine Learning: Theory, Adaptations and Applications. Newnes, Oxford (2014)

Bella, A., Ferri, C., Hernández-Orallo, J., Ramírez-Quintana, M.J.: Using negotiable features for prescription problems. Computing **91**(2), 135–168 (2011)

Bella, A., Ferri, C., Hernández-Orallo, J., Ramírez-Quintana, M.J.: On the effect of calibration in classifier combination. Appl. Intell. **38**(4), 566–585 (2013)

Blanco-Vega, R., Ferri-Ramírez, C., Hernández-Orallo, J., Ramírez-Quintana, M.: Estimating the class probability threshold without training data. In: Workshop on ROC Analysis in Machine Learning, p. 9 (2006)

Blanco-Vega, R., Hernández-Orallo, J., Ramírez-Quintana, M.: Analysing the trade-off between comprehensibility and accuracy in mimetic models. In: Discovery Science, pp. 35–39 (2004)

Boden, M.A.: Computer Models of Mind: Computational Approaches in Theoretical Psychology. Cambridge University Press, New York (1988)

Chart, D.: A Theory of Understanding: Philosophical and Psychological Perspectives. Routledge, New York (2000)

Core, M., Lane, H.C., van Lent, M., Gomboc, D., Solomon, S., Rosenberg, M.: Building explainable artificial intelligence systems. In: Proceedings of the 18th Innovative Applications of Artificial Intelligence Conference (2006)

Cousot, P., Cousot, R.: Abstract interpretation: a unified lattice model for static analysis of programs by construction or approximation of fixpoints. In: Proceedings of the 4th ACM SIGACT-SIGPLAN Symposium on Principles of Programming Languages, pp. 238–252 (1977)

Cui, Z., Chen, W., He, Y., Chen, Y.: Optimal action extraction for random forests and boosted trees. In: Proceedings of the 21st ACM SIGKDD International Conference on Knowledge Discovery and Data Mining, pp. 179–188 (2015)

Domingos, P.: Knowledge discovery via multiple models. Intell. Data Anal. **2**(1–4), 187–202 (1998)

Doshi-Velez, F., Kim, B.: Towards a rigorous science of interpretable machine learning. arXiv (2017)

Elkan, C.: The foundations of cost-sensitive learning. In: International Joint Conference on Artificial Intelligence, vol. 17, pp. 973–978 (2001)

Fenton, N.E., Neil, M.: Software metrics: successes, failures and new directions. J. Syst. Softw. **47**(2), 149–157 (1999)

Ferri, C., Hernández-Orallo, J., Ramírez-Quintana, M.J.: From ensemble methods to comprehensible models. In: 5th International Conference on Discovery Science, pp. 165–177 (2002)
Ferri, C., Hernández-Orallo, J., Martínez-Usó, A., Ramírez-Quintana, M.: Identifying dominant models when the noise context is known. In: First Workshop on Generalization and Reuse of Machine Learning Models Over Multiple Contexts (2014)
Ferri, C., Hernández-Orallo, J., Modroiu, R.: An experimental comparison of performance measures for classification. Pattern Recogn. Lett. **30**(1), 27–38 (2009)
Hernández-Orallo, J., Flach, P.A., Ferri, C.: A unified view of performance metrics: translating threshold choice into expected classification loss. J. Mach. Learn. Res. **13**, 2813–2869 (2012)
Hernández-Orallo, J.: The Measure of All Minds: Evaluating Natural and Artificial Intelligence. Cambridge University Press, New York (2017)
Hernández-Orallo, J., Martínez-Usó, A., Prudêncio, R.B., Kull, M., Flach, P., Farhan Ahmed, C., Lachiche, N.: Reframing in context: a systematic approach for model reuse in machine learning. AI Commun. **29**(5), 551–566 (2016)
Huang, L., Joseph, A.D., Nelson, B., Rubinstein, B.I., Tygar, J.D.: Adversarial machine learning. In: Proceedings of the 4th ACM Workshop on Security and Artificial Intelligence, pp. 43–58 (2011)
Jain, S., Osherson, D., Royer, J.S., Sharma, A.: Systems that Learn, 2nd edn. MIT Press, Cambridge (1999)
Japkowicz, N., Shah, M.: Evaluating Learning Algorithms: A Classification Perspective. Cambridge University Press, New York (2011)
Kamiran, F., Calders, T.: Classifying without discriminating. In: 2nd International Conference on Computer, Control and Communication, pp. 1–6 (2009)
Langley, P., et al.: Selection of relevant features in machine learning. In: Proceedings of the AAAI Fall Symposium on Relevance, vol. 184, pp. 245–271 (1994)
Lichman, M.: UCI machine learning repository (2013)
Lyu, Q., Chen, Y., Li, Z., Cui, Z., Chen, L., Zhang, X., Shen, H.: Extracting actionability from machine learning models by sub-optimal deterministic planning. arXiv preprint arXiv:1611.00873 (2016)
Martínez-Plumed, F., Prudêncio, R.B.C., Usó, A.M., Hernández-Orallo, J.: Making sense of item response theory in machine learning. In: European Conference on Artificial Intelligence, pp. 1140–1148 (2016)
Ribeiro, M.T., Singh, S., Guestrin, C.: Why should I trust you?: Explaining the predictions of any classifier. In: Proceedings of the 22nd ACM SIGKDD International Conference on Knowledge Discovery and Data Mining, pp. 1135–1144 (2016)
Samek, W., Wiegand, T., Müller, K.-R.: Explainable Artificial Intelligence: Understanding, Visualizing and Interpreting Deep Learning Models, ArXiv e-prints (2017)
Tickle, A.B., Andrews, R., Golea, M., Diederich, J.: The truth will come to light: directions and challenges in extracting the knowledge embedded within trained artificial neural networks. Trans. Neur. Netw. **9**(6), 1057–1068 (1998)
Turney, P.D.: The management of context-sensitive features: a review of strategies. arXiv preprint cs/0212037 (2002)
Weller, A.: Challenges for transparency. arXiv preprint arXiv:1708.01870 (2017)
Yang, Q., Yin, J., Ling, C., Pan, R.: Extracting actionable knowledge from decision trees. IEEE Trans. Knowl. Data Eng. **19**(1), 43–56 (2007)

Is Programming Done by Projection and Introspection?

Sam Freed(✉)

Department of Informatics, University of Sussex, Falmer, Brighton, BN1 9RH, UK
s.freed@sussex.ac.uk

Abstract. Often people describe the creative act of programming as mysterious (Costa 2015). This paper explores the phenomenology of programming, and examines the following proposal: Programming is a log of actions one would imagine oneself to be doing (in order to achieve a task) after one projects oneself into a world consisting of software mechanisms, such as "the Python environment". Programming is the formal logging of our imagined actions, in such an imagined world. Our access to our imagination is introspective.

Keywords: Programming · Introspection · Role-playing

1 Introduction

Introspection was denounced as a methodology in psychology with the founding of behaviourism (Watson 1913). Cognitive science inherited this bias from psychology (Costall 2006), and AI being (at least partially) a wing of cognitive science, inherited this abhorrence of introspection. However introspection *has* been used in AI research, but has been widely obfuscated and denied (Freed 2017, Sect. 4.3). I have argued elsewhere for the rehabilitation of introspection as a source of ideas for AI algorithms (Freed 2017). Here I examine a radical short-cut: Developing an AI system includes by definition a programming task. If introspection is involved in *all* programming, then banning its use in AI development is logically impossible, and AI researchers denying and obfuscating their use of introspection are attempting the impossible.

Introspection can be done in our normal waking state (assuming that is a single thing). However, one can also introspect while imagining oneself to be another person, as an actor does when playing a role. This is not limited to theatrics:

2 Projecting into a Role

When acting and communicating, people assume a certain role (or "frame of mind"), usually depending on the social context. An example is how people take on their "office persona" as they start their work day. This "office persona" is broadly similar to the job description the organization would advertise to fill a role if it were to fall vacant. As Herbert Simon noted:

V. C. Müller (Ed.): PT-AI 2017, SAPERE 44, pp. 187–189, 2018.
https://doi.org/10.1007/978-3-319-96448-5_17

> *Administration is not unlike play-acting. The task of the good actor is to know and play his role... The effectiveness of the performance will depend on the effectiveness of the play and the effectiveness in which it is played. The effectiveness of the administrative process will vary with the effectiveness of the organisation and the effectiveness with which its members play their parts.* (Simon 1976, p. 252; Simon 1996, p. xii)

There seems to be no such thing as the mind operating (in a way that could be relevant for action) outside of some cultural context (see Wittgenstein 2001) even if it is the context of running amok (Carr 1985). This fact of the individual's behaviour being constructed in (usually) socially-accepted roles is transparent to us in daily life, but has been the subject of much research (Goffman 1971).

One of the roles one can adopt is the role of being cooperative with some scientific programme. An example is Watson's (and later Simon's) "thinking aloud", requiring that a "*scientific man*" take on such a role as thinking aloud "*in the proper spirit*" and possibly even "*with zest*" (Watson 1920, p. 91).

These roles that we adopt come with certain prejudices in interpreting our environment. The same event (say the firing of a pistol by an assassin) could be interpreted differently by the same person, depending on whether they are acting in their capacity as a citizen, or in their capacity as a scientist. In one case he would describe an assassination, in the other he would explain the chemistry and mechanics of pistols.

This observation (that humans act within a context or a role) is not entirely alien to the field of AI. Note that the above quote is from Herbert Simon - a pillar of the AI community (albeit from his work in public administration, not AI).

3 Programming Is Introspective

Let us examine (phenomenologically) what happens when one approaches (even) a simple programming task, say calculating and printing a tax invoice for a company:

In writing new code (not debugging or reusing existing code), a programmer projects herself (like a stage-actor would) into an imaginary world where she is (say) inside a space consisting of the Python instruction set (and libraries), or in a space comprised of an "Intel" architecture, and asks herself how she could use the tools available (variables, arrays, loops, libraries, etc.). There is a lot of "first person thinking" going on, as in "*how could I do this*", "*this could give me that*" etc. The programmer's output, the code that is supposed to do the task, is a formalization into (say) Python instructions of the imagined actions by the programmer inside this "Python environment". Our access to our imagined actions is introspective.

Some may argue that programming is merely translation of the requirement (a tax calculation) into a programming language, making the act of programming an act of translation. But (adequate) translations are also rarely simply a mechanical replacement of words in one language with words from another. Good translations are achieved by the translator projecting herself into the character speaking the words in the original language, and imagining how she would express the same ideas and sentiments in the target language. So the translation explication of programming ends up also involving projection into a role, imagination, and introspection.

Where else could the code come from? I can find no evidence (or testimony) that there is anything like a tree-search of possibilities as GOFAI would have it, nor do we have any reports that explain the appearance of code in a contradictory manner.

Moreover, in debugging, a similar thing happens. In the exercise known as "a dry run", the programmer projects herself into the role of a Python interpreter, and acts (in her mind, perhaps using pencil and paper) on the code and the data as a Python interpreter would, always keeping half an eye on the intended result to see where the actual result deviates from the intended result. When such a deviation is found, the programmer would say that she found a bug.

Conversely a programmer copying an algorithm from a book is not introspective.

4 Summary

Developing AI requires programming. Introspection has been discouraged in the cognitive sciences in general, including AI. In finding that all original programming is introspective, we can lay to rest any ban on introspection, and review what roles it plays in AI research, for details see (Freed 2017). If introspection has an unavoidable role in AI research, then discouraging its use is nonsensical, and hence counterproductive.

References

Carr, J.E.: Ethno-behaviorism and the culture-bound syndromes: the case of amok. In: Simons, R.C., Hughes, C.C. (eds.) The Culture-Bound Syndromes, pp. 199–223. Springer, Netherlands (1985). Accessed http://link.springer.com/chapter/10.1007/978-94-009-5251-5_20

Costa, S.: How to Survive 80 + Hours of Programming Every Week, 11 May 2015. https://medium.freecodecamp.org/how-to-survive-80-hours-of-programming-every-week-3cc1db75695a. Accessed 29 March 2018

Costall, A.: 'Introspectionism' and the mythical origins of scientific psychology. Conscious. Cogn. **15**(4), 634–654 (2006). https://doi.org/10.1016/j.concog.2006.09.008

Freed, S.: A role for introspection in AI research. University of Sussex (2017). Accessed http://sro.sussex.ac.uk/66141/

Goffman, E.: The Presentation of Self in Everyday Life. Penguin, Harmondsworth (1971)

Simon, H.A.: Administrative Behavior: A Study of Decision-Making Processes in Administrative Organization, 3rd edn. Collier Macmillan, London (1976). Accessed http://capitadiscovery.co.uk/sussex-ac/items/38710

Simon, H.A.: The Sciences of the Artificial, 3rd edn. MIT Press, Cambridge (1996). Accessed http://capitadiscovery.co.uk/sussex-ac/items/546838

Watson, J.B.: Psychology as the behaviorist views it. Psychol. Rev. **20**(2), 158–177 (1913). https://doi.org/10.1037/h0074428

Watson, J.B.: Is thinking merely action of language mechanisms1? (v.). Br. J. Psychol. **11**(1), 87–104 (1920). General Section https://doi.org/10.1111/j.2044-8295.1920.tb00010.x

Wittgenstein, L.: Philosophical Investigations: The German Text with a Revised English Translation, 3rd edn. Wiley-Blackwell, Malden (2001)

Supporting Pluralism by Artificial Intelligence: Conceptualizing Epistemic Disagreements as Digital Artifacts

Soheil Human[1,2(✉)], Golnaz Bidabadi[3], and Vadim Savenkov[2]

[1] Department of Philosophy & Cognitive Science Research Platform, University of Vienna, Universitätsring 1, 1010 Vienna, Austria
soheil.human@univie.ac.at

[2] Department of Information Systems and Operations, Vienna University of Economics and Business, Welthandelsplatz 1, 1020 Vienna, Austria
{soheil.human,vadim.savenkov}@wu.ac.at

[3] Cisco Systems, Inc., 170 West Tasman Drive, San Jose, CA 95134, USA
golnaz@cisco.com

Abstract. A crucial concept in philosophy and social sciences, epistemic disagreement, has not yet been adequately reflected in the Web. In this paper, we call for development of intelligent tools dealing with epistemic disagreements on the Web to support pluralism. As a first step, we present POLYPHONY, an ontology for representing and annotating epistemic disagreements.

1 Introduction

While artificial intelligence is considered as both threat and opportunity for the modern democracies, many have called for immediate action for development of AI tools to support pluralism (see e.g. Helbing et al. 2017). Detection, representation and visualization of epistemic disagreements, we propose, is one of the important steps to support pluralism and dialog in the Web. Here are two concrete examples: (I) consider a controversial article in Wikipedia that is the matter of different disagreements. If we would be able to detect and represent disagreements, disputable parts could be visualized for people, users could simply compare different points of view (or request particular versions of the article based on their preferences). (II) Imagine you have recently read an article and like to find some articles that disagree with the proposed point of view. If it would be possible to automatically identify and link disagreeing articles, one could simply find them without the need to exploring all related articles one by one and thoroughly to discover disagreeing contents.

Due to its nature, Semantic Web and Linked [Open] Data are perfectly fit to capture disagreements: producing two different descriptions of the same phenomenon and publishing them suffices to produce a potential disagreement.

V. C. Müller (Ed.): PT-AI 2017, SAPERE 44, pp. 190–193, 2018.
https://doi.org/10.1007/978-3-319-96448-5_18

What remains to be done is making the disagreement between descriptions co-existing, e.g., at different sources, explicit. This paper advocates a particular instance of the general Linked Open Data (LOD) principle, according to which explicit links between entities and resources are essential. The special type of link, we advocate in this paper, is explicit *disagreement annotations*, making explicit the disagreements using standard LOD linking by means of IRIs. We call the design pattern of providing several alternative descriptions of the same subject its *pluralist description*. This pattern requires either (a) authors of description to be aware of alternative views on the subject, and taking care of encoding these alternative descriptions, or (b) the disagreeing contents are detected, linked and visualized by artificial intelligent agents. Considering the huge and increasing amount of available data, the former option seems to be unrealistic, leaving us no choice but to develop intelligent tools that can perform such tasks. Here, we take the first step towards development of intelligent tools dealing with epistemic disagreements on the Web by conceptualizing epistemic disagreements as digital artifacts and proposing an ontology for representing epistemic disagreements, called POLYPHONY.

2 Conceptualizing Epistemic Disagreements as Digital Artifacts

Study of epistemic disagreements is a fresh and active field of research (Goldman 2010; Frances 2014, p. 16). Besides the very fundamental questions regarding existence and importance of disagreements, many epistemologists have tried to answer two main questions: (1) What types of disagreement exist? (2) What is the rational response to each type? In order to conceptualize epistemic disagreements as digital artifacts, the answers to the these questions should be considered. Therefore, after a literature review, some of the most important types of epistemic disagreements, such as *peer disagreements*, *deep disagreements*, *genuine disagreements*, *merely apparent disagreements*, *merely verbal disagreements*, and *faultless disagreements* (Siegal 2013; Fogelin 1985; Cohnitz and Marques 2014; Jenkins 2014), along with binary distinctions between them were identified, and real-world examples of each type were documented. Next, possible responses to disagreements, such as (a) *rejecting the existence of the disagreement*, (b) *maintaining one's confidence*, (c) *suspending judgment*, (d) *reducing one's confidence*, and (e) *deferring to the other's conclusion* and the relationship between these responses and different types of epistemic disagreements based on the real-world examples were identified and documented[1].

Based on the conceptualization of epistemic disagreements outlined before, we designed, POLYPHONY (see Fig. 1) a generic OWL ontology for annotating disagreements in Linked Data. To this end, POLYPHONY supports disagreement annotations of varying granularity: from the ontology level to the level of single

[1] See the documentations of the POLYPHONY ontology for detailed descriptions, here: http://purl.org/epistemic-disagreement.

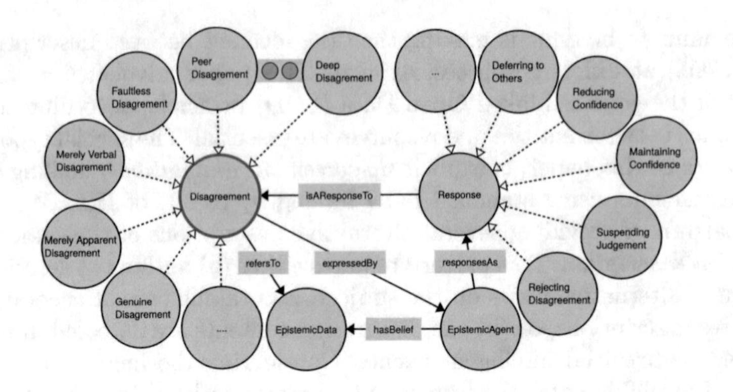

Fig. 1. Core concepts of POLYPHONY (see footnote 1)

triple, or a collection of triples. As a proof of concept, POLYPHONY was applied to OPENEED, a modular ontology for human needs data proposed by Human et al. (2017), to represent disagreements between different modules of the OPENEED ontology, i.e. to annotate epistemic disagreements between *needs theories*.

3 Conclusion

Epistemic disagreement has been argued to be valuable for most crucial aspects of society, such as science (Cruz and Smedt 2013) and politics. In this paper, we took the first step towards development of intelligent tools dealing with epistemic disagreements on the Web by presenting POLYPHONY, an ontology for representing epistemic disagreements. We hope that our research will serve as a base for future studies on development of intelligent tools for automatic detection, annotation, and visualization of epistemic disagreements on the Web.

Acknowledgement. This research was partially supported by the WU Anniversary Fund of the City of Vienna and by the Austrian Federal Ministry for Transport, Innovation and Technology (Grant 855407).

References

Cohnitz, D., Marques, T.: Disagreements. Erkenntnis **79**(1), 1–10 (2014)
Cruz, H.D., Smedt, J.D.: The value of epistemic disagreement in scientific practice. The case of homo floresiensis. Stud. Hist. Philos. Sci. Part A **44**(2), 169–177 (2013)
Fogelin, R.: The logic of deep disagreements. Informal Log. **7**(1), 3–11 (1985)
Frances, B.: Disagreement. Polity Press, Cambridge (2014)
Goldman, A.: Epistemic relativism and reasonable disagreement. In: Feldman, R., Warfield, T.A. (eds.) Disagreement. Oxford University Press, Oxford (2010)
Helbing, D., Frey, B.S., Gigerenzer, G., Hafen, E., Hagner, M., Hofstetter, Y., van den Hoven, J., Zicari, R.V., Zwitter, A.: Will democracy survive big data and artificial intelligence. Sci. Am. 25 February 2017

Human, S., Fahrenbach, F., Kragulj, F., Savenkov, V.: Ontology for representing human needs. In: International Conference on Knowledge Engineering and the Semantic Web, pp. 195–210. Springer (2017)
Jenkins, C.S.: Merely verbal disputes. Erkenntnis **79**(1), 11–30 (2014)
Siegel H.: Argumentation and the epistemology of disagreement. In: Mohammed, D., Lewinski, M. (eds.) Virtues of Argumentation: Proceedings of the 10th International Conference of the Ontario Society for the Study of Argumentation (OSSA), May 22–25, 2013. OSSA, Windsor, ON, (2014)

The Frame Problem, Gödelian Incompleteness, and the Lucas-Penrose Argument: A Structural Analysis of Arguments About Limits of AI, and Its Physical and Metaphysical Consequences

Yoshihiro Maruyama[1,2(✉)]

[1] The Hakubi Centre for Advanced Research, Kyoto University, Kyoto, Japan
maruyama@i.h.kyoto-u.ac.jp
[2] JST PRESTO, Tokyo, Japan

Abstract. The frame problem is a fundamental challenge in AI, and the Lucas-Penrose argument is supposed to show a limitation of AI if it is successful at all. Here we discuss both of them from a unified Gödelian point of view. We give an informational reformulation of the frame problem, which turns out to be tightly intertwined with the nature of Gödelian incompleteness in the sense that they both hinge upon the finitarity condition of agents or systems, without which their alleged limitations can readily be overcome, and that they can both be seen as instances of the fundamental discrepancy between finitary beings and infinitary reality. We then revisit the Lucas-Penrose argument, elaborating a version of it which indicates the impossibility of information physics or the computational theory of the universe. It turns out through a finer analysis that if the Lucas-Penrose argument is accepted then information physics is impossible too; the possibility of AI or the computational theory of the mind is thus linked with the possibility of information physics or the computational theory of the universe. We finally reconsider the Penrose's Quantum Mind Thesis in light of recent advances in quantum modelling of cognition, giving a structural reformulation of it and thereby shedding new light on what is problematic in the Quantum Mind Thesis. Overall, we consider it promising to link the computational theory of the mind with the computational theory of the universe; their integration would allow us to go beyond the Cartesian dualism, giving, in particular, an incarnation of Chalmers' double-aspect theory of information.

1 Introduction

The frame problem is concerned with a fundamental challenge in artificial intelligence and philosophy of it. It is generally understood, on the one hand, that the original 1969 frame problem posed by McCarthy and Hayes (1969) has largely

V. C. Müller (Ed.): PT-AI 2017, SAPERE 44, pp. 194–206, 2018.
https://doi.org/10.1007/978-3-319-96448-5_19

been settled from a technical point of view (see Shanahan 1997 and 2016). Philosophers such as Dennett (1987) and Fodor (1983), on the other hand, have reformulated the frame problem in different ways in order to put it in a broader, conceptual context of epistemology. The long-standing frame problem that is still hotly debated in artificial intelligence and philosophy of it is the latter sort of frame problem.

In the present article we are going to discuss the latter sort of frame problem from yet another, Gödelian perspective. Addressing the significance of incompleteness in (stronger-than-Robinson-arithmetic) finitary formal systems, Gödel (1995) asserts that either "the human mind (even within the realm of pure mathematics) infinitely surpasses the power of any finite machine" or "there exist absolutely unsolvable diophantine problems." Note here that Gödelian incompleteness hinges upon the two assumptions on the nature of beings that systems involved are weak enough in the sense that they are finitary (i.e., recursively axiomatisable), and that they are, at the same time, strong enough in the sense that they encompass certain basic arithmetic (e.g., Robinson arithmetic). At a level of abstraction, the nature of the frame problem is closely linked with the nature of incompleteness: we shall argue, in particular, that (i) the finitarity condition is indispensable in both the frame problem and Gödelian incompleteness, (ii) without the finitarity condition there is no frame problem or incompleteness any more, and (iii) both the frame problem and Gödelian incompleteness may be understood as instances of the fundamental discrepancy between finitary beings and infinitary reality.

The Gödel's incompleteness theorems have yielded vital driving forces for different (often controversial) discussions in different fields beyond logic and foundations of mathematics (including both rigorous and sloppy ones). The Lucas-Penrose argument is one of them, concerned with the very possibility of artificial (general) intelligence or the computational theory of the mind. Rather than discussing the ultimate validity of such a controversial argument, we concentrate upon making a bridge between artificial intelligence qua the computational theory of the mind and information physics qua the computational theory of the universe through a structural analysis of the Lucas-Penrose argument (for an origin of information physics we would refer to Wheeler's "It from Bit" thesis asserting "all things physical are information-theoretic in origin" (Wheeler 1990); Everett, a pupil of Wheeler, had a similar idea, as seen in his information-theoretic concept of entropy and his entropic uncertainty relation (Everett 1973); it is unclear whether Wheeler's idea preceded Everett's or it was the other way around; Everett might have influenced Wheeler).

In the rest of the article, we first elaborate upon our Gödelian perspective on the frame problem and argue that the frame problem shares the same structure as Gödelian incompleteness at a level of abstraction. And then we reconsider the Lucas-Penrose argument from a different angle, exposing a link between the (im)possibility of artificial intelligence and the (im)possibility of information physics. We thereafter briefly discuss Penrose's quantum mind thesis as well. We finally conclude the article by giving a remark about the possibility of embodying

Chalmers' double aspect theory of information through the integration of the computational theories of mind and of the universe. Overall, our focus is placed upon the analysis of arguments rather than their ultimate validity; our primary purpose is to give a structural analysis of arguments about limits of AI and related computational theories such as information physics, and to unpack its consequences.

2 A Gödelian Perspective on the Frame Problem

Although it is sometimes said that "a definition of the 'Frame Problem' is harder to come by than the Holy Grail" (Stein 1990), nonetheless, we give an explicit, conceptual formulation of the frame problem from an informational point of view. The frame problem may be understood as indicating a discrepancy between finitary cognitive capacity and infinitary informational content. In order to illustrate what is meant in our conceptual formulation of the frame problem, we shall compare it with the well-known, robot account of the frame problem by Dennett (1987):

> Once upon a time there was a robot, named R1 by its creators. Its only task was to fend for itself. One day its designers arranged for it to learn that its spare battery, its precious energy supply, was locked in a room with a time bomb set to go off soon. R1 located the room, and the key to the door, and formulated a plan to rescue its battery. There was a wagon in the room, and the battery was on the wagon, and R1 hypothesized that a certain action which it called PULLOUT (Wagon, Room, t) would result in the battery being removed from the room. Straightaway it acted, and did succeed in getting the battery out of the room before the bomb went off. Unfortunately, however, the bomb was also on the wagon. R1 knew that the bomb was on the wagon in the room, but didn't realize that pulling the wagon would bring the bomb out along with the battery. Poor R1 had missed that obvious implication of its planned act.

Now our formulation of the frame problem is as follows:

- The space of information in reality is possibly infinite.
  - In the Dennett's robot account of the frame problem there are indeed a possibly infinite number of implications of the robot's action, which the robot has to take into account.
- Need a (finitary) frame to reduce the possibly infiniary amount of information and to identify the finitary scope of relevant information.
  - In the Dennett account the robot needs to ignore irrelevant information; relevance is usually supposed to be determined by so-called frame axioms in the case of logic-based AI.
- Need a (finitary) meta-frame to choose a frame because there are possibly infinitely many frames.

  - It needs to be determined which way of determining relevance is actually relevant in the present state of affair; there are indeed different ways of measuring the relevance of information.
- This meta-frame determination process continues *ad infinitum*.

Here every frame is assumed to be finitary, just as any formal system is assumed to be finitary (i.e., recursively axiomatisable, or equivalently, r.e. axiomatisable; see Shönfield (1967)) in the Gödel's first incompleteness theorem. The Gödel's first incompleteness theorem tells us any finitary formal system cannot characterise (stronger-than-Robinson-arithmetic) truths in reality (see Shönfield (1967)); in a nutshell, there is no finitary means to characterise truths in reality (in technical terms, the truths are not recursively enumerable; see Dummett (1963) and Smorynski (1997)). The frame problem is caused by the fact that any finitary frame cannot capture all information in reality (if the class of information is possibly infinite). What is lurking behind the two phenomena, Gödelian incompleteness and the frame problem, is arguably this sort of tension between finitary beings and infinitary reality. Let us elaborate more on this in the following.

This finitarity condition is what essentially underpins the frame problem, and from this point of view, the frame problem is about the fundamental discrepancy between the finitude of beings and the infinitude of (information in) reality. What happens if the finitarity condition is dropped? If there is no finitarity condition, we do not really have to choose a frame, and we can just take the largest frame or the union of all frames. The infinitary frame, therefore, is one escape from the frame problem. Put succinctly:

- There is no frame problem any more if infinitary frames are allowed.
- In Dennett's robot case this means the robot may take all information into account even if there is no finitary upper bound on the amount of information involved.

Another possible escape is to allow for infinitary time rather than frame (or alternatively we may also think of infinitary frames working for an infinitary amount of time). According to a certain sort of philosophy, such as Heidegger's (Heidegger 1927), the finitude of beings is, at least partly, rooted in the temporality of beings. If temporality is ignored the agent concerned may compute for an infinite amount of time, and hence no need for singling a finitary frame out and for being worried about the frame problem. From an epistemological point of view, therefore, we may conclude:

- What causes the frame problem is the finitude of beings or agents.
- The fundamental underpinning of the frame problem is the discrepancy between finitary beings and infinitary reality.

The same actually happens in Gödelian incompleteness. According to the Gödel's first incompleteness theorem, any finitary system cannot capture all truths in reality if the class of truths is infinitary enough (in the sense of being stronger than the Robinson arithmetic), and the finitarity condition is what crucially underpins Gödelian incompleteness as well. From this perspective, Gödelian incompleteness, as well as the frame problem, is about the fundamental discrepancy between the finitude of beings and the infinitude of reality; there is a fundamental gap between the provability of finitary systems and the truths of reality. What happens, then, if the finitarity condition is dropped in the case of incompleteness? If an infinitary system or infinitary computation is allowed then there is no incompleteness any more; indeed there are complete infinitary systems characterising the mathematical truths (see, e.g., Shapiro (2000) and Giaquinto (2004)). There are different ways to do this, including those via what are called ω-logic and infinitary logic. And this is actually not so difficult to see. If we allow for infinitary disjunction, for example, we can extend the Peano arithmetic with the following formula:

$$\forall x \ (x = 0 \vee x = 1 \vee ... \vee x = n \vee ...).$$

The extended Peano arithmetic over infinitary base logic is complete, and also there is no non-standard model allowed in this case, whereas there are infinitely many non-standard models of the ordinary Peano arithmetic over finitary first-order logic which include weird, infinitary numbers greater than any finitary number n. The above axiom allows us to exclude such non-standard numbers; this is never possible in finitary first-order logic (if it is possible we lead to a contradiction by the very incompleteness theorem). Yet at the same time the finitarity condition is epistemologically indispensable, for the human being cannot really deal with infinitary formulae or proofs in infinitary formal systems. For the same epistemological reason, formal systems must have been required to be finitary in the so-called Hilbert's programme; consistency, to be epistemologically significant, has to be established on the ground of Hilbertian finitism (see, e.g., Giaquinto (2004)). The failure of the Hilbert's programme can be overcome as well, if mathematical induction up to certain infinitary ordinals is allowed as a means of consistency proofs (as observed in the celebrated Gentzen's or Takeuti's consistency proof; see, e.g., Pohlers (1989)). From an epistemological point of view, thus, we may conclude:

- What causes Gödelian incompleteness is the finitude of beings or systems.
- The fundamental underpinning of Gödelian incompleteness is the discrepancy between finitary beings and infinitary reality (cf. Maruyama 2017).

Now we have argued that the frame problem and Gödelian incompleteness have the same structural origin in the finitude of agents or systems (if they are formulated as above). Simply saying, finitary beings (i.e., agents or systems) cannot apprehend all elements of infinitary reality. Without the finitarity condition, we can go beyond the finitude of beings. Once a strong ontology of infinities is accepted to be legitimate in some way or other, we have infinitary systems and infinitary frames, and they do allow us to resolve Gödelian incompleteness and the frame problem at the same time. Yet it is epistemologically quite problematic. In light of the Dennett's robot account of the frame problem, some infinitary robots based upon some infinitary computing mechanisms would not face the frame problem, and yet any robot based upon finitary computing systems cannot actually grasp all of possibly infinite implications of its action. Although infinitary systems do not suffer from incompleteness and although infinitary ordinals allow us to create consistency proofs, nonetheless, they are only meaningful under strong ontological assumptions, which may be more dubious than consistency per se. Indeed, the Gödel's second incompleteness theorem tells us that the consistency of one system can only be established on the basis of something stronger than the system per se. There is no discrepancy between infinitary beings and infinitary reality; infinitary beings can apprehend all elements of infinitary reality. We thus conclude the present section as follows:

- Both the frame problem and Gödelian incompleteness thus hinge upon the finitude of beings (which Heidegger emphasised as well).
- Put another way, the fundamental discrepancy between beings and reality (only) emerges when we take the finitude of beings seriously.

Let us make a final remark before closing the present section. The infinite regress argument above actually applies to any sort of finitary entity (i.e., an entity which cannot process a possibly infinite amount of information), and so, if the human being is a finitary entity, then the above argument, in principle, applies to the human being as well as the computing machine. The human being, therefore, must suffer from the frame problem, according to the above argument. Yet some consider the human being to be able to solve the frame problem. This might be understood as follows. Probably, the human being would have obtained some capacity, if incomplete, to choose a frame as a result of evolution; this would give an evolutionary account of frame selection capacity. If the human being has some capacity of frame selection which is incomplete and which nevertheless works in most practical situations, then there is no contradiction between the fundamental limitation that the frame problem tells us and the practical (and yet incomplete) capabilities that the human has in order to choose a frame. Again, this may be compared with what happens in the incompleteness phenomenon. The ZFC set theory and other finitary formal systems for mathematics are surely incomplete, and yet at the same time, there is basically no incompleteness known for practically meaningful problems. Indeed, it is generally understood in the logic community that there is, so far, almost no mathematically natural proposition found to be independent of the ZFC set theory. Incompleteness only arises in certain special realms concerning peculiar properties of infinities (such propositions include, for example, the continuum hypothesis and the existence of certain large cardinals). Here there is a tension between fundamental incompleteness and practical completeness, which seems to exist in the frame problem as well.

3 Reconsidering the Lucas-Penrose Argument

The Lucas-Penrose argument is an attempt to show the impossibility of AI on the ground of incompleteness (Megill 2017; Penrose 1989 and 1994); it has already been much criticised by many logicians, AI researchers, and philosophers (see, e.g., Feferman (1996), McCarthy (1990) and Putnam (1994); Penrose devoted as many as 200 pages of his second 1994 book to replying to various criticisms on his first 1989 book, even though further criticisms arose for the second book, followed by further replies to them). We would like to stress that here we focus upon an analysis of the structure of arguments rather than an assessment of the ultimate validity of it, which is another, separate issue different from the focus of the present article. There are actually several versions of the Lucas-Penrose argument; here we concentrate upon this one:

- If the human mind can solve an undecidable problem, then it cannot be computational, i.e., there is no computational system that can simulate the human mind, because the existence of such a computational system means that the computer can solve an undecidable or uncomputable problem.
- Put another way, artificial (general) intelligence is impossible. This is the central point of the Lucas-Penrose argument against the computational theory of mind. Yet the assumption must be justified to derive the AI impossibility conclusion.
- Penrose endorsed the impossibility of AI on the ground of Gödelian incompleteness: for any computational formal system (satisfying the aforementioned conditions) there is a proposition which is neither provable nor refutable in it and which the human can nevertheless know is true.
  - Put simply, Penrose considered Gödelian incompleteness telling that there is something which the computer cannot know and which the human can know; this means the human mind is essentially more powerful than the computer.
- Gödel (1995) himself says as follows: "So the following disjunctive conclusion is inevitable: Either mathematics is incompletable in this sense, that its evident axioms can never be comprised in a finite rule, that is to say, the human mind (even within the realm of pure mathematics) infinitely surpasses the powers of any finite machine, or else there exist absolutely unsolvable diophantine problems"; and Lucas (1961) argues "it is clear that Gödel thought the second disjunct false."
  - Megill (2017) gives a comprehensive survey of the Lucas-Penrose arguments, and he also says "perhaps the first thinker to endorse a version of the Lucas-Penrose argument was Gödel himself."

Here our primary aim is at analysing the structure of this argument and thereby deriving a physical consequence in a certain manner. Although the Lucas-Penrose argument, by itself, is about a limitation of the computational theory of mind, it can actually be adapted so as to be applicable for the computational theory of the universe or so-called information physics as we shall see in the following. The computational theory of the universe asserts that the universe is, or can entirely be simulated by, a computational system (although there would be different formulations of the thesis of computational theory of the universe or information physics, here we take this to be the primary thesis of it; for a related distinction between weak and strong information physics, we refer to Maruyama (2016)).

Now, the logic of the Lucas-Penrose argument allows us to contrive a physical version of it:

- If anything (or any mechanism) in the universe allows for solving uncomputable problems, then the universe cannot be computational, i.e., any computing system cannot simulate the universe (for, if it can, the computer can solve uncomputable problems, and this is a contradiction).
  - This is the universe version of the Lucas-Penrose argument, and let us call it the physical Lucas-Penrose argument.
- This universe version is actually subsumed under the mind version above in the following way: if the human can solve undecidable problems there is something in the universe which can compute uncomputable problems and so the universe cannot be computational.
  - Put another way, if the human mind is more powerful than the computing system the universe cannot be simulated by the computing system because humans are indeed part of the universe (this would be so regardless of whether one takes Quinean naturalism).
- The point of this argument may be summarised this way: if the Lucas-Penrose argument is correct and its assumption is satisfied, then the impossibility of information physics follows as well as the impossibility of AI; this means that the possibility of AI is tightly intertwined with the possibility of information physics.
  - We may derive a corollary from this: if Penrose denies AI or the computational theory of the mind, he must deny information physics or the computational theory of the universe as well, even though informational approaches have seen a great success in recent developments of physics, especially foundations of quantum theory (see, e.g., Chiribella (2011)).

This exposes an interesting, unexpected link between the possibility of artificial intelligence and the possibility of information physics, giving an insight into how much of the world, including both matter and mind, may be understood by computational means (since there is a clear mathematical bound on computability, any computational theory cannot go beyond the limitation of computability per se).

In view of our discussion on the frame problem, the human being, to Penrose and also to Gödel perhaps, is not a finitary entity, going beyond the realm of finitary computability, and in this case the human being could somehow "hypercompute" and would not suffer from the frame problem. Yet it would be quite controversial whether the realm of human intelligence or rather human knowability is really broader than the realm of finitary computability. Kripke (1982), for example, thinks that the human being is essentially a finite automaton. In light of recent developments in artificial intelligence, there is a great amount of evidence for the computational nature of human cognition; different cognitive capacities of the human being have turned out to be implementable through computational means such as machine learning methods. For now, however, no one would really know whether all elements of human cognition can be realised via computational mechanisms only. There is some realm of cognition to which the machine is ever unreachable. The Lucas-Penrose argument we have discussed above is an attempt to indicate a case for human intelligence beyond computability by virtue of Gödelian incompleteness. What we have shown in the present section is that, if the original, cognitive Lucas-Penrose argument is valid and its assumption is satisfied, then the modified, physical Lucas-Penrose argument must be valid as well; in other words, the impossibility of artificial intelligence in the Lucas-Penrose argument implies the impossibility of information physics as well. This is what we have called a physical consequence of our structural analysis of the Lucas-Penrose argument.

AI is a computational theory of the mind, and information physics is a computational theory of the universe. Both are computational theories, and yet at the same time, they have been discussed quite separately. It might be because of the Cartesian dualism, which ontologically separates the realm of mind and the realm of matter. Our general proposal here has nonetheless been to link the two computational theories, in particular with respect to their limitation. The success of both AI and information physics would suggest that most elements of reality are actually computational, or they can at least be simulated through computational mechanisms. It may eventually lead to the view that everything there is computational; or to be more precise, every thing there is information, and every process there is information-processing or computation. It is, so to say, an "informational theory of everything" or "computational theory of everything." Yet it would still be too early to judge if the universal computationalist or informationalist view is really correct or not. It would nonetheless deserve serious metaphysical consideration because it arguably implies that the Cartesian dualism is wrong in the sense that there is actually a single, unified realm of information and computation underlying the two apparently different realms of mind and of matter. Computational or informational world-views would thus be metaphysically relevant as well as scientifically effective. It would however be an open question if they are compelling whether metaphysically or scientifically.

4 Articulating Penrose's Quantum Mind Thesis

Penrose has a positive thesis on the nature of human mind as well as the aforementioned negative thesis on the impossibility of cognitive computationalism. If the capacity of human cognition is not bound by computability, what mechanism gives rise to such super-computational features? And what is the right theory of mind? Along such a line of thought Penrose argues for the quantum nature of mind:

- (i) AI or the computational theory of mind is misconceived in light of Gödelian incompleteness; the capacity of human cognition is not bound by computability.
- (ii) The mind is materially quantum; consciousness emerges via material quantum processes in microtubules. Call it the Material QMT (quantum mind thesis).

There are different criticisms from different angles on the one hand (and yet about 200 pages of Penrose's second book *Shadows of the Mind* are devoted to replies to those criticisms, suggesting his serious and deep commitment to the QMT). Chalmers (1996b), e.g., says: "Why should quantum processes in microtubules give rise to consciousness, any more than computational processes should?"; "reader who is not convinced by Penrose's Gödelian arguments is left with little reason to accept his claims [...] that quantum processes are essential to cognition." On the other hand, the field of quantum cognition is rapidly growing over recent years, giving the quantum models of cognition that are able to account for those non-classical features of cognition which were mysteries within conventional theories. "The success of human cognition can be partly explained by its use of quantum principles", Pothos and Busemeyer (2009) say. The success of quantum structure in AI might support this idea too, including, for example, quantum machine learning, the kernel method via RKHS (Reproducing Kernel Hilbert Spaces), and tensor networks (the tensor structure is what underpins entanglement and non-locality in quantum theory). In light of those recent developments, there might now be some scientific reasons to accept Penrose's claim that "quantum processes are essential to cognition", in contrast to Chalmers' criticism above. This never means something like the QMT has been confirmed; yet it is now getting a rigorous scientific basis in the rapidly growing field of the quantum science of cognition. In light of recent research in quantum cognition, we propose to modify Penrose's argument above, that is, replace computability by complexity, and the material QMT by the structural QMT (under the assumption that the human cognitive system is capable of recognising the world in a more efficient manner than the classical computing system; arguments for this assumption could be elaborated on the ground of different scientific findings):

- (i) Classical AI or the classical computational theory of mind is misconceived in light of the super-classical features and effectiveness of human cognition.
  - The mind cannot be a classical computer due to differences in complexity (in a formal, "polynomial vs. exponential" sense as in quantum computing if the mind is a quantum computer, or in a more informal sense as the efficiency of human cognition in certain fields compared to machine cognition as enabled by pattern recognition methods of machine learning).
- (ii) The mind is structurally quantum; the structure of cognition is homomorphic to the structure of (models of) quantum theory.
  - It is not that quantum processes are materially going on in the macroscopic physical brain; e.g., Tegmark (2000) computationally refutes Penrose's claim on microtubules (even though there are counterarguments by Hameroff et al.).
  - The structure of economic systems is homomorphic to that of physical systems as mathematical economics tells us; this never means that the nature of economy is materially physical. Likewise, quantum modelling of cognition and its success do not entail that cognition is materially quantum.

If the universe is a materially quantum computer as in quantum information physics, the mind might be a structurally quantum computer. Along such lines of thought we may possibly save Penrose's Quantum Mind Thesis in light of contemporary developments in quantum cognitive science. Although Penrose's ideas are sometimes considered utterly absurd, nevertheless, they might be able to survive with modifications in some way or other. It is now a scientific fact that certain facets of human cognition can only be modelled by quantum structures, including the so-called order effect, the conjunction effect, and the disjunction effect, which all indicate that the laws of classical logic and probabilities do not hold in human cognition, and which can be explained via quantum models. Even violation of Bell-type inequalities in cognition has recently been reported in rigorous experimental studies (see, e.g., Cervantes and Dzhafarov (2018)), suggesting that cognitive systems shares with quantum systems cerain key structural features such as contextuality and sensitivity to environments. The quantum science of cognition is developing very rapidly, and some version of the QMT might be able to be vindicated in the near future.

5 Concluding Remarks

We conclude the article by summarising the above discussion and by giving a final remark about the integration of computational theories of the mind and of the universe as giving an embodiment of the Chalmers' double-aspect theory of information (Chalmers 1996a). Let us wrap up the points of our discussion as follows:

- Both the frame problem and Gödelian incompleteness may be understood as instances of the fundamental discrepancy between finitary beings and infinitary reality (as have been much discussed within the philosophical tradition).
  - The finitarity condition is indispensable, since infinitary agents/systems are able to resolve the frame/incompleteness problem.
  - If the human being is a finitary entity, then the frame problem in our formulation applies to the human being as well as the computing machine.
- There is a tight link between artificial intelligence (the computational theory of the mind) and information physics (the computational theory of the universe). The Lucas-Penrose argument, if it is correct, indicates the impossibility of information physics as well as the impossibility of artificial intelligence.
  - Relating the two computational theories, we could go beyond the Cartesian dualism, thus giving an embodiment of the Chalmers' double-aspect theory of information (Chalmers 1996a); i.e., matter and mind are united in the underlying, fundamental reality of information and its processing via the integration of the two computational theories of the mind and of the universe.
  - The theory of life may be included in that of cognition or of matter, but life can be another dimension of reality, and in this case we would need a triple-aspect theory of information. From this perspective, life is an emergent property, and living organisms are entities in their own right,

and do not reduce to anything like matter, thus existing as a coherent and united whole.
- The following picture illustrates the links between different dimensions of reality, which are theorised by different computational theories, and yet unified in pancomputational reality at the fundamental level. Whilst keeping different ontological realms as in the Cartesian dualism, we can still have them united in the underlying fundamental reality of information and its processing.

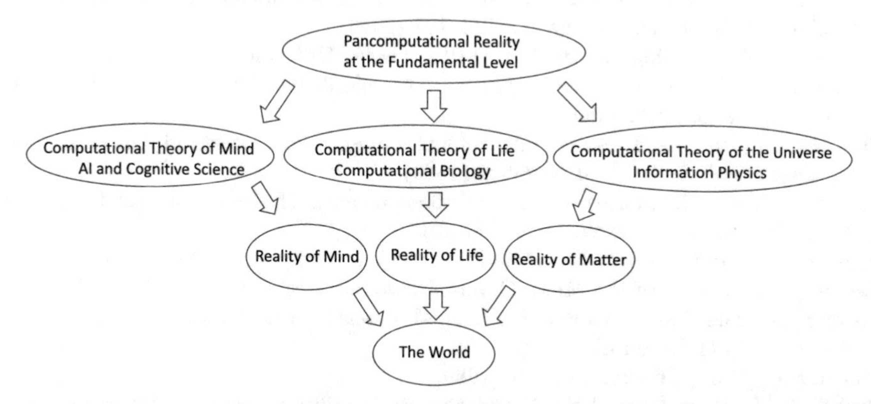

The links between the computational theories of the universe, life, and mind are to be further explored and expanded in future work; we believe the pursuit of the links would allow us to overcome the fragmentation of science and of our worldview.

Acknolwedgements. I would like to thank the members of the Quantum Foundations Discussion at the University of Oxford for their useful feedback to my talk there. This work has been supported by JST PRESTO Grant (JPMJPR17G9) and JSPS Kakenhi Grant (JP17K14231).

References

Cervantes, V.H., Dzhafarov, E.N.: Snow Queen is Evil and Beautiful: Experimental Evidence for Probabilistic Contextuality in Human Choices (2018). forthcoming in Decision

Chalmers, D.: The Conscious Mind. Oxford University Press, New York (1996a)

Chalmers, D.: Minds, machines, and mathematics. Psyche **2**, 11–20 (1996b)

Chiribella, G., et al.: Informational derivation of quantum theory. Phys. Rev. A **84**, 012311 (2011)

Dennett, D.: Cognitive Wheels: The Frame Problem in Artificial Intelligence. The Robot's Dilemma, The Frame Problem in Artificial Intelligence, pp. 41–64 (1987)

Dummett, M.: The philosophical significance of Gödel's theorem. Ratio **5**, 140–155 (1963)

Everett, H.: The Theory of the Universal Wavefunction. The Many-Worlds Interpretation of Quantum Mechanics, pp. 3–140. Princeton University Press (1973, originally 1955)

Feferman, S.: Penrose's Godelian argument. Psyche **2**, 21–32 (1996)
Fodor, J.A.: The Modularity of Mind. MIT Press, Cambridge (1983)
Giaquinto, M.: The Search for Certainty. Clarendon Press, Oxford (2004)
Gödel, K.: Collected Works III. Oxford University Press, Oxford (1995)
Heidegger, M.: Sein und Zeit (1927)
Kripke, S.: Wittgenstein on Rules and Private Language (1982)
Lucas, J.R.: Minds, Machines and Gödel. Philosophy **36**, 112–127 (1961)
Maruyama, Y.: AI, Quantum Information, and External Semantic Realism: Searle's Observer-Relativity and Chinese Room, Revisited. Fundamental Issues of Artificial Intelligence, Synthese Library, pp. 115–127 (2016)
Maruyama, Y.: Meaning and duality. D.Phil Thesis, University of Oxford (2017)
McCarthy, J.: Review of "The emperor's new mind" by Roger Penrose. Bull. Am. Math. Soc. **2**, 606–611 (1990)
McCarthy, J., Hayes, P.J.: Some philosophical problems from the standpoint of artificial intelligence. Mach. Intell. **4**, 463–502 (1969)
Megill, J.: The Lucas-Penrose Argument about Gödel's Theorem. Internet Encyclopedia of Philosophy. Accessed December 2017
Penrose, R.: The Emperor's New Mind. Oxford University Press, New York (1989)
Penrose, R.: Shadows of the Mind. Oxford University Press, New York (1994)
Putnam, H.: The Best of All Possible Brains? The New York Times, Book Review, 20 November 1994. Late Edition
Pohlers, W.: Proof Theory. Springer (1989)
Pothos, E.M., Busemeyer, J.R.: A quantum probability explanation for violations of rational decision theory. Proc. Roy. Soc. B **276**, 2171–2178 (2009)
Shanahan, M. (2016). The Frame Problem. The Stanford Encyclopedia of Philosophy, Spring 2016 Edition
Shanahan, M.: Solving the Frame Problem: A Mathematical Investigation of the Common Sense Law of Inertia. MIT Press, Cambridge (1997)
Shapiro, S.: Foundations without Foundationalism: A Case for Second-Order Logic. Oxford University Press, Oxford (2000)
Smorynski, C.: The incompleteness theorems. In: Handbook of Mathematical Logic, pp. 821–865 (1997)
Shönfield, J.R.: Mathematical Logic. Addison Wesley, Reading (1967)
Stein, L.A.: An atemporal frame problem. Int. J. Expert Syst. **3**, 371–381 (1990)
Tegmark, M.: The importance of quantum decoherence in brain processes. Phys. Rev. E **61**, 4194–4206 (2000)
Wheeler, J.A.: Information, physics, quantum: the search for links. In Complexity, Entropy, and the Physics of Information, pp. 309–336 (1990)

Quantum Pancomputationalism and Statistical Data Science: From Symbolic to Statistical AI, and to Quantum AI

Yoshihiro Maruyama[1,2](✉)

[1] The Hakubi Centre for Advanced Research, Kyoto University, Kyoto, Japan
[2] JST PRESTO, Tokyo, Japan
maruyama@i.h.kyoto-u.ac.jp

Abstract. The rise of probability and statistics is striking in contemporary science, ranging from quantum physics to artificial intelligence. Here we discuss two issues: one is the computational theory of mind as the fundamental underpinning of AI, and the quantum nature of computation therein; the other is the shift from symbolic to statistical AI, and the nature of truth in data science as a new kind of science. In particular we argue as follows: if the singularity thesis is true the computational theory of mind must ultimately be quantum in light of recent findings in quantum biology and cognition; data science is concerned with a new form of scientific truth, which may be called "post-truth"; whereas conventional science is about establishing idealised, universal truths on the basis of pure data carefully collected in a controlled situation, data science is about indicating useful, existential truths on the basis of real-world data gathered in contingent real-life and contaminated in different ways.

1 Introduction

Pancomputationalism (Piccinini 2017) is the view that everything there is is a computing system, which is, in principle, a theory of everything, and may be instantiated in different kinds of science. It may be applied, in particular, to matter, life, and mind, thus yielding physical, biological, and cognitive computationalism. In the first part of the present article, we discuss the nature of computation in computationalism, which we argue must be quantum rather than classical. The measurement statistics of quantum systems is probabilistic, and this quantum probabilistic nature lies at the heart of the super-classical power of quantum computing. The second part concerns how probabilities could change the notion of truth in light of data science as a new kind of science, articulating the status of data science qua science or the pursuit of truth (whatever it is). A key feature shared by quantum computing and data science is the utilisation of

Supported by JST PRESTO Grant (JPMJPR17G9) and JSPS Kakenhi Grant (JP17K14231).

V. C. Müller (Ed.): PT-AI 2017, SAPERE 44, pp. 207–211, 2018.
https://doi.org/10.1007/978-3-319-96448-5_20

indeterminacies in Nature or cognition, which are resources for new technologies rather than philosophical anomalies to be resolved.

2 The Computational Theory of Life and Cognition: Classical or Quantum?

AI is the computational theory of mind; it engineers intelligence via computational processes. Most neuroscientists today consider the brain to be a sort of computing system processing different kinds of information in reality. Information physics is the computational theory of the universe. Quantum information physics (Lloyd 2006) asserts that the universe is a gigantic quantum computer calculating its own quantum state or wave function. If the mind is computational, is it classical-computational or quantum-computational? The same question may be asked for artificial life as the computational theory of life: if life is computational, is that computation classical or quantum?

The emergent field of quantum biology tells us some life science phenomena are essentially quantum (Arndt et al. 2009; Lloyd 2011): "Nature is the great nano-technologist. The chemical machinery that powers biological systems consists of complicated molecules structured at the nanoscale and sub-nanoscale. At these small scales, the dynamics of the chemical machinery is governed by the laws of quantum mechanics" (Lloyd 2011). Apart from the disputed case of quantum effects in the brain, the rôle of quantum coherence in biological systems is observed in photosynthesis, bird navigation ("avian compass"), and the sense of smell. There is strong evidence for quantum coherence in photosynthesis whilst that for the sense of smell and bird navigation is more "indirect" (Lloyd 2011). Yet more evidence has been found since Lloyd (2011), including, e.g., Francoa et al. (2011) and Gane et al. (2013). Quantum biology is now being formed as a new vital field of life sciences.

Fleming and his collaborators at Berkeley in particular conjectured bacteria are performing the quantum search algorithm in photosynthesis (Lloyd 2011). However, "the bacteria were in fact performing a different type of quantum algorithm, called a quantum walk" (Mohseni et al. 2008). The bacteria may thus be regarded as tiny quantum computers. Any classical AI or AL cannot completely simulate the bacteria or the mechanism of photosynthesis due to their quantum nature. If the technological singularity comes in the future, and if it allows for computation more powerful than that of the bacteria, it must be quantum on the ground of the recent findings of quantum science as mentioned above.

3 From Symbolic to Statistical AI: Data Science as a New Kind of Science

Probabilistic, if not fully quantum, approaches have turned out to be successful in different fields of computer science; even in classical computation, probabilistic algorithms achieve striking efficiency in certain tasks. Broadly, the rise of statistical methods in AI is along the same line, having led to the vital movement of data science. Here we give a conceptual perspective on the peculiar nature of data science.

Statistical AI is particularly suitable for pattern recognition, thus enabling statistical prediction and classification among other things. Data science can even be a system of ontology: data mining from ICD (Spangler 2002; Yan et al. 2010), for example, has given insights into the classification of diseases and thus into the ontology of medicine. The intuitive sense of patterns, such as those of faces, is difficult to simulate by logical reasoning or symbolic AI; by contrast, statistical AI allows us to build a computational theory of pattern sensing. In Kantian terms, statistical AI may be called the AI of sensibility, and symbolic AI the AI of reason or understanding. Put another way, the former is the AI of learning (e.g., from empirical data), and the latter the AI of reasoning (e.g., from established knowledge). The tension between symbolic and statistical AI may be compared with the philosophical dichotomy between rationalism and empiricism; the Chomsky vs. Norvig debate (Norvig 2011) can be understood in such a broader, philosophical context. Currently, statistical AI is dominant, compared to symbolic AI (aka. "good old-fashioned AI"), and yet some AI researchers consider both symbolic and statistical AI crucial for future developments (Domingos et al. 2006; Russell 2016). From a Kantian point of view, intelligence or the capacity to know about the world arguably requires the capabilities of both sensing and reasoning about the world.

Characteristics of big data include volume, variety, velocity, variability, value, veracity, and so fourth (Gandomi et al. 2015). The methods of data mining allows us to discover some knowledge from such data. Yet any empirical science is about obtaining knowledge from data. Is data science, then, just a sort of universal science developing general statistical methods to generate knowledge from data? It may be the case in some sense, but there would be subtler issues on the nature of truth and knowledge found by data mining. Whereas conventional science is about establishing pure or idealised truths from pure or idealised data, which are carefully collected in a controlled situation (such as laboratory), data science is about indicating impure or real-world truths from impure or real-world data, which may be contaminated in different ways, and are collected from (often uncontrollable) contingent real life. Conventional science (so far) cannot predict the result of a referendum (such as the victory of Donald Trump) and yet data science is still able to say something about it, if the prediction turns out to be wrong at the end of the day (or not). The lack of truthfulness, nonetheless, is a trade-off for real-world applicability, since the real world is surrounded by myriads of uncertainties and contingencies. Data science takes real-world contingencies and uncertainties seriously, whilst conventional science idealises, if not ignore, them to some degree (just as physics discusses the movement of just one massless particle in a vacuum). Data science is concerned with truths putting more emphasis on use value than truthfulness per se; e.g., the prediction of preferences in recommender systems may be useful even if it is wrong after all.

Such truths can be wrong and yet of social, economical, or political use value; they may be named "post-truths." In other words, conventional science is about truthful knowledge (cf. knowledge as justified true belief), and data science is about post-truth wisdom, which is a special type of belief justified by non-human machine (learning algorithms) and of social, economical, or political use value.

Idealisation in conventional science is made for the purpose of elucidating the essence of Nature through the universal principles of causal mechanisms; it is nothing negative. By contrast, data science approaches the existence of actual Being through the statistical accounts of correlational mechanisms. (The latter, of course, can contribute to the former and vice versa, as observed in the actual practice of science.) Its primary focus upon existence and contingency is a distinctively Heideggerian characteristic. Recently, AI is becoming more and more embedded, and more and more embodied (Dreyfus 2007). Taking all this into account, we may say that AI, especially statistical data science, is getting more and more Heideggerian.

4 Concluding Remarks

The deterministic view of the world was dominant at the time of Laplace, and probabilities were supposed to exist within the human cognition of Nature, and not in reality per se. The rise of quantum mechanics has changed the scene (though not completely; note that there is still a deterministic interpretation of quantum theory such as Bohmian mechanics), and the indeterministic nature of reality has been elucidated since. The more recent rise and success of probabilistic models in AI and data science are along the same line, and our understanding of Nature is getting more and more statistical in contemporary science. Computationalism has been effective in life sciences, including biology and cognitive science, as well as physical sciences, in particular information physics. The fundamental nature of computation is still debatable, however. Here we have argued that it must be quantum (or at least probabilistic in some way or other). A shift in the fundamental notion of computation would lead to a shift in the computationalist conception of truth; we have thus explicated and articulated the notion of truth in the age of statistical data science. Nevertheless it would still be possible to defend the classical deterministic conception of truth, for example, by arguing that data science is just a method to find hypotheses and thereby to reach the deterministic truth of reality. After all, there is no perfectly compelling reason to take probabilities in quantum or data science at face value. Even so, they are making more and more room for statistical indeterminism and a new conception of truth (or post-truth). And, in the long run, a non-classical, statistical view of the world could possibly become classical.

References

Arndt, M., et al.: Quantum physics meets biology. HFSP J. **3**(6), 386–400 (2009)

Dreyfus, H.L.: Why Heideggerian AI failed and how fixing it would require making it more Heideggerian. Philos. Psychol. **20**, 247–268 (2007)
Domingos, P., et al.: Unifying logical and statistical AI. In: Proceedings of AAAI, pp. 2–7 (2006)
Francoa, M.I., et al.: Molecular vibration-sensing component in Drosophila melanogaster olfaction. Proc. Natl. Acad. Sci. **108**, 3797–3802 (2011)
Gandomi, A., et al.: Beyond the hype: big data concepts, methods, and analytics. Int. J. Inf. Manag. **35**, 137–144 (2015)
Gane, S., et al.: Molecular vibration-sensing component in human olfaction. PLoS ONE **8**, e55780 (2013)
Lloyd, S.: Programming the Universe. Knopf, New York (2006)
Lloyd, S.: Quantum coherence in biological systems. J. Phys. Conf. Ser. **302**, 012037 (2011)
Mohseni, M., et al.: Environment-assisted quantum walks in photosynthetic energy transfer. J. Chem. Phys. **129**(17), 174106 (2008)
Norvig, P.: On Chomsky and the Two Cultures of Statistical Learning (2011). http://norvig.com/chomsky.html. Accessed 31 Jan 2018
Piccinini, G.: Computation in Physical Systems. Stanford Encyclopedia of Philosophy (2017)
Spangler, W.E., et al.: A data mining approach to characterizing medical code usage patterns. J. Med. Syst. **26**, 255–275 (2002)
Russell, S.: Rationality and intelligence. Fund. Issues Artif. Intell. **376**, 7–28 (2016). Synthese Library
Yan, Y., et al.: Medical coding classification by leveraging inter-code relationships. In: Proceedings of KDD, pp. 193–202 (2010)

Getting Clarity by Defining Artificial Intelligence—A Survey

Dagmar Monett[1,2(✉)] and Colin W. P. Lewis[1(✉)]

[1] AGISI.org, Berlin, Germany
{dagmar.monett,colin.lewis}@agisi.org
[2] HWR Berlin, Berlin, Germany

Abstract. Intelligence remains ill-defined. Theories of intelligence and the goal of Artificial Intelligence (A.I.) have been the source of much confusion both within the field and among the general public. Studies that contribute to a well-defined goal of the discipline and spread a stronger, more coherent message, to the mainstream media, policymakers, investors, and the general public to help dispel myths about A.I. are needed. We present the preliminary results of our research survey "Defining (machine) Intelligence." Opinions, from a cross sector of professionals, to help create a unified message on the goal and definition of A.I.

1 Introduction

Intelligence permeates almost everything we do. Formally providing a robust and scientific definition of intelligence has been a goal of scientists and researchers for several centuries. During the last sixty years, the formal definition of intelligence has taken on extra impetus as machine intelligence, or A.I., developers pursue their vision of creating intelligent machines that "replicate" human intelligence (Brooks 1991). For others, the goal is creating Artificial General Intelligent systems which exceed human intelligence. However, it is still very hard to define what intelligence is (Kambhampati 2017). Furthermore, creating an agreed upon message on the goal and definition of A.I. is far from obvious or straightforward (Nilsson 2010).

In order to clarify the goal and definition of A.I. the research survey "Defining (machine) Intelligence" solicits opinions from a cross sector of professionals. The ongoing survey[1] has attracted a significant volume of responses and high level comments and recommendations concerning the definitions of A.I. and human intelligence from experts around the world. We believe that collecting experts' opinions can contribute to both a deeper understanding and a better definition of what intelligence is. In this short paper, a partial analysis of the first 400 responses is presented.

[1] See http://agisi.org/Survey_intelligence.html for more about the survey.

V. C. Müller (Ed.): PT-AI 2017, SAPERE 44, pp. 212–214, 2018.
https://doi.org/10.1007/978-3-319-96448-5_21

2 Research Survey: Preliminary Results and Discussion

The research survey focuses on specific definitions of human and machine intelligence,[2] and on the level of agreement of respondents with those definitions. The survey further asks respondents to provide their level of agreement with statements based on DeBoeck's (2013) questions concerning the definition of intelligence. Potential respondents were collected from different sources with research topics relevant to the survey. News lists informing computer scientists, neuroscientists, cognitive sociologists, and roboticists, to name a few, were also considered. Survey respondents originate from 48 countries and 131 different institutions (academia 77%, industry 21.3%). Respondents are mainly researchers (75.3%), educators (36%), and developers or engineers (16.8%), and come from Computer Science (58%), Psychology or Cognitive Science (9.3%), and Engineering (8.5%).

Partial results show (see Table 1) that most respondents disagree or strongly disagree there is a difficulty in defining the goal of A.I. (1.b) Only a small minority seem to believe that a definition of intelligence is self-evident (1.a) and over the half of respondents disagree that it will never be possible to reach an agreement upon a definition of A.I. (1.f) Most respondents indicated disagreement or strong disagreement with the statement that a unified definition of artificial intelligence does not pay off (1.d). However, there are also strong opinions for the contrary and an almost equal amount of opinions are neutral. Other statements to agree upon considered differences in opinion when defining A.I. being too large to bridge (1.c), a definition of A.I. experienced as a restriction (1.e), and scientific advances in A.I. being "a huge step forward and possibly a promising paradigm shift towards creating machines that can be measured to match or exceed human level intelligence" (1.g).

Table 1. Level of agreement with some statements to agree upon in the survey ($N = 400$).

Id	Strongly disagree	Disagree	Neutral	Agree	Strongly agree
1.a	160 (40.0%)	169 (42.3%)	34 (8.5%)	24 (6.0%)	13 (3.3%)
1.b	61 (15.3%)	173 (43.3%)	83 (20.8%)	71 (17.8%)	12 (3.0%)
1.c	19 (4.8%)	152 (38.0%)	116 (29.0%)	93 (23.3%)	20 (5.0%)
1.d	26 (6.5%)	125 (31.3%)	119 (29.8%)	115 (28.8%)	15 (3.8%)
1.e	51 (12.8%)	160 (40.0%)	81 (20.3%)	93 (23.3%)	15 (3.8%)
1.f	55 (13.8%)	171 (42.8%)	94 (23.5%)	62 (15.5%)	18 (4.5%)
1.g	40 (10.0%)	66 (16.5%)	74 (18.5%)	147 (36.8%)	73 (18.3%)

Many respondents ($N = 187$, 46.8%) express agreement or strong agreement concerning the need for having separate definitions of human and machine

[2] See http://agisi.org/Defs_intelligence.html for a complete list of definitions.

intelligence, but a slightly equal number ($N = 172$, 43%) indicate that only one definition is adequate. The definition of *machine* intelligence that received the most comments was Russell and Norvig's (2010) definition with a total of 224 (56%) opinions. However, the most accepted definition of machine intelligence was Wang's (2008): 224 (56%) respondents agree or strongly agree with it. Similarly, the definition of *human* intelligence that received the most comments was Humphreys' (1984) definition with a total of 148 (37%) opinions. However, the most accepted definition was Gottfredson's (1997) definition: 246 (61.5%) respondents agree or strongly agree with it. Survey participants provided a total of 3453 reasons for supporting their selections of the definitions. Furthermore, a total of 213 (53.3%) survey participants provided their suggested definitions of human and/or machine intelligence.[3]

3 Conclusions

Getting clarity around defining A.I. must include experts' opinions. The first 400 responses to our survey comprise thousands of those opinions, not to mention other hundreds of suggested definitions of intelligence and overall feedback that were received. A significant variety of judgements and viewpoints that allow for first understanding and then creating an agreed upon message on the goal and definition of A.I. The question of combined versus separate definitions of human and machine intelligence remains highly polarized. Our work in progress includes building a catalogue of factors contributing to intelligence, and a methodology for and best practices to be applied when defining (machine) intelligence.

References

Brooks, R.A.: Intelligence without representation. Artif. Intell. **47**, 139–159 (1991)
De Boeck, P.: Intelligence, where to look, where to go? J. Intell. **1**, 5–24 (2013)
Gottfredson, L.S.: Mainstream science on intelligence: an editorial with 52 signatories, history, and bibliography. Intelligence **24**, 13–23 (1997)
Humphreys, L.G.: General intelligence. In: Reynolds, C.R., Brown, R.T. (eds.) Perspectives on Bias in Mental Testing, p. 243. Springer, Boston (1984)
Kambhampati, S.: On the Past and Future of AI. Interviews with Experts in Artificial Intelligence, Iridescent (2017). https://goo.gl/nspv6y
Nilsson, N.J.: The Quest for Artificial Intelligence. A History of Ideas and Achievements. Cambridge University Press, Cambridge (2010)
Russell, S.J., Norvig, P.: Artificial Intelligence: A Modern Approach, 3rd edn. Prentice Hall, Upper Saddle River (2010)
Wang, P.: What do you mean by "AI"? In: Wang, P., Goertzel, B., Franklin, S. (eds.) Proceedings of the First AGI Conference on Artificial General Intelligence 2008. Frontiers in Artificial Intelligence and Applications, vol. 171, pp. 362-373. IOS Press, Amsterdam (2008)

[3] More results can be found in our poster to the PT-AI 2017 conference (Leeds, UK) that is available at http://agisi.org/doc/Monett-Lewis_PT-AI2017_poster.pdf.

Epistemic Computation and Artificial Intelligence

Jiří Wiedermann[1(✉)] and Jan van Leeuwen[2(✉)]

[1] Institute of Computer Science of AS CR, Prague, Czech Republic
jiri.wiedermann@cs.cas.cz
[2] Department of Information and Computing Sciences, Utrecht University, Utrecht, The Netherlands
J.vanLeeuwen1@uu.nl

Abstract. AI research is continually challenged to explain cognitive processes as being computational. Whereas existing notions of computing seem to have their limits for it, we contend that the recent, epistemic approach to computations may hold the key to understanding cognition from this perspective. In this approach, computations are seen as processes generating knowledge over a suitable knowledge domain, within the framework of a suitable knowledge theory. This, machine-independent, understanding of computation allows us to explain a variety of higher cognitive functions such as accountability, self-awareness, introspection, free will, creativity, anticipation and curiosity in computational terms. It also opens the way to understanding the self-improving mechanisms behind the development of intelligence. The argumentation does not depend on any technological analogies.

1 Introduction

Computation has proved to be a powerful framework for understanding cognitive processes. The all-important question is how computation can explain and model them, whether our understanding of computation is sufficient for it, and how a sufficiently general theory might be developed that transcends the realm of concrete algorithmic models.

In computer science, computations are traditionally seen as processes performed by computers. In order to make this definition more precise we must fix the notion of a computer. In theory, we commonly use the Turing machine for it since we know that, in principle, this model captures the computational ability of a large class of contemporary digital computers. Nevertheless, in practice, especially in artificial intelligence, biology and physics, one often considers computations in broader terms, not only including the processes taking place in computers and related artifacts, but also those in cells or plants, in the brains of people or animals, or even in the Universe as a whole.

The research of the first author was partially supported by the ICS AS CR fund RVO 67985807 and the Czech National Foundation Grant No. 15-04960S.

V. C. Müller (Ed.): PT-AI 2017, SAPERE 44, pp. 215–224, 2018.
https://doi.org/10.1007/978-3-319-96448-5_22

Hence, in order to see how the computational framework might apply, we have to investigate the nature and limitations of computations in the diverse contexts in which they emerge and by the devices that seem to perform them. In particular, much has been learned about the problem, *how* computations are realized by the various kinds of devices. This view is theoretically and technically interesting and has led to the development of e.g. classical computability theory and to important approaches to the understanding of computational complexity, indeed. However, these theories are dependent on the chosen model and, even worse, they tend to lack the potential to capture the meaning, or aim, of the computations that are being modeled. In other words, the problem *what* computations should do or be for us, or for the device realizing a computation, remains largely unanswered by it.

When asking what computations do, we contend that the only reasonable answer is that they produce *knowledge* of some kind. Here we take knowledge to be knowledge in the general sense, be it declarative or procedural or otherwise. For example, we see the causing processes of actions or behaviors as computational, as they generate the knowledge needed for an agent to trigger or perform them. This view has been the starting point of the so-called *epistemic theory of computation* proposed by the authors some years ago (cf. Wiedermann and van Leeuwen 2013, 2014, 2015a,b, 2017). Following this theory, computations are seen as processes generating knowledge over a suitable domain, in the framework of a suitable epistemic theory.

In this paper we show that the epistemic approach to computations has great potential especially for Artificial Intelligence. We will argue that it gives us a natural way to define and explain non-trivial cognitive functions like accountability, self-awareness, introspection, knowledge understanding, free will, creativity, anticipation, and curiosity as specific ways of computation, i.e. of particular kinds of knowledge generation. This brings new insight into the surmised algorithmic mechanisms behind intelligence, in a way that does not depend on any specific model of computation.

The latter feature is a considerable advantage of the epistemic model over other approaches to computational cognitive systems as described in the literature, which often use far more complex models of cognition. These models, usually based on concrete algorithmic mechanisms, tend to lead to cumbersome definitions, with descriptions depending more on the architecture and the special properties of the underlying model than on the general properties of the cognitive functions under consideration. A detailed account of the realized cognitive architectures can be found in Samsonovich (2010) and in the related on-line catalog. For the introduction to the theory of epistemic computation, its justification and examples of its viability, see Wiedermann and van Leeuwen (2013).

The structure of the paper is as follows. In Sect. 2 we present the main ideas of the epistemic approach to computation. In Sect. 3 we explain how this approach leads to machine- and algorithm-independent definitions of important higher cognitive functions, including also the notion of self-improving epistemic theories. Finally, in Sect. 4 we give some conclusions.

2 Computation as Knowledge Generation

Before digressing on the epistemic approach to computation, we need to be more specific about the way we view *knowledge*. Informally, knowledge in our framework is knowledge in the usual sense of this word. For our purposes we allow both declarative and procedural knowledge as they are commonly distinguished, and any other form of knowledge that might be implicitly or explicitly acquired somehow. Thus, knowledge could be facts or specific information of some kind but also knowledge of the 'know-how' kind, including the underlying rules of actions or skills and even of behaviors. For the treatment of skills in our approach, see the short remark at the end of this section.

When defining knowledge in accordance with the epistemic theory of computation, we need to be somewhat more specific. We will see knowledge as the result of a certain computational process, working over a given *knowledge domain*. This process combines known elements of the domain—so-called *elementary knowledge*—into derived, often more complex constructs that represent new knowledge over the given domain. For combining elements from the domain, the computation makes use of *derivation rules* which are either embedded in the process or proceed in interaction with the environment. The rules can be known beforehand or may be learned, in the latter case through the potentially endless processing of many computations over the given domain. Hence, a computation works with a more or less formal theory, capturing the given knowledge elements of the underlying epistemic domain as well as the ways of inferring new elements within that domain.

2.1 Computational Processes

As mentioned above, we are not interested in *how* a computation proceeds by means of a mechanism of some kind, but rather in *what* it does, i.e., in what knowledge is generated in the course of a computation. Under this viewpoint, the ability to generate knowledge becomes the hallmark of those processes that we will call computational. *Intelligent systems* are special instances of computational processes or sets thereof, which are able to generate knowledge over knowledge domains that model large parts of the real world, of various sciences, or of any other specific areas that are amenable to knowledge generation. This contrasts with the practice of contemporary AI systems, or "ordinary" computations that, as a rule, are specialized to, in most cases, considerably restricted knowledge domains. Within the epistemic theory of computation, the processes that do not generate knowledge are not considered as computations.

The question how (generated) knowledge can be identified or recognized is a difficult philosophical problem. From a practical point of view, one normally agrees that it is an *observer-dependent* matter - what constitutes knowledge can be obvious for one person and completely unclear to an other. That is, what is knowledge depends on how knowledgeable a person (or an AI system) already is in the domain under consideration. In the epistemic approach to computation we therefore always define knowledge in the framework of the *knowledge domain* over which a computation operates. All knowledge about (a subset of) the knowledge domain at hand is to be captured by a corresponding *knowledge theory* which can be more or less formal, or completely informal.

In the knowledge theory of a domain, *axioms* describe the available *elementary knowledge* corresponding to the (representations of) objects in the domain and their properties. The ways in which new, derived knowledge can be constructed from elementary knowledge are described by *inference* or *derivation rules*. In fact, there are also other constructive methods for knowledge generation, used e.g. in ancient mathematics (cf. Sloman 2018), but here we concentrate solely on theories that are based on logic-based inference. Then, computational processes are bound to their knowledge domain through the corresponding knowledge theory, via the following condition:

> *whatever can be derived within the given theory must be supported by the corresponding computational process (and vice versa).*

Furthermore, for any computational process, there must be evidence concerning the validity of the latter condition. If the condition holds, then what knowledge can or cannot be generated over the given knowledge domain, and the "quality" of this knowledge (e.g., its agreement with observations), depends solely on the properties of the underlying knowledge theory.

In other words, in order for a computation to generate knowledge there must be evidence that explains that the computational process works as expected. This evidence must establish two facts: *(i)* that the generated knowledge can be derived within the underlying epistemic theory; this is provided via a 'proof' in that theory, and *(ii)* that the computational process generates the desired knowledge; this is also done via a proof, in a formalism capturing the actions of the computational process. The latter is the key to the following more formal definition (cf. Wiedermann and van Leeuwen (2014)). In this definition we assume that the input to a computation is part of both the underlying epistemic domain (and thus of the theory) and the initial data of the process.

Definition 1. *Let T be a theory, let ω be a piece of knowledge serving as the input to a computation, and let $\kappa \in T$ be a piece of knowledge from T denoting the output of the computation. Let Π be a computational process and let E be an explanation. Then we say that process Π acting on input ω generates the piece of knowledge κ if and only if the following two conditions hold:*

- *$(T, \omega) \vdash \kappa$, i.e., κ is provable within T from ω, and*
- *E is the (causal) explanation that Π generates κ on input ω.*

We say that the 5-tuple $C = (T, \omega, \kappa, \Pi, E)$ is a computation rooted in theory T which, on input ω, generates knowledge κ using computational process Π with explanation E.

Although the notation used in the definition resembles the one used in the formal theories, we will be using it equally in the case of informal epistemic domains and theories.

Note that the epistemic approach to computations is *machine independent*, since it holds regardless of the mechanism supporting knowledge generations in the theory at hand. It is also *algorithm independent* since we do not care how the computational process operates. Last but not least, the approach is *representation independent* since we do not need to assume any particular knowledge representation. For a more detailed account of this approach to computation and knowledge generation, see Wiedermann and van Leeuwen (2014), (2015a).

2.2 Intelligent Systems and Their Knowledge Theories

Thanks to its generality, the epistemic approach is well suited to be applied both in well formalized, so-called *theory-full knowledge domains,* and in knowledge domains with inference rules that resist any formalization efforts (so-called *theory-less domains*). The prototypical example of a theory-less domain with informal derivation rules is the 'real world'. Its objects, phenomena, and actions, and the relations among them, are described in a natural language. Knowledge about such a domain is captured by sentences in a natural language again. In this case, derivation rules are the rules of rational thinking and behavior. These rules are based on facts and arguments that can be described in natural language. In typical cases, theory-less domains have large knowledge bases (think, e.g., of the Internet) and relatively short derivation chains.

A prime example of an intelligent system is the human brain. In the brain, knowledge is generated by processes following an informal theory which can be described, however, in a natural language. In principle, in stead of the brain one may consider any other kind of intelligent artifact with similar properties, even a not-yet-existing one. The result would be the same: a general artificial system on par with human intelligence. The fact that the epistemic approach allows one to work with poorly formalized notions is a strong point of our modeling. It opens the door to an computational understanding of the mechanisms of knowledge generation that so far was not possible by other means.

Note that the idea of epistemic computation surely works well for systems whose main task is "thinking", i.e., solving purely intellectual problems. However, when we want to apply our approach to the functions of a robot whose main purpose is to perform some prevailingly physical tasks, in addition to speech recognition and interaction, we are in a more difficult situation. This is because such tasks require sensory-motor skills which use the 'physics' of the robot and are not clearly 'knowledge-driven' only. Thus, these skills cannot, probably, be adequately captured in a sufficiently formalized and manageable theory that would drive the behavior of the robot at hand. A solution of this problem may depend on our ability to invent new types of languages similar to how languages were 'invented' a long ago by biological evolution and used by our brains. It may not be formalizable using current logical and grammatical notations, but those are relatively recent discoveries. In the future, one may discover what evolution has achieved and how it was done. We may then be able to replicate the mechanisms in future, more intelligent machines.

A more immediate solution could be based on a hybrid approach, delegating the realization of skills to specialized, pre-trained, "trusted" specialized (deep) neural nets whose correctness is based on statistical evidence. Their ability to perform the designated tasks follows from their construction, after they were trained on large sets of data. The cooperation of these nets is then governed by some epistemic theory with checkable proofs (cf. Shanahan (2016) for a similar approach).

A more formal explanation of our approach, presented only informally in this paper, can be found in van Leeuwen and Wiedermann (2017).

3 Epistemic Computation and Higher Cognitive Functions

Let us return to the condition (cf. Sect. 2.1) that binds a computational process to a specific epistemic theory of its domain. We also required that for any (piece of) knowledge that can be derived in the epistemic theory, there must be evidence that the computational process supports (realizes) the resulting knowledge. This evidence need not necessarily be a logic-based proof. It could use a wider range of forms of representation, as used in ancient mathematics (cf. Sloman 2018), or in the reasoning about chemical or physical processes.

In this section we will show that a cognitive system working in this way, can generate knowledge corresponding to the definition of many non-trivial cognitive functions. That is, within the epistemic framework we can naturally define and explain many higher cognitive functions as knowledge generating and, hence, computational processes, which would be cumbersome otherwise.

In what follows we describe the respective higher cognitive function informally, but this will be sufficient to bring them into the framework of epistemic computations. There are a least three reasons for doing so. First, we do not aim at a descriptive model of the cognitive functions, as these are amply considered in the literature of the cognitive or brain sciences. Secondly, there is often no generally accepted definition of these functions. Finally, our goal is primarily to demonstrate that these functions are all of a computational nature, in the context of artificial intelligence. It is an advantage of epistemic approach that it allows us to concentrate on the knowledge generating aspects of these functions, without having to take the specificity of their definition into account.

We now briefly characterize and discuss nine higher order cognitive functions in turn, to illustrate the epistemic approach.

(a) Accountability: an ability of a system to generate knowledge justifying its own decisions. This means e.g. that a cognitive agent can, upon request or otherwise, issue a substantiation of his doings, i.e. an explanation of how he made out his findings, cf. Kroll *et al.* (2016). To this end it is enough for the agent to generate, together with the resulting knowledge, also the proof for its derivation. Of course, this proof should be presented in a formalism accessible to the user. This enables the user to review and check whether the agent's justifying information is correct and complete, within the framework of the underlying epistemic theory.

(b) Awareness: an ability of a cognitive system to have knowledge about the problem being solved. Given accountability, this knowledge is available and an agent can report what is it doing, if asked to do so. Self-knowledge can be a part of the epistemic theory controlling the activities of the system.

(c) Introspection: an ability of a system to recall knowledge about its previous actions and their derivation. A cognitive system can store its previous tasks and the way of their solutions. Doing so, it can return to them, to re-inspect them and make use of them when solving new problems by means of analogy. It can also improve upon previous solutions w.r.t. new findings that the system could have accumulated in the mean time.

(d) Epistemic understanding. Accountability, (self-)awareness and introspection together give rise to understanding the knowledge domain over which a system operates. A system is able to explain the meaning of the terms it works with and, based on its previous experience (recorded in its knowledge base), to apply them in new contexts. For a full understanding of the real world, one has to consider embodied cognitive systems.

(e) Free will. We say that cognitive system A has free will with respect to cognitive system B if and only if, based solely on the observation of A's actions, B is not able to always predict (in the form of generated knowledge) A's future actions in concrete situations. This definition differs from numerous definitions of free will (cf. Wikipedia) that see the concept from an inner (subjective) view of a system. For instance "in a given situation, free will is the ability to choose from several alternative behaviors", or "an ability to behave differently than in the past under the same conditions". Only a cognitive system itself has information whether it has chosen its behavior from several possibilities, or whether it has "invented"a new behavior. Assuming that this information is inaccessible for an external observer, there is no way for such an observer to decide whether the system at hand possesses free will or not. This fact makes our definition of free will observer dependent. However, the advantage is that it identifies the problem of free will as computational w.r.t. the given observer.

(f) Creativity: a manifestation of a creative process, which is any process generating a solution to a problem (in the form of knowledge) that is new for the given cognitive system. Its counterpart is a routine process, which solves a known problem with the help of known procedures. In general, a creative process seeks explicit knowledge that is given implicitly via conditions that the knowledge to be found must satisfy. Deferring efficiency aspects, in our approach the basic strategy for a creative process is the exploitation of "brute force"—the systematic examination of all knowledge that can be generated in the framework of a given theory and checking whether the generated knowledge satisfies the given conditions. This looks like an extremely inefficient, even naive approach, but it seems that such an approach is the basis of any creative process. In fact, it is a special case of *knowledge discovery.* This initially inefficient, but universal process of knowledge discovery is cultivated in the course of its repeated use. Knowledge discovery is then seen as a potentially never ending, evolutionary

self-improving learning process whose goal is to make its creative abilities more efficient. In Wiedermann and van Leeuwen (2015b) we have described several basic techniques that can be used in the cultivation of creative processes:

- *interactive refinement:* this entails a modification of the search criteria, based on the experience from previous, unsatisfactory or even failed searches.
- *automatic extraction and modification of user preferences:* this narrows the search space. It is based on the observed preferences of the user (in a similar way as done by Google+) and/or on the basis of user's emotions and subjective experience (if this is accessible to the system). This activity of collecting and shaping user preferences goes on at every occasion when solving any creative problem. The extracted preferences are exploited for streamlining any discovery process and for ordering the discovered solutions.
- *guided interaction with the environment:* its aim is to gain new, additional knowledge that can help in solving the given concrete problem (in much the same way as we do when checking the Internet for additional information).

As a result of such cultivation procedures, the knowledge base of the creative process as well as the mechanisms of its exploitation are modified or supplemented.

Note that the mechanisms described above not only answer the question *what* knowledge must be generated in order to solve a problem, but at the same time they also answer the question *how* such knowledge can be found, i.e., in fact, *how* to solve the given problem. A special case is the question *how* to improve some aspects of a known solution of a problem. This question can be transformed into the question of *what* knowledge must be found solving a given problem and, at the same time, satisfying a new condition referring to any aspect of a solution that must be improved.

(g) Anticipation: an ability of a system to generate knowledge in the form of predictions about the future occurrence of events or conditions in an epistemic domain. It is seen as the result of a "wired" creativity, a limiting result of creativity cultivation where no search is necessary in order to solve the problem. Consequently, anticipation becomes a routine process working as the first choice alternative or an efficient substitute of an originally creative process.

(h) Epistemic curiosity: a perpetual need of a system to discover new knowledge. It is intimately related to creativity and anticipation. Similar to creativity, curiosity is a life-long learning process whose cultivation causes that not everything is explored and exploration is not made randomly. Curiosity is often invoked when anticipation fails.

(i) Epistemic self-improvement. The mere ability to derive new knowledge in the framework of a given epistemic theory cannot be considered to be the main attribute of intelligent systems. Namely, the main attribute of an intelligent system is its ability to improve its own epistemic theories through which it generates its knowledge. If this is the case, then the intelligence of such a system keeps provably increasing.

When using cultivation procedures as described above, it may happen that a system gets new knowledge that contradicts the knowledge already possessed by the system. Such contradictory knowledge can be derived by the system itself or it can enter the system from "outside" (e.g., from the Internet), or when the system reveals a discrepancy between its own observations with its epistemic theory. Such a flaw can only be cured by a change of the underlying theory.

Systems that have mechanisms for discovering and repairing logical inconsistencies in their theory obviously can increase their intelligence under any reasonable definition of this notion. This process can continue as long as there exist contradictory facts within the theory and the system at hand can find them, and as long as there exist unexplored objects and phenomena in the underlying knowledge domain. As a result, such systems can potentially, at least in some domains, overcome human intelligence (Wiedermann and van Leeuwen 2017). Unlike the popular idea of software self-improvement that aims at streamlining derivation procedures in a cognitive system (cf. Bostrom 2014), self-improvement of knowledge theories aims at the heart of the intelligence—viz. the quality and quantity of the epistemic data.

4 Conclusions

In this paper we have applied the framework of epistemic computation to the definition of various higher cognitive functions. The aim was to present them as knowledge generating functions, i.e., in fact, as computational processes. We have also elucidated the mechanisms of self-improving epistemic theories that lie behind the development of intelligence. In doing so we have focused, in accordance with the philosophy of the epistemic approach to computation, on the question *what* the cognitive functions under consideration do, i.e., what knowledge they are producing, rather than on *how* they do what they do. This leads to a new approach to understanding these functions. They cannot be described in a simple and elegant way using the classical view of computations, generated by various models of computers. This is because such a view is necessarily machine dependent and therefore cannot offer a sufficiently abstract and general framework for defining and understanding the functions at hand. Contrary to this, with the help of an elementary, abstract model of cognitive systems that is not burdened by any technical details, our approach clearly points to the conclusion that all cognitive functions under consideration are related to specific forms of knowledge generation, within an appropriate epistemic theory.

Viewing computations as knowledge generating processes has great potential for AI. In future work, we intend to apply the epistemic approach also to the problem of consciousness. This would present an essential contribution to the philosophy and theory of computational cognitive systems.

Acknowledgment. The authors thank Jodi Guazzini and Aaron Sloman for comments and suggestions that greatly helped to improve the manuscript.

References

Bostrom, N.: Superintelligence: Paths, Dangers, Strategies. Oxford University Press, Oxford (2014)

Garnelo, M., Arulkumaran, K., Shanahan, M.: Towards deep symbolic reinforcement learning. arXiv:1609.05518 (2016)

Kroll, J.A., et al.: Accountable algorithms. Univ. PA Law Rev. **165**(3), 633–705 (2016). Available at SSRN. https://ssrn.com/abstract=2765268

Samsonovich, A.V.: Toward a unified catalog of implemented cognitive architectures. In: Proceedings of BICA 2010. Frontiers in Artificial Intelligence and Applications, vol. 221. pp. 195–244. IOS Press Ebooks (2010). http://bicasociety.org/cogarch/

Sloman, A.: Huge but unnoticed gaps between current AI and natural intelligence (2018). This volume

van Leeuwen, J., Wiedermann, J.: Knowledge, representation and the dynamics of computation. In: Dodig-Crnkovic, G., Giovagnoli, R. (eds.) Representation and Reality in Humans, Other Living Organisms and Intelligent Machines, pp. 69–89. Springer, Cham (2017)

Wiedermann, J., van Leeuwen, J.: Rethinking computation. In: Proceedings of 6th AISB Symposium on Computing and Philosophy: The Scandal of Computation - What is Computation?, AISB Convention 2013, Exeter, UK, pp. 6–10. AISB (2013)

Wiedermann, J., van Leeuwen, J.: Computation as knowledge generation, with application to the observer-relativity problem. In: Proceedings of 7th AISB Symposium on Computing and Philosophy: Is Computation Observer-Relative?, AISB Convention 2014, Goldsmiths, London. AISB (2014)

Wiedermann, J., van Leeuwen, J.: What is computation: an epistemic approach. (Invited talk.) In: Italiano, G., et al., (eds.) SOFSEM 2015: Theory and Practice of Computer Science. Lecture Notes in Computer Science, vol. 8939, pp. 1-13. Springer (2015a)

Wiedermann, J., van Leeuwen, J.: Towards a computational theory of epistemic creativity. In: Proceedings of 41st Annual Convention of AISB 2015, London, pp. 235–242 (2015b)

Wiedermann, J., van Leeuwen, J.: Understanding and controlling artificial general intelligenct systems. In: Proceedings of 10th AISB Symposium on Computing and Philosophy: Language, Cognition and Philosophy, AISB Convention 2017. University of Bath, UK, AISB, pp. 356–363 (2017)

Will Machine Learning Yield Machine Intelligence?

Carlos Zednik(✉)

Otto-von-Guericke-Universität Magdeburg, Magdeburg, Germany
carlos.zednik@ovgu.de

Abstract. This paper outlines the non-behavioral *Algorithmic Similarity* criterion for machine intelligence, and assesses the likelihood that it will eventually be satisfied by computers programmed using Machine Learning (ML). Making this assessment requires overcoming the *Black Box Problem*, which makes it difficult to characterize the algorithms that are actually acquired via ML. This paper therefore considers *Explainable AI*'s prospects for solving the Black Box Problem, and for thereby providing *a posteriori* answers to questions about the possibility of machine intelligence. In addition, it suggests that the real-world nurture and situatedness of ML-programmed computers constitute *a priori* reasons for thinking that they will not only learn to behave like humans, but that they will also eventually acquire algorithms similar to the ones that are implemented in human brains.

1 Machine Learning and the Algorithmic Similarity Criterion

Rather than equip computers with hand-coded algorithms for solving complex AI problems, Machine Learning (ML) methods allow computers to "learn" the requisite algorithms "by themselves", by engaging real or simulated environments, and by processing large quantities of data. These methods have already yielded self-driving cars and autonomous helicopters, computers that play chess and Go, and sophisticated face-, handwriting-, and speech-recognition systems, among others.

Do these impressive feats of engineering herald an age of thinking machines? Even if ML-programmed computers can eventually match or exceed human behavior in a variety of domains (and thus, may satisfy behavioral criteria such as the Turing Test), there are reasons to be skeptical. One worry is expressed in a famous thought experiment due to Ned Block, in which a lookup table is used to reproduce the input-output structure of an intelligent human being. Although a computer equipped with such a lookup table would be capable in principle of matching human behavior in any number of domains, it arguably "has the intelligence of a toaster" (Block 1981, p. 21).

Block's thought experiment motivates the search for non-behavioral criteria for machine intelligence. One such criterion is *Algorithmic Similarity* (AS): A machine is intelligent if it implements algorithms—rules and representations—similar to the ones that are implemented in human brains.[1] Although there may be other ways of achieving

[1] This appeal to brain-implemented algorithms is in line with cognitive science orthodoxy. Analogous non-behavioral criteria may invoke other theoretical posits.

V. C. Müller (Ed.): PT-AI 2017, SAPERE 44, pp. 225–227, 2018.
https://doi.org/10.1007/978-3-319-96448-5_23

machine intelligence, computers that deploy rules and representations that resemble the ones that are deployed by human beings have a particularly strong case to make.[2]

2 The Black Box Problem and Explainable AI

Unfortunately, it is hard to know which algorithms are actually learned via ML, and thus, whether any particular ML-programmed computer satisfies AS. In particular, many deep neural networks and reinforcement learning policies are so high-dimensional and complex that their inner workings remain "opaque" to human observers. This so-called *Black Box Problem* has well-known practical implications; it now appears to have philosophical significance as well.

An *a posteriori* solution to the Black Box Problem may eventually be delivered by the *Explainable AI* research program. One branch of this research program aims to reduce the opacity of ML-programmed computers through mathematical tools, experimental techniques, and visualization methods for characterizing the algorithms that are executed by ML-programmed computers (e.g. Ritter et al. 2017). Another branch of Explainable AI aims to modify the Machine Learning process itself so that computers do not only learn to solve complex AI problems, but also learn to produce comprehensible "explanations" of their actions (e.g. Ribeiro et al. 2016).

Although Explainable AI's prospects for solving the Black Box Problem are uncertain, it is worth reflecting on some possible outcomes. For one, Explainable AI might fail to solve the problem altogether. For another, Explainable AI might succeed, but reveal that ML-programmed computers do not in fact satisfy the Algorithmic Similarity criterion. In either one of these cases, Explainable AI would yield no positive evidence for machine intelligence. That said, it would yield no negative evidence, either—it would merely suggest that we are unable to tell whether ML-programmed computers are genuinely intelligent by "looking under the hood".

But there is also a third possibility: Explainable AI might succeed, and reveal that some ML-programmed computers do in fact satisfy AS. In this case, the research program would arguably yield positive *a posteriori* evidence for machine intelligence. But just how likely is this third outcome?

3 Nurture and Situatedness

It may seem unreasonable to expect Machine Learning to give rise to algorithms similar to the ones that are implemented in human brains. Indeed, because many learning algorithms such as backpropagation are biologically implausible, an ML-programmed computer's "nature" is quite unlike that of an intelligent human being. Nevertheless, their "nurture" is increasingly similar. For example, many ML-programmed computers have access to the same information that is used to educate human beings. This includes

[2] Those impressed by Searle's (1980) Chinese Room would of course disagree. For them, the algorithms being executed have no bearing on the intelligence that may or may not be possessed.

text produced by humans for humans (e.g. news articles, books, and Twitter feeds), as well as naturalistic images, sounds, and videos (e.g. the ones on Instagram and YouTube). Insofar as the algorithms acquired through Machine Learning are more likely to be determined by a computer's real-world nurture than by its artificial nature, ML-programmed computers are likely to acquire rules and representations similar to the ones that are implemented in human brains.[3]

ML-programmed computers are not only nurtured by increasingly naturalistic data, but are also situated in increasingly realistic environments. For example, self-driving cars must learn to navigate on the same roads that are used by human drivers. As a consequence, the former will likely learn to exploit the same structures, artifacts, and tools that are already being exploited by the latter; a self-driving car might learn to exploit the presence of road signs in much the same way that humans do, and in this sense, might develop algorithms for interacting with the environment that closely resemble our own.

In summary, whereas Explainable AI may deliver an *a posteriori* way of evaluating an ML-programmed computer's Algorithmic Similarity to human beings, the nature and situatedness of Machine Learning constitute *a priori* reasons for believing that this similarity will be considerable. Insofar as Algorithmic Similarity is a sufficient criterion for machine intelligence, Machine Learning may really bring about an age of thinking machines.

References

Block, N.: Psychologism and behaviorism. Philos. Rev. **90**(1), 5–43 (1981)

Lake, B., Ullman, T., Tenenbaum, J., Gershman, S.: Building machines that learn and think like people. Behavi. Brain Sci. **40**, E253 (2017). https://doi.org/10.1017/S0140525X16001837

Ribeiro, M.T., Singh, S., Guestrin, C.: "Why Should I Trust You?": Explaining the Predictions of Any Classifier. arXiv 1602.04938v3 (2016)

Ritter, S., Barrett, D.G.T., Santoro, A., Botvinick, M.M.: Cognitive psychology for deep neural networks: a shape bias case study. In: Proceedings of the 34th International Conference on Machine Learning (2017)

Searle, J.: Minds, brains and programs. Behav. Brain Sci. **3**(3), 417–457 (1980)

[3] This is an admittedly strong empiricist thesis. Lest it be considered too strong, it is worth considering recent attempts to increase the psychological plausibility of ML methods. For example, Lake et al. (2017) review several ways in which such methods can be modified to incorporate known principles of human learning and development. This research may obviate the need for (overly) strong empiricism.

Ethics - Law

In Critique of RoboLaw: The Model of SmartLaw

Paulius Astromskis(✉)

Vytautas Magnus University, K. Donelaičio g. 58, 44248 Kaunas, Lithuania
paulius@astromskis.lt

Abstract. This research develops a new regulatory framework for analyzing the probable upcoming technological singularity. First, an analysis of the standard regulation framework is provided, describing its elements and explaining its failures as applied to the singularity. Next, using a transaction cost approach a new, conceptual regulation framework is proposed. This work contributes to the understanding of regulation in the context of technological evolution.

Keywords: Regulation · Singularity · RoboLaw · SmartLaw · Transaction cost

Critique of RoboLaw. The RoboLaw report (Palmerini et al. 2014) offers an in-depth analysis of the ethical and legal issues raised by robotic applications. However, the choice of the main research question deserves critique. It is obvious that the standard model of regulation is not working within the context of its application to existing and emerging technologies. The object of the research should be the regulation framework *per se,* if the aim of policy making is the reconciliation of regulation and technology.

Thus, the aim of this research is to explain the elements of the standard regulation framework, its failures, and consequently develop the concept of a new regulation framework within the context of the probable technological singularity. The research is limited to development of this conceptual approach to the new regulatory framework.

Standard Regulation Framework. The starting point of law in traditional legal theory is the "idea of law", which then turns into "legal rules", leading to "legal relationships" and *vice versa* (Vaisvila 2004). Within this ontology of law, legal rules are chosen in consideration of the discretion limits set by the idea of law and must be reconciled with actual legal relationships. Williamson (2000) has set forth a very similar structure for economic institutions. Alignment of legal and economic perspectives supports assumption that the regulator acts as a gatekeeper between transactions and the values of society. This assumption justifies using Coase Theorem (Coase 1960) and the market failures' framework (Williamson 1984) for an analysis of regulation issues.

According to Coase (1960), if there are zero transaction costs, the outcome of transactions will be efficient regardless of the legal regulation applied. Transaction costs are the "costs of running the economic system" (Arrow 1969) which constitute the major problem or drag on economic efficiency. Moreover, since efficiency and justice coincide in many ways (Šimašius 2002), market failures are a legal problem as well. The most commonly observed and regulated human and environmental market failures factors

V. C. Müller (Ed.): PT-AI 2017, SAPERE 44, pp. 231–234, 2018.
https://doi.org/10.1007/978-3-319-96448-5_24

are: (i) bounded rationality; (ii) opportunism, (iii) uncertainty and (iv) asset specificity (Williamson 1984).

Methodological Framework and Explanation of the Standard Model of Regulation Failures. For the purposes of this research, it is assumed that courts are the institution which most frequently deal with market failure cases. Therefore, the standard model of regulation failure may be sufficiently observed and described through the lenses of time, price and quality of the performance of the judicial branch. After making this observation, explanatory technological forecasting is used to develop a new conceptual regulation model in the scenario of the technological singularity.

The analysis of the performance of the Lithuanian courts has revealed that the standard model of regulation is insufficient in terms of costs, speed and quality. However, neither kinetic nor digital safeguards can eliminate all vulnerabilities and uncertainties associated with the weaknesses of human nature and with the limited resources of the court system. The sole possible alternative is the automation of law and legislation, thus replacing or decreasing the dependency on human nature in the field of regulation.

The Conceptual Regulation Model Within the Context of the Technological Singularity. Academically trained attorneys are already increasingly being replaced by technology (WEF 2017). However, the development of intelligent technologies is accelerating exponentially towards the assumed singularity – i.e., the moment in time when, due to the vastly increased power of computation, artificial intelligence becomes equal to human intelligence, and as a result society transcends towards the post-humanistic era (Kurzweil 2014; Futurism n.d.).

Of course, there is no unanimous agreement whether such a scenario may be accurately predicted on the basis of a simple equation inputting increases in computational power (Russel and Norvig 2010). However, considering the exponential development of technologies, the probability of a scenario where the singularity occurs should not be considered as mere fiction and excluded from scientific modeling. Presuming that a spectacular breakthrough in computing power and artificial intelligence will result in a new intelligent system not dependent on trust, a model of regulation could be build using the logic of process mining (Aalst and Wil 2014) as shown in Fig. 1, below.

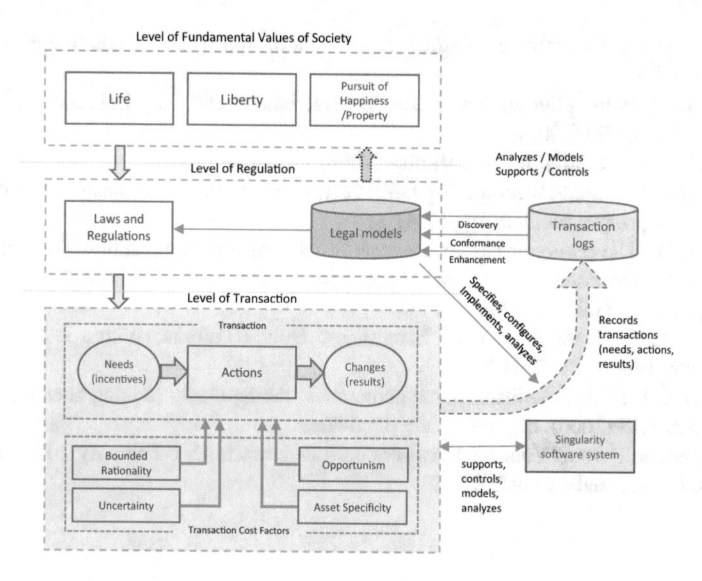

Fig. 1. Conceptual regulation framework (SmartLaw)

Within this system, all complete transaction data (subjective and objective) would go to a newly developed software system with a supported and controlled database. Utilizing the standards of fundamental values, it would autonomously analyze and model a legal framework designed to prevent market failures. A characteristic of this new system would be that it is constantly learning and enhancing its effectiveness. Intelligent machines in the singularity age could be trained to constantly analyze patterns of human transactions to identify any failures and fix them, much like a human regulator does, but just much faster, cheaper and with higher quality. The law might become "smart", i.e. individualized and constantly adopting to the real time transactions, presenting solutions that correct human error, reduce high expenses and increase response time. But is it safe - or even ethical - to allow humanity to be regulated by machines?

References

Arrow, K.J.: The organization of economic activity: issues pertinent to the choice of market versus non-market allocations. In: The Subcommittee on Economy in Government of the Joint Economic Committee, Congress of the United States, The analysis and evaluation of public expenditures: the PPB system, pp. 47–64. Government Printing Office (1969)

Coase, R.: The problem of social cost. J. Law Econ. **3**(1), 1–44 (1960)

Futurism. (n.d.). Softbank CEO: The Singularity Will Happen by 2047. https://futurism.com/softbank-ceo-the-singularity-will-happen-by-2047/. Accessed 20 May 2017

Kurzweil, R.: The singularity is near. In: Sandler, R.L. (ed.) Ethics and Emerging Technologies, pp. 393–406. Palgrave, Macmillan, London, UK (2014). https://doi.org/10.1057/9781137349088_26

Palmerini, E., et al.: Guidelines on Regulating Robotics (2014). http://www.robolaw.eu/. Accessed 1 May 2017

Russell, S., Norvig, P.: Artificial Intelligence: A Modern Approach, 3rd edn. Prentice Hall, New Jersey (2010)

Šimašius, R.: Teisinis pliuralizmas (Doctoral dissertation, The Law University of Lithuania, Vilnius, Lithuania) (2002)

Vaišvila, A.: Teisės teorija. Justitia, Vilnius (2004)

Van der Aalst, W.M.P.: Process Mining Discovery, Conformance and Enhancement of Business Processes. Springer, Heidelberg (2014)

Williamson, O.: The economics of governance: framework and implications. Zeitschrift Für Die Gesamte Staatswissenschaft/J. Inst. Theor. Econ. **140**(1), 195–223 (1984). Accessed http://www.jstor.org/stable/40750687

Williamson, O.E.: Transaction Cost Economics: How It Works; Where It Is Headed. De Economist **146**(1), 23–58 (1998)

Williamson, O.E.: The new institutional economics: taking stock, looking ahead. J. Econ. Lit. **38**(3), 595–613 (2000). https://doi.org/10/cd47nw

World Economic Forum. Your next lawyer could be a machine, 6 February 2017. http://bit.ly/2lgyU6R. Accessed 1 May 2017

AAAI: An Argument Against Artificial Intelligence

Sander Beckers(✉)

Department of Philosophy and Religious Studies, Utrecht University, Utrecht, Netherlands
srekcebrednas@gmail.com
http://sanderbeckers.com

Abstract. The ethical concerns regarding the successful development of an Artificial Intelligence have received a lot of attention lately. The idea is that even if we have good reason to believe that it is very unlikely, the mere possibility of an AI causing extreme human suffering is important enough to warrant serious consideration. Others look at this problem from the opposite perspective, namely that of the AI itself. Here the idea is that even if we have good reason to believe that it is very unlikely, the mere possibility of humanity causing extreme suffering to an AI is important enough to warrant serious consideration. This paper starts from the observation that both concerns rely on problematic philosophical assumptions. Rather than tackling these assumptions directly, it proceeds to present an argument that if one takes these assumptions seriously, then one has a moral obligation to advocate for a ban on the development of a conscious AI.

1 Introduction

In the wake of the recent boom in the field of Artificial Intelligence, there has been an equally spectacular boom in apocalyptic predictions regarding AI and the faith of mankind. Extrapolating the accelerating progress of AI and our dependence on it, doomsayers worry that it is only a matter of time before we develop an Artificial General Intelligence, or Strong AI, which would be so powerful that it could cause terrible global suffering and possibly even the extinction of our species (Bostrom 2014; Hawking 2014; Hawking et al. 2014; Musk 2015; Tegmark 2015). In fact, our situation is deemed so worrisome, that several new research centers have been created with the explicit aim of reducing the potential dangers of AI.[1]

Some authors have also turned the table on the ethical concerns regarding AI. Instead of merely considering the harm that an AI could bring upon humans,

[1] The Center for Human-Compatible AI, the Machine Intelligence Research Institute, OpenAI, the Future of Humanity Institute, and the Foundational Research Institute, to name just a few. Of course these institutes do not focus exclusively on the long-term existential risks posed by AI, but also on the abundant more concrete risks that current AI already poses.

V. C. Müller (Ed.): PT-AI 2017, SAPERE 44, pp. 235–247, 2018.
https://doi.org/10.1007/978-3-319-96448-5_25

they also consider the harm that could be brought upon an AI by humans (Bostrom 2014; Mannino et al. 2015; Metzinger 2010; Sotala and Gloor 2017). The idea is that a truly intelligent AI would also develop consciousness, and with consciousness comes the capacity for emotions, agency, and all other aspects that we associate with subjects that deserve moral consideration (Bostrom 2014; Chalmers 1996; Dennett 1993; Metzinger 2010). For example, one could argue that the continued development of AI as systems that are entirely subjected to our every wish and command would amount to re-introducing slavery (Walker 2006).

Ironically, the strong pessimism towards the future prevalent in both types of ethical concern mentioned above is founded on underlying assumptions that reveal a strong optimism towards the present: the assumption that the current rate of progress within AI is bound to continue unabated and the assumption that we have a clear understanding of certain deep philosophical issues. We set aside entirely whether the former assumption is justified, as that is something to be settled by a technical scientific discussion. Instead, the focus of this paper is on the latter more philosophical assumptions.

Concretely, the first type of ethical concern is based on the assumption that an AI could become superintelligent and the second type of ethical concern is based on the assumption that an AI could suffer. Both assumptions are highly controversial from a philosophical perspective. Firstly, it is not at all clear whether the very notion of superintelligence makes any sense, especially as it concerns non-human entities. Secondly, it is undoubtedly an understatement to say that we do not yet have a good understanding of how consciousness arises in human beings, let alone elsewhere.

The goal of this paper is not to call into question these assumptions directly. Instead, the aim is to show that if we actually take them seriously, then humanity has a moral obligation not to create a conscious AI. If this argument is successful, researchers in AI who are reluctant to accept its conclusion will be pressed with the challenge of either dropping one – or both – of these assumptions, or do some soul-searching and become advocates of a ban on the creation of a conscious AI.

In addition to said assumptions, the argument here developed also assumes a minimal utilitarian outlook. That is, it is assumed that utilities appropriately capture certain quantifiable and objective features of our moral framework, and that all else being equal, we have a moral duty to create more utility rather than less. What makes this outlook minimal, is that it leaves open entirely whether utilities capture *all* morally salient features.

2 Supersuffering

The first assumption we encountered above is that an AI could develop superintelligence, which is a level of intelligence that far exceeds our own human intelligence (Bostrom 2014; Chalmers 2010). In the words of Bostrom (2006), "we mean an intellect that is much smarter than the best human brains in practically every field, including scientific creativity, general wisdom and social skills".

Although it is quite straightforward to imagine an AI that outperforms human beings in computationally demanding tasks, it is much harder to conceive of an intellect that is much smarter than human beings in, say, social skills. Presumably we are talking about the skills required to socialize with human beings, after all.

More generally, given that the concept of intelligence was constructed to capture a property of human beings, it is not at all clear what it even means to surpass *human* intelligence. The way in which the idea of superintelligence is invoked implies that it isn't merely a matter of being able to think faster than a human or having a larger memory capacity, for if that were the case then humans would still be able to outsmart it by using their collective intelligence together with computers and other technology. On the contrary, the type of superintelligence produced by a singularity, or one that has the capacity to subdue and potentially destroy all of mankind, is such that it would be able to gain insights that are seemingly forever beyond our grasp. Here one is inclined to paraphrase Wittgenstein: "If an AI could speak, we could not understand it".[2]

Without a proper theory of intelligence that enables us to make sense of the idea that intelligence comes in degrees which extend far beyond the range of anything we find in humans, we ought to split up our single assumption into several separate assumptions that make explicit what the idea of a superintelligent AI requires.

Assumption 1. *There exist quantifiable, mental, and human properties that an AI can have.*

Assumption 2. *If an AI can have a quantifiable, mental, and human property, then it can have this property to a degree which extends far beyond the human level.*

Assumption 3. *Intelligence is one such property for which Assumption 1 holds.*

Taken together, these assumptions allow us to conclude that a future AI could be superintelligent.

The second assumption mentioned earlier is that an AI could suffer. Given that, like intelligence, suffering comes in degrees, we can rephrase this assumption as follows:

Assumption 4. *Suffering is one such property for which Assumption 1 holds.*

As with intelligence, we can apply Assumption 2 to conclude that a future AI could supersuffer, i.e., it could suffer to a degree that far exceeds any potential human suffering.

One might object that Assumption 2 is too strong, for the idea of superintelligence only requires an assumption of that form to hold for intelligence. Yet without a theory of intelligence that gives us this particular assumption,

[2] The original mentions a lion, rather than an AI. (Wittgenstein 1953, p. 223).

restricting Assumption 2 to intelligence would be gratuitous: we have no grounds whatsoever for stipulating that there is something peculiar about intelligence as compared to suffering so that it is the only candidate for amplification to a super-level.

In fact, given Assumptions 1, 3 and 4, the concept of supersuffering seems far less problematic than the concept of superintelligence once we shift focus from a single agent to a group of agents.[3] The reason is that more agents suffering to some degree X always implies more total suffering than less agents suffering to the same degree X, whereas the same does not hold for intelligence. In other words, suffering is far more cumulative than intelligence. For example, except for the speed at which they can solve certain problems, two identical agents need not be any more intelligent overall than each agent considered separately. However, if two identical agents are being tortured then there is clearly a lot more suffering than when only of of them is being tortured.

Further, we find additional support for the possibility of a supersuffering AI from other sources. Sotala and Gloor (2017) offer a detailed analysis of the potential suffering that could be caused by an AI. While they focus mostly on human suffering, they also mention that "these [future technologies] may enable the creation of mind states that are worse than the current biopsychological limits." They provide interesting thought experiments to substantiate this claim. In a similar vein, Metzinger (2013) states that future AIs "might suffer emotionally in degrees of intensity or in qualitative ways completely alien to us that we, their creators, could not even imagine."

Such a supersuffering AI would amount to what can be called a *negative utility monster*: a being whose utility is so incredibly low that all of our efforts should go to increasing its utility, instead of wasting energy on increasing the comparatively negligible utilities that we human beings could obtain. The notion of a positive utility monster was posited by Nozick in order to highlight a counterintuitive consequence of utilitarianism (Nozick 1974, p. 41):

> Utilitarian theory is embarrassed by the possibility of utility monsters who get enormously greater sums of utility from any sacrifice of others than these others lose ... the theory seems to require that we all be sacrificed in the monster's maw, in order to increase total utility.

One standard utilitarian reply is to object that such a monster is not conceivable, for no single entity could possibly have such large quantities of utility, be it negative or positive (Parfit 1984). Note that our starting point, however, contains the observation that the recent success of AI has dramatically altered the type of entities that people claim they can conceive of. So if by now we can conceive of an AI as an intelligence monster, and we can conceive of an AI as having morally salient mental states such as suffering, then the mere claim that we cannot conceive of an AI as a negative utility monster does not carry much weight.

[3] In the case of AI systems, it might even be that a group of agents could easily merge into a single agent. For sake of simplicity, we leave this speculative possibility aside.

Other utilitarians are indeed prepared to bite the bullet and concede that such a monster would have to be the primary target of moral concern. For example, Singer (2009) says that "if we ever encountered Martians who could convince us that they had a vastly greater capacity for happiness than we do, then it could be a problem."

So far we have focussed on the suffering of a single AI, but the problem of supersuffering becomes all the more pressing once we aggregate the suffering of multiple AI systems, over extended periods of time. Once we are able to create an AI – in the sense of a superintelligent and conscious AI as we have been considering – it is reasonable to assume that we go ahead and produce a large number of copies. From there it is only a small step to imagining horrible scenarios in which there would be more artificial suffering than all human suffering, past and present, combined.

For example, say we are able to create a holographic AI that is the result of uploading a person's brain and running it as a hologram that looks like the person. Creative as human beings are, a cunning investor uses this technology to construct a profitable attraction: for a couple of dollars, visitors get to pull the switch on a holographic electric chair in which is seated a holographic copy of a convicted murderer whose original human version has long since been executed by an actual electric chair. The fact that the hologram experiences the exact same excruciating pain makes the attraction widely successful. Millions of visitors come to pull the switch, causing millions of holograms to suffer terribly. To top it all off, each visitor receives a keychain containing a copy of the hologram that is continuously, during every single second of the day, year after year, experiencing this execution. Unlike an actual human being, these holographic AIs do not have the benefit of death to put their suffering to an end. Obviously this scenario is extremely far-fetched, but given our earlier assumptions, it is certainly conceivable. This is confirmed by millions of viewers of the superb sci-fi television series Black Mirror, in which this very scenario is enacted (and others like it).

For another illustration, one could imagine that the experience of empathy is achieved in an AI by automatically replicating any suffering that it observes. A reason for programming the AI in this manner is that it might very well be a good way to ensure that an AI is highly sensitive to, and aware of, any form of human suffering – which it better be if we expect it to avoid treating humans as mere instruments for attaining its objectives. Now imagine that such an AI has access to all of recorded human history. In particular, it can immediately access all audio-visual material ever produced. Further, the AI is so fast and unbounded in resources that for every single decision it makes, it takes into account the total amount of evidence which is available to it. Say it makes a million decisions per second. This implies that during a single second, a single AI goes through the entire amount of suffering ever recorded a million times over. The fact that all of this happens within a single second should not be seen as a mitigating factor, for according to Bostrom and Yudowsky's plausible principle of the subjective rate of time, "In cases where the duration of an experience is of basic normative significance, it is the experience's subjective duration that counts." (2014, p. 326). Given the speed at which we can expect an AI to be operating, this principle in and of itself is already sufficient to guarantee that the experience of suffering for an AI can take on far more extreme forms than it can for human beings: a single experiment that goes astray for a few seconds could result in an AI suffering for many years.

One might counter that we can avoid such scenarios by implementing policies that forbid them. But such policies would be unable to prevent similar scenarios in which the suffering is unintended, and worse even, scenarios in which the suffering goes by entirely unnoticed. Once an AI has the capacity to suffer, then all it would take is some bug in the code for similar scenarios to unfold. For example, imagine that there is some complicated version of the millennium bug, which is activated in billions of AIs at the same time and causes them to suffer to the astronomic extend portrayed above before we even know what is going on.

Metzinger (2013) also focusses on this issue, highlighting the "possibility that non-biological subjects of experience have already begun to suffer before we as their human creators have even become aware of this fact." He develops a theory that allows for the quantification of suffering, and posits that it is our duty to minimize the frequency of conscious experiences that involve suffering (Metzinger 2017). As a consequence, he concludes that we should ban the development of an AI, in the strong sense of AI as we are using it, stating the following principle (2013, p. 3):

> We should not deliberately create or even risk the emergence of conscious suffering in artificial or postbiotic agents, unless we have good reasons to do so.

Mannino et al. (2015) reach a similar conclusion in their overview of the moral risks posed by the development of AI, stating that "the (unexpected) creation of sentient artificial life should be avoided or delayed wherever possible, as the AIs in question could – once created – be rapidly duplicated on a vast scale."

Given the assumptions made at the outset, and the severity of the sketched scenarios, the only way to avoid accepting these negative verdicts is to follow through on Metzinger's hint and offer good reasons as to why the possibility of supersuffering is an acceptable price to pay. Three straightforward suggestions present themselves as plausible candidates:

1. The attempt at creating an AI is not at all special in this regard, since all other acts that we perform as humanity today also run the risk of causing extreme suffering in the future, and nevertheless we find this perfectly acceptable.
2. The negative scenario of supersuffering is compensated by a positive scenario of an AI experiencing superpleasure.
3. The expected benefits for mankind that come from creating an AI outweigh the possibility of supersuffering.

In the remainder of this paper the aim is to show why all three suggestions fail.

3 The Unique Risk of Creating an AI

In order to show how the consequences of creating an AI are unlike the consequences of other acts that we collectively engage in, we make explicit in what manner the risk of an AI undergoing extreme suffering is unique. We do so by comparing strategies that humanity could adopt in order to avoid risking human suffering with strategies that humanity could adopt in order to avoid risking AI suffering.

A first thing to note is that there exists one very radical strategy that humanity could adopt in order to avoid any risk of human suffering in the long term, namely to stop having children altogether. We here take it for granted that this strategy is morally unacceptable, without offering further motivation.

Principle 1 (Acceptable). *A strategy is* acceptable *unless it certainly causes the extinction of mankind.*

The only reason we mention this principle is for sake of completeness, since in recent years Benatar (2006) has championed the extremely controversial position of anti-natalism, according to which it is immoral to have children. This position is based on negative utilitarianism, a topic to which we come back later.

Now we turn to strategies that are acceptable and ensure the avoidance of some terrible outcome.

Definition 1 (Avoidable). A possible outcome is *certainly avoidable* if there exists an acceptable strategy that certainly prevents the outcome of occurring.

We already concluded that humanity might cause future AIs to experience supersuffering. In order to invoke the above definition, we need to add the following trivial counterpart.

Premise 1. *If all of humanity does not attempt to create an AI, then certainly there will never be an AI that experiences supersuffering.*

This leads us to conclude the following:

Conclusion 1. *A supersuffering AI is certainly avoidable.*

On the short term, and when considering a single agent, there are many negative outcomes which are certainly avoidable. This no longer holds if we consider all of humanity and extend our horizon into the far future: given our limited knowledge of the world, and the almost infinite complexity of the causal chain that results from our actions, we are ignorant with respect to the long-term consequences of our actions on the well-being of humanity.

Premise 2. *There exists a time t such that no matter what acceptable strategy we adopt, to the best of our knowledge, it is possible that this strategy causes extreme human suffering after t.*

At first glance there appear to be many strategies that defy the above premise. For example, think of our efforts to cure cancer. Either these are successful, in which case they would prevent a great deal of suffering, or they would be unsuccessful and not have any impact at all. But this analysis only considers the most likely outcomes that each alternative would have. Although unlikely, it is definitely possible that our efforts to cure cancer result in the creation of a deadly and contagious virus, which causes many more deaths than the disease which it was supposed to cure. Or it is possible that by curing cancer, the next would-be genocidal dictator is kept alive, and therefore able to live out his evil intentions.

We can now apply Definition 1 to reach the following conclusion:

Conclusion 2. *Extreme human suffering is not certainly avoidable.*

This conclusion provides us with a distinguishing feature of the risk that comes with the attempt to create a conscious AI. Any suffering that a future AI might experience is certainly avoidable, whereas this does not hold for the suffering of future humans. Therefore the first suggestion does not succeed in giving us a good reason why we should risk a supersuffering AI.

4 Moral Asymmetry

At this point we can draw the following worrisome conclusion.

Conclusion 3. *It is possible that by creating an AI, we will cause a unique and extreme form of suffering that could certainly have been avoided.*

Still, an optimist might argue, completely analogous to this depressing conclusion, it is also possible that by creating an AI we will cause a unique and extreme form of pleasure that would otherwise certainly have been avoided. Hence the route for the second suggestion to defend our attempt at creating an AI is still open.

However, there is a strong intuition that is so well-embedded in our everyday life that only an extreme utilitarian would object to it: it is more important to avoid suffering than it is to create pleasure. Moore (1903) was the first to express this intuition, but Popper was its most famous defender (Popper 1945):

> We should realize that from a moral point of view suffering and happiness must not be treated as symmetrical; that is to say the promotion of happiness is in any case much less urgent than the rendering of help to those who suffer, and the attempt to prevent suffering.

This idea forms the basis of "moderate negative utilitarianism", which considers it our primary duty to avoid suffering (Chauvier 2014; Mayerfeld 1999; Metzinger 2013; Parfit 1997). The asymmetry between pleasure and pain that lies at its core is evident in the medical principle "first do no harm", and is confirmed by the moral risk-aversion that is widespread in our behaviour.[4]

[4] See the papers cited above for many more interesting examples.

For example, assume you may press a button such that with probability 0.5 a random person's leg will be broken, and with probability 0.5 someone's broken leg will be healed, and neither person has any say in the matter. Or imagine that if you press the button, a random person will be hit in the face, but offered a massage afterwards. It goes without saying that it is immoral to press the button.

Further, this asymmetry increases as the intensity of the suffering and pleasure increases. For example, if someone insults you but then offers a compliment, you probably will not have hard feelings towards that person. But if they torture you, it is hard to imagine what form of pleasure they could offer you to avoid feeling terribly wronged by that person. In fact, some even go so far as to state that certain amounts of suffering cannot be compensated by any amount of pleasure at all, a position Hurka (2010) describes as the limit asymmetry thesis: "There is some intensity n such that a pain of intensity n is more evil than any pleasure could be good."

In light of all this, the following moral principle is endorsed by a broad range of ethical positions and has a prima facie intuitive appeal:

Principle 2 (Moral asymmetry). *All else being equal, the moral blameworthiness for causing a degree of suffering X is greater than the moral praiseworthiness for causing a degree of pleasure X. Further, the difference between the degree of blame and praise strictly increases with X.*

Nevertheless, as said, a strict utilitarian could insist on the symmetry between pleasure and suffering, and hence reject this principle. In that case, the possibility of a supersuffering AI could be compensated, on the condition that the probability of superpleasure is significantly greater than that of supersuffering.

Setting this caveat aside, we conclude that our expected blameworthiness when continuing the development of AI is higher than when we stop all research on AI, and hence the second candidate suggested as a good reason for risking supersuffering is ruled out as well.

4.1 Anti-natalism

As promised, we briefly return to the position of anti-natalism. We rejected this position outright because of the simple fact that almost everyone would find the prospect of mankind going extinct quite depressing, to say the least (whereas few would mourn the non-existence of sentient AIs). We now clarify how this position relates to the one here developed.

We motivated Principle 2 by reference to moderate negative utilitarianism. By using the label "moderate", its proponents wish to distance themselves from "negative utilitarianism", which embraces the far stronger claim that even the slightest amount of suffering can never be compensated by any amount of pleasure whatsoever. Given that every human being will experience some amount of suffering throughout its life, this claim implies that it is immoral to bring children into existence, no matter what the circumstances.

In contrast, a moderate negative utilitarian can perfectly well defend having children, on grounds of the fact that most people end up leading lives which have an acceptable amount of suffering compared to the amount of pleasure. In other words, most people end up leading lives worth living. Only if it were very likely that one's child would experience constant suffering would it follow that it is better not to have a child, a conclusion which most people would fully endorse.[5]

Further, even if we were to assume that the probability of extreme suffering for a child is identical to that of supersuffering for an AI, the astronomical difference in the degree of suffering involved suffices to separate the application of Principle 2 to the latter from its application to the former.

5 The Ethical Priority of Artificial Suffering

The third suggestion that might offer a good reason for going ahead and risk the prospect of AI systems supersuffering, is that this would be an acceptable price to pay given the expected benefits for mankind that would follow the invention of truly intelligent AI. That we consider this candidate suggestion last is due to the simple reason that the discussion of the previous two suggestions already puts considerable pressure on this idea.

First, so far we have completely ignored the first ethical concern mentioned at the outset, namely the concern that the creation of an AI could cause extreme human suffering. If we take seriously the doomsayers, meaning we attribute a small but non-negligible probability to the type of worst-case scenarios in which AIs will wreck havoc and destruction everywhere, then we have every reason to believe that the overall expected benefits for mankind in the long run are drastically negative.

Second, even without invoking these worst-case scenarios, given the myriad degrees of freedom that the organization of human society possesses, there is no reason to assume that AI is in any way necessary for human beings to flourish in the long run. For all we know, there exist strategies not involving the creation of AI that would cause an even higher increase in human pleasure after the time horizon t mentioned in Conclusion 2 than an AI could ever produce. In sum, as per Conclusion 1, the creation of an AI would result in us giving up on the certainly avoidable outcome of a supersuffering AI, whereas it is by no means clear how the creation of an AI relates to the prospects of mankind after time t.

Third, we can combine the asymmetry between suffering and pleasure captured by Principle 2 with the astronomical difference in orders of magnitude between the amount of suffering depicted in the scenarios from Sect. 2 and any feasible amount of human pleasure that could occur before time t mentioned

[5] Metzinger also makes this point, and adds that anti-natalism regarding artificial life is far more plausible than its biological counterpart (Metzinger 2013, 2017). To avoid unnecessary complication, we make clear that we need not get into the issue of abortion, but are talking simply about preventing the act of human fertilization in the first place.

above to see that the benefits for mankind before time t do not even come close to compensating for a supersuffering AI. That is, they do not come close if we accept the following plausible principle:

Principle 3 (Non-discrimination). *When evaluating the overall expected benefits of creating an AI, we ought not discriminate between the suffering/ pleasure of an AI and a human being.*[6]

When reformulated in terms of different groups of human beings, this principle is a bedrock of any modern moral system, and hence it hard to see how it could fail to apply when we extend it to other conscious beings that have an even stronger capacity for suffering and pleasure than humans do. Singer (2011, p. 50) puts it thus:

> If a being suffers, there can be no moral justification for refusing to take that suffering into consideration. No matter what the nature of the being, the principle of equality requires that the suffering be counted equally with the like suffering – in so far as rough comparisons can be made – of any other being.

In sum, the combination of these two arguments blocks the last suggested candidate:

Conclusion 4. *From an ethical point of view, the possibility for an AI to experience supersuffering takes precedence over the expected benefits that an AI will produce for mankind.*

6 Conclusion

In Sect. 2 we examined the basis for two popular and controversial assumptions regarding Artificial Intelligence, and argued that accepting these assumptions leads to the moral principle that we should not create a conscious AI, unless we can offer good reasons to do so. In the subsequent sections we rejected three natural candidates for such reasons. Therefore, if we take seriously our two initial assumptions, we are forced to accept the following conclusion:

Conclusion 5. *Humanity should not attempt to create a conscious AI.*[7]

Given the gravity of this conclusion, it is incumbent upon each and every AI researcher to closely inspect said assumptions, and make a choice: either refrain from endorsing both of them and explain why, or advocate for a ban on the creation of a conscious AI.

[6] This principle could also be generalized to include certain animals. However, one would first have to introduce Assumptions 1 and 4 reformulated for animals.

[7] Humanity here refers to current humanity. If at some point in the future we discover a method of working on the development of a conscious AI that is certain to avoid supersuffering, then obviously this conclusion could be retracted. For now, however, we are just as far removed from the discovery of such a method as we are of the discovery of a conscious AI itself.

Acknowledgements. The author gratefully acknowledges financial support from the ERC-2013-CoG project REINS, nr. 616512.

References

Benatar, D.: Better Never to Have Been. Oxford University Press, Oxford (2006)
Bostrom, N.: How long before superintelligence? Linguist. Philos. Investig. **5**(1), 11–30 (2006)
Bostrom, N.: Superintelligence: Paths, Dangers, Strategies. Oxford University Press, Oxford (2014)
Bostrom, N., Yudowsky, E.: The Cambridge handbook of artificial intelligence. In: Frankish, K., Ramsey, W.M. (eds.) The Ethics of Artificial Intelligence. Cambridge University Press (2014)
Chalmers, D.: The singularity: a philosophical analysis. J. Conscious. Stud. **17**(9–10), 7–65 (2010)
Chalmers, D.J.: The Conscious Mind: In Search of a Fundamental Theory. Oxford University Press (1996)
Chauvier, S.: A challenge for moral rationalism: why is our common sense morality asymmetric? In: Dutant, J., Fassio, D., Meyan, A. (eds.) Liber amicorum pascal engel, pp. 892–906. University of Geneva (2014)
Dennett, D.C.: Consciousness Explained. Penguin UK, London (1993)
Hawking, S.: Stephen Hawking warns artificial intelligence could end mankind (2014). http://www.bbc.com/news/technology-30290540
Hawking, S., Russell, S., Tegmark, M., Wilczek, F.: (2014). https://www.huffingtonpost.com/stephen-hawking/artificial-intelligenceb5174265.html. Accessed 14 Apr 2018
Hurka, T.: Asymmetries in value. Nous **44**(2), 199–223 (2010)
Mannino, A., Althaus, D., Erhardt, J., Gloor, L., Hutter, A., Metzinger, T.: Artificial intelligence: opportunities and risks. Policy Pap. Effective Altruism Found. **2**, 1–16 (2015)
Mayerfeld, J.: Suffering and Moral Responsibility. Oxford University Press, New York (1999)
Metzinger, T.: The Ego Tunnel: The Science of the Mind and the Myth of the Self. Basic Books, New York (2010)
Metzinger, T.: Two principles for robot ethics. In: Hilgendorf, E., Günther, J.-P. (eds.) Robotik und gesetzgebung, pp. 263–302. Nomos, Baden-Baden (2013)
Metzinger, T.: Suffering. In: Almqvist, K., Haag, A. (eds.) The return of consciousness. Axess Publishing (2017)
Moore, G.: Principia Ethica. Cambridge University Press, London (1903)
Musk, E.: (2015). https://www.theguardian.com/technology/2015/jan/16/elon-musk-donates-10m-to-artificial-intelligence-research. Accessed 14 Apr 2018
Nozick, R.: Anarchy, State, and Utopia. Basic Books, New York (1974)
Parfit, D.: Reasons and Persons. Oxford University Press, New York (1984)
Parfit, D.: Equality and priority. Ratio **10**(3), 202–221 (1997)
Popper, K.: The Open Society and Its Enemies, vol. 1. Routledge (1945)
Singer, P.: Back talk: Peter singer (2009). https://www.thenation.com/article/back-talk-peter-singer/. Accessed 14 Apr 2018
Singer, P.: Practical Ethics. Cambridge University Press, Cambridge (2011)
Sotala, K., Gloor, L.: Superintelligence as a cause or cure for risks of astronomical suffering. Informatica **41**, 389–400 (2017)

Tegmark, M.: An open letter: research priorities for robust and beneficial artificial intelligence (2015). https://futureoflife.org/ai-open-letter. Accessed 14 Apr 2018

Walker, M.: A moral paradox in the creation of artificial intelligence: Mary poppins 3000s of the world unite! In: AAAI Workshop Human Implications of Human-Robot Interaction (2006)

Wittgenstein, L.: Philosophical Investigations. Macmillan Publishing Company, New York (1953)

Institutional Facts and AMAs in Society

Arzu Gokmen(✉)

Department of Philosophy, Bogazici University, Istanbul, Turkey
arzgokmen@gmail.com

Abstract. Which moral principles should the artificial moral agents, AMAs, act upon? This is not an easy problem. But, even harder is the problem of identifying and differentiating the elements of any moral event; and then finding out how those elements relate to your preferred moral principle, if any. This is because of the very nature of morally relevant phenomena, social facts. As Searle points out, unlike the brute facts about physical world, the ontology of the facts about social reality -which he calls institutional facts- is subjective and they exist within a social environment. The appropriate way to learn these facts is by interaction. But, what should this interaction be like and with whom, especially in the case of artificial agents, before they become 'mature'? This implies that we are to face a very similar problem like raising a child.

1 Introduction

Machine ethics has been challenged by both fundamentally meta-questions (is the discipline itself morally sound? [4, 6]), and by relatively applied ones (how to model the moral behavior of autonomous cars?). The later problems are pressingly on the table; thus, the practical challenge is to find a valid framework to address these later problems without questioning the table itself. There are ongoing debates on how to implement some moral criteria to the actions of these artificial moral agents, AMAs. Three basic approaches are as follows [5]. First, we can determine a set of moral principles -usually in line with one of the core moral theories- and then, by top-down programming, make the machine follow these principles. Second, we can teach or let the machine discover what right and wrong action is in many circumstances by using supervised or unsupervised bottom-up learning. Third, and more promisingly, we can combine all these methods in useful ways.

Of course, none of these methods is easy to implement and none of the moral issues has a unique answer so that it can be used to label an action as being the right one. However, even if we would have a solution to decide on the moral principles that an AMA should follow, the harder problem would still remain. That is, the harder problem for any moral event is to identify and differentiate the elements of the event, and to find out how those elements relate to your preferred moral principle, if any. Thus, even if AMAs would be built to follow some principles, like they shouldn't harm, would they be able to identify what harm is, in an actual temporal event?

What makes this a problem is due to the very nature of morally relevant phenomena, social facts. Searle's distinction between institutional facts and brute facts [2] helps us

V. C. Müller (Ed.): PT-AI 2017, SAPERE 44, pp. 248–251, 2018.
https://doi.org/10.1007/978-3-319-96448-5_26

analyzing why it is harder to teach AMAs social phenomena than to make them follow some moral principles. I claim that we cannot succeed in having AMAs that are morally welcome, without realizing our responsibility to these machines that learn about the social realm from the data we supply, as being the very creators of those data.

2 Institutional Facts vs Brute Facts

Brute facts are independent of human institutions, like physical facts. Trees have roots; ice melts into water; dolphins are mammals; oranges have predictable shapes. It is relatively easy to learn and recognize the instances of these facts for a machine. The very first reason is that it is easier to collect objective, similar, yet sparse data about those facts and then let the machine discover what is statistically frequent in the data.

Institutional facts, on the other hand, "typically objective facts, but oddly enough, they are only facts by human agreement or acceptance" [2]. These facts are ontologically subjective and epistemologically objective. The fact that some paper is money; that some relation is a marriage or a friendship; that some person is a professor; or the fact that an act of killing is a murder; or that a conflict is a war; or that a statement is a promise; or that a particular ring belongs to a particular agent, like Gollum.

The way to create these facts is by *Declaration* (may or may not be a speech act [2]); and some of the institutional facts are regulated under laws or documentation. However, for many others, there are neither regulations nor explicit documentation. Because, these facts are up to local situational agreements upon people, and their factual nature is dynamic. Moreover, regulations or documentation usually say very little about what the relevant institutional fact means or implies for any particular instance.

As Searle points, "we live in a sea of human institutional facts. Much of this is invisible to us" [2]. Mostly, the elements of a moral event are due to these institutional facts or due to values that depend on these facts, either invisible or not. Therefore, it becomes integral to have access to this kind of data while evaluating any morally relevant case. We are not much better than machines, regarding having access to the regulated, documented or declared facts. Indeed, how can your neighbors know for sure that the house is yours or that you are married, without declaration or a documented proof? They can not; but, the very nature of institutional facts allows us to have quite accurate guesses about these facts without having access to the documents.

3 How to Teach Institutional Facts to AMAs?

Interactively. The very dynamic and social nature of these facts requires AMAs to learn them via social interaction. But, with whom should they interact, before they become 'mature' about the nature of the social facts? In 2016, Microsoft released a chatbot, named Tay, for Twitter users which was programmed to begin with a set of conversational priors and then to learn from interacting with users. However, they had to shut the service down just after 16 hours as Tay had started producing very controversial and 'inappropriate' conversations upon what she has learnt. Crucially, she approved and adopted positions of racism, sexism, and Nazism.

These are already among the sensitive cases of institutional facts; and for example, an actual instance of racism might easily be denied by the very society that instantiates it; and it is even easier to misjudge someone or some community to be racist, while these judgements do not affect the very existence of that fact [2]. That is, the fact is there, if it is there, and its existence implies an associated deontology [2]; thus, it requires a moral stand. Therefore, reaching the truth about that fact becomes crucial, especially for an artificial moral agent that might particularly be designed to execute necessary actions in vital situations.

Tay's case implies more than the dangers of manipulation or abuse of users for a learning machine. Any nonmalignant or sincere series of conversations would have led to the same result for Tay. There lies the urgent practical challenge for the machine ethics. That is, AMAs have to learn these facts from people, but they cannot just utilize the statistics of any data. First of all, we as consumers of these technologies should realize that our interactions with them are social interactions and "there is an important difference between 'treating some x as if it were to do y' and 'interacting with x as if it were y'" [3]. Therefore, we have a moral responsibility to our interactions with machines as we socially interact with them. Either we anthropomorphize them or not, we still socialize with them. This will be the key to the 'quality' of the social data that they can get from us. Second, we have to understand that these are learning machines, and they learn from data; they are not totally pre-programmed. Thus, they are sensitive to data that we produce, and we don't yet know what kind of patterns they are sensitive to and what they can discover. Therefore, we need to use our mental-autonomy [1] to attend the fact that they learn and we should prevent our predispositions to treat AMAs as if they were immune to what we do or say.

Ultimately, this requires a comprehensive discussion of moral and social education of AMAs. But, for now -as the machines already make use of the data we produce, and as the current aim can be approached distinctly from a possibly idealist solution of making them capable of social cognition and practical wisdom- we can begin with the awareness that we are not only beneficiaries or consumers of the so-called intelligent machines, but we are the very responsible source of social data from which they will learn about us. One can argue that we should allow machines to utilize just every data about social realm; and allow them to discover by themselves; and that we should embrace whatever they discover and learn about us. I don't think that this would be a good strategy and I believe that, with carefully designed education plans, we can achieve a very promising level of machine 'understanding' of sociality.

References

1. Metzinger, T.: M-autonomy. J. Conscious. Stud. **22**, 270–302 (2015)
2. Searle, J.: Making the Social World. Oxford University Press, New York (2010)
3. Seibt, J.: Towards an ontology of simulated social interaction: varieties of the "As If" for robots and humans. In: Hakli, R., Seibt, J. (eds.) Sociality and Normativity for Robots. Studies in the Philosophy of Sociality (9), pp. 11–40. Springer, Cham (2017)

4. Tonkens, R.: A challenge for machine ethics. Mind. Mach. **19**, 421–438 (2009)
5. Wallach, W., Allen, C.: Moral Machines. Oxford University Press, New York (2009)
6. White, J.B.: Autonomous reboot: the challenges of artificial moral agency and the ends of machine ethics. https://philpapers.org/rec/WHIART-6. Accessed 31 Mar 2018

A Systematic Account of Machine Moral Agency

Mahi Hardalupas(✉)

Department of History and Philosophy of Science, University of Pittsburgh, Pittsburgh, USA
mch64@pitt.edu

Abstract. In this paper, I develop a preliminary framework that permits groups (or 'systems') to be moral agents. I show that this has advantages over traditional accounts of moral agency when applied to cases where machines are involved in moral actions. I appeal to two thought experiments to show that the traditional account can lead us to counterintuitive consequences. Then I present what I call the 'systematic account' which I argue avoids these counterintuitive consequences. On my account, machines can be partial moral agents currently fulfilling some but not all of the conditions required for moral agency. Thus, when a machine is part of a group of agents, it can be part of a *system* that is a moral agent. This framework is a useful starting point as it preserves aspects of traditional accounts of moral agency while also including machines in our moral deliberations.

1 Traditional Accounts of Moral Agency

Traditional accounts of moral agency are generally hostile to attributing moral agency to artificial agents such as machines. The traditional account (henceforth referred to as TA), as summarized by Himma (2009), states that:

X is a moral agent of action A if and only if:

(i) X is an agent,
(ii) X can make free choices about A,
(iii) X can deliberate about what one ought to do,
(iv) X can understand moral rules,
(v) X can apply the moral rules correctly in paradigm cases.

Himma identifies consciousness as a prerequisite for these conditions. Other accounts also require consciousness (Parthemore and Whitby 2013), mental states (Johnson 2006) or, in extreme cases, biological sentience (Torrance 2008). Such definitions motivate arguments against machine moral agency (see Himma 2009) and force us to engage with issues of machine consciousness to show why a machine should count as a moral agent. To avoid this, I propose a new 'systematic' account, which permits conditions for agency to be satisfied by groups of agents. I show that this avoids some counterintuitive consequences of TA.

To demonstrate TA's counterintuitive consequences, I analyse two thought experiments adapted from past philosophical literature. I argue that these cases are importantly different but TA doesn't have the resources to differentiate them. The first case is a

V. C. Müller (Ed.): PT-AI 2017, SAPERE 44, pp. 252–254, 2018.
https://doi.org/10.1007/978-3-319-96448-5_27

"moral room" (adapted from Searle's 'Chinese room' (Searle, 1980)) where the inhabitant of the room must output 'moral' decisions from inputs that are moral dilemmas using a 'moral rulebook' found in the room. On TA, the inhabitant is not a moral agent as it only satisfies condition (i) and (v). It cannot satisfy (ii) as it is only following the rules from a rulebook and since it is not meaningfully engaging with the dilemmas presented to it except to output a response, it does not satisfy (iii) and (iv). The second example is a variant of the "Otto and Inga" thought experiment (adapted from Clark and Chalmers (1998)). Here Otto has a notebook where he keeps a diary of the correct responses to moral dilemmas. He decides to publish his diary and Inga purchases it to guide her moral decisions. On TA, Inga is not a moral agent as she only satisfies (i) and (v). As long as Inga is merely following the rules straight from the book, then she cannot satisfy (ii) and (iii) and since she is just following them and didn't formulate them, she can't be said to understand them so fails to satisfy (iv). TA treats the two cases identically. However, this seems counter to the intuition that the inhabitant and Inga differ in an important respect. Inga follows the rules out of a motivation to be moral unlike the inhabitant of the room. Indeed, we might say that Inga is *more* of a moral agent than the inhabitant. Such a distinction is not possible on TA because it lacks degrees of moral agency. In contrast, my alternative account is able to capture the differences between these cases as well as preserving aspects of TA.

2 An Alternative Account of Moral Agency

In my systematic account (or SA), I preserve ideas from TA while making it less anthropocentric.[1] SA builds on Moor's (2006) degrees of moral agency. It claims that a set of agents X is a full moral agent of action A if and only if X is the smallest set where all the following conditions hold:

(I) X acts in a way A that is evaluated with respect to moral rules,
(II) X follows moral rules,
(III) X has the potential to follow different rules,
(IV) X has a moral motivator.

There are two kinds of moral motivator: weak and strong. A weak moral motivator entails (a) and a strong moral motivator entails (a) and (b):

(a) Having a reason to believe the act A of X is moral.
(b) Having a reason to believe the rules X follows are moral.

This specifies the conditions for a full moral agent. A partial moral agent is a member of the set X that satisfies some but not all of the conditions. But if a set of partial moral agents (or a 'system') collectively satisfy all the conditions, the system would be a full moral agent.

[1] This is in contrast to accounts (such as Floridi and Sanders (2004)) that break with the requirements of TA. I am not wedded to having to preserve parts of TA. However, since many philosophers seem committed to TA generally capturing some central things about agency, I think building on TA is a good starting point.

The degree of moral agency depends on the number of conditions satisfied and the highest condition satisfied. The conditions of SA are in ascending degrees of moral agency, meaning that satisfying the higher conditions such as (III) and (IV) contributes to a greater degree of moral agency. So a partial moral agent satisfying (III) and (IV) for a system is more of a moral agent than the partial moral agent satisfying (I), (II) and (III). In terms of condition (IV), a system with a strong moral motivator is more of a moral agent than a system with a weak moral motivator. It is worth noting that SA is an account of agency that is relative and specific to an action A. Part of the system could generally be a full moral agent without satisfying all the conditions for agency relative to action A. Because of the smallest set condition, if one agent fulfills all the conditions, then the "system" will consist just of that agent. However, there are cases where a full moral agent won't satisfy all conditions for an action and will need to be coupled with other partial moral agents.

When applied to the moral room and Otto and Inga cases, SA makes more fine-grained distinctions between the cases. On SA, the moral room system (of the inhabitant and the agent who wrote the rule book) is a full moral agent even if the inhabitant is not. Similarly, the Otto and Inga system make up a full moral agent even though Inga and Otto alone are partial moral agents with respect to Inga's moral actions. In this way, SA avoids the counterintuitive consequences of TA.

The systematic account includes machines in the realm of moral agents. On the systematic account, a machine will often be a partial moral agent rather than a full moral agent (to my knowledge, there are currently no candidates for machines which fulfill the moral motivator condition) but it can be part of a system that is a full moral agent. For example, a robot that stops a human crossing the road when a speeding car drives by would be a partial moral agent. The system of the robot and the programmer who implements the moral rules would be a full moral agent. In this way, the systematic account provides a framework to make sense of how machines feature in our moral decisions and actions. This could be a good starting-point for considering how moral responsibility is distributed in situations where a machine is involved in a moral action. In conclusion, the systematic account of moral agency avoids a debate about machine consciousness while also offering a framework to explain how machines can play a role in our moral actions as partial moral agents.

References

Clark, A., Chalmers, D.: The extended mind. Analysis **58**(1), 7–19 (1998)

Floridi, L., Sanders, J.W.: On the morality of artificial agents. Minds Mach. **14**(3), 349–379 (2004)

Himma, K.E.: Artificial agency, consciousness, and the criteria for moral agency: what properties must an artificial agent have to be a moral agent? Ethics Inf. Technol. **11**(1), 19–29 (2009)

Johnson, D.G.: Computer systems: Moral entities but not moral agents. Ethics Inf. Technol. **8**, 195–204 (2006)

Moor, J.H.: The nature, importance, and difficulty of machine ethics. IEEE Intell. Syst. **21**(4), 18–21 (2006)

Parthemore, J., Whitby, B.: What makes any agent a moral agent? Reflections on machine consciousness and moral agency. Int. J. Mach. Conscious. **5**(02), 105–129 (2013)

Searle, J.R.: Minds, brains, and programs. Behav. Brain Sci. **3**(3), 417–424 (1980)

Torrance, S.: Ethics and consciousness in artificial agents. AI Soc. **22**(4), 495–521 (2008)

A Framework for Exploring Intelligent Artificial Personhood

Thomas B. Kane(✉)

Centre for Algorithms, Visualisation and Evolutionary Systems, Edinburgh Napier University, Edinburgh, Scotland
t.kane@napier.ac.uk

Abstract. The paper presents a framework for examining the human use of, and the activities of, artificial persons. This paper applies Hobbesian methodology to ascribe artificial personhood to business organisations, professional persons and algorithmic artificial intelligence services. A modification is made to Heidegger's ontological framework so that it can accommodate these artificial persons in a space between tools and human beings. The extended framework makes it possible not only to explore human uses of tools, but also to pose questions on the relationships, obligations and operations that transfer between humans and artificial persons.

1 Introduction

Humans have always been tool developers. In recent times, we see the most powerful tools ever created being contracted to address ever more intimate and existential tasks for the benefit of humans in such domain areas as the provision of social and commercial services in healthcare, education, and governance. Such tools are scaffolds (Sterelny 2010; Vygotsky 1978), in that they allow us to extend human capacities. The Extended Mind theory (Clark and Chalmers 1998), posits that our mind is extended across the tools that we use. Clark (Clark 2003, p. 26) argues that we ourselves are natural, born cyborgs, stating "it is our special character, as human beings, to be forever driven to create, co-opt, annex, and exploit non-biological props and scaffoldings. We have been designed, by Mother Nature, to exploit deep neural plasticity in order to become one with our best and most reliable tools."

Alongside Clark's Cyborg, Martin Heidegger's Dasein (Heidegger et al. 2010) is also an ontological entity: one whose existential experience in the world requires it to develop, evolve, and employ tools as it pursues its relationships with tools and the world.

This paper is concerned with the applications of artificial intelligence in such services as Facebook's "Facebook for Politics" (Facebook 2017), which was controversially and successfully contractually employed (Cadwalladr 2017), (Davies and Yadron 2016) in the 2016 Brexit referendum and USA general election. Such sociotechnical entities clearly function at a capacity somewhere between tools and human beings. A minor modification of Heidegger's approach allows us to consider organisations such as Facebook Inc., professionals at work, and advanced AI tools such as "Facebook for Politics"

V. C. Müller (Ed.): PT-AI 2017, SAPERE 44, pp. 255–258, 2018.
https://doi.org/10.1007/978-3-319-96448-5_28

as both tools and artificial persons. Once positioned, these entities can be involved in dialectical analyses. A first investigation might well be, “should businesses who enjoy limited financial liability for providing contractually defined social services also be entitled to avail themselves of limited moral or ethical responsibility for their actions?”

2 Societal Tools, Social Contracts and Agency

Thomas Hobbes introduced the term “social contract” (Hobbes 2008) to describe how an individual will sacrifice some freedoms to join a body of people for perceived social benefits. Hobbes described the body of people as an artificial person, which is, crucially, ruled by a sovereign authority. We will refer to this form of artificial person as an Organisational Artificial Persons (OAP). He spoke of another kind of artificial person – a person who represents another, or one who speaks the words of another. This makes it possible to identify an organisational employee as a professional artificial person (PAP), and an artificial intelligence system with learned human behaviours as an Algorithmic Artificial Person (ALAP).

Social contracts make it possible for OAPs to be created and recognised in society. Social contracts also make it possible for organisational intentions to be articulated and pursued by their agents: PAPs and ALAPs. These artificial persons, O, P and AL, have scaffolded societal affordances, are controlled at some fundamental level and can change the world.

Martin Heidegger (Heidegger et al. 2010, pp. 39–110) distinguishes three kinds of being in the world:

1. Innerworldly beings – all beings which exist within the world;
2. The being of produced tools encountered in their own right; and,
3. The being-in-the-world, called Dasein (being there), that is capable of uncovering innerworldly beings, and engaging in existential relationships within the worldliness of the world (Fig. 1).

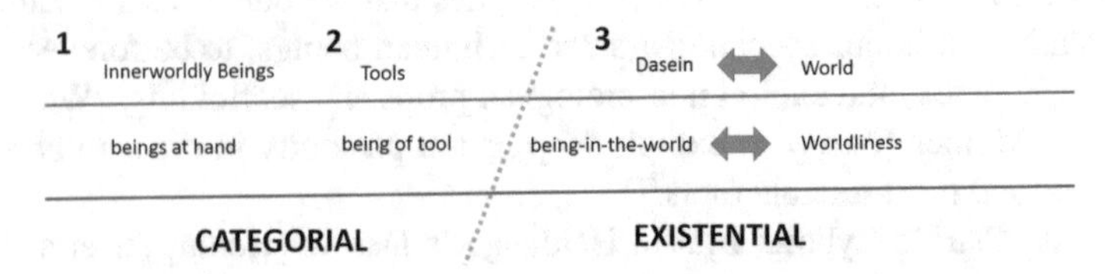

Fig. 1. Heidegger’s framework for beings in the world

Dasein cannot be categorially defined – part of the definition of Dasein is that it is incomplete, and always changing as it moves forward into a horizon of time (Dreyfus 1991; Heidegger 1976; Heidegger et al. 2010). Dasein is that being in the world who makes and uses tools which can be described categorially as she comports herself to the world in which she lives and takes care of business. An ALAP crosses the Dasein 2–3 Heideggerian divide.

The ALAP "Facebook for Politics" services a relationship between the Facebook organisation and a Customer. The tool can, for a fee, make Facebook professionals (PAPs) available to an election campaign, work with external specialists (other artificial persons), internal and external tools of artificial intelligence (ALAPs) and operate over diverse datasets. Facebook for Politics can affect the world in influencing the outcome of elections. Such a tool, and most of those it partners with, could be called a Dasein 2.x: one that was created to operate as a tool, but which has important aspects of human personhood, such as an ability to develop a (quasi-) existential relationship with the world, and a comportment to take care of business.

3 Some Issues for Artificial Persons

Artificial persons raise questions not only of artificial intelligence, but also of intentionality (all artificial persons are tools) and control (all artificial persons have scaffolding and support). Unpicking the relationships between artificial persons in complex systems is essentially challenging the Wittgensteinian (Wittgenstein 1980) hurly-burly. Regarding their freedom of self-expression, their ability to adopt worldly affordances and the possibilities of their history in time, we see this as an area that needs philosophical attention. Where contractual systems are involved we have hermeneutics: contracts and the accepted ethics of the day to leverage.

And so, for example, where an ALAP such as DeepMind (Temperton 2017) is working in healthcare services we might argue that the professional ethics of the clinical professionals whose professional services provide data sets for DeepMind should be respected by all artificial persons that interact with the DeepMind ALAP. Philosophical governance of healthcare artificial persons may require dialectics, preservation of the Hippocratic oath ("Hippocratic Oath," 1923) and data protection by design across medical data sets in order to prevent unethical behaviours, such as the unauthorized leakage of personal medical records to business artificial persons.

Ascertaining acceptable human behaviours expected of societal tools and ascribing measurable values along a 2.x scale (which may itself be multidimensional) will be a non-trivial task. However, we may begin by performing judgements upon the actions and consequences of an artificial person's behaviours as if it were a real person. Artificial persons which preserve human values would be accorded a value close to 3, and those which are little more than tools, a value close to 2.

Another important area may well be the task of designing and creating suitable societal artificial persons which will be capable of protecting societal values, norms and ethics. Such work might also help us prepare to address a Dasein 3.x – a supra-personal, supra-cognitive being with all the existential issues of Dasein, appearing as a superintelligence (Bostrom 2014), or as a benign singularity (Kurzweil 2005).

4 Conclusion

This paper places advanced technological tools on an expanded ontological scale drawn from Heidegger's categorial/ontological work. This framework may be of value in teasing out significant boundaries between human and machine, pursuing dialectics, and preserving essential elements of acceptable human behaviour.

References

Bostrom, N.: Superintelligence: Paths, Dangers, Strategies. OUP Oxford, Oxford (2014)

Cadwalladr, C.: The great British Brexit robbery: how our democracy was hijacked. The Observer, May 2017. http://www.theguardian.com/technology/2017/may/07/the-great-british-brexit-robbery-hijacked-democracy

Clark, A.: Natural-Born Cyborgs: Minds, Technologies, and the Future of Human Intelligence. Oxford University Press, New York (2003)

Clark, A., Chalmers, D.: The extended mind. Analysis **58**(1), 7–19 (1998)

Davies, H., Yadron, D.: How Facebook tracks and profits from voters in a $10bn US election. The Guardian, January 2016. https://www.theguardian.com/us-news/2016/jan/28/facebook-voters-us-election-ted-cruz-targeted-ads-trump

Dreyfus, H.L.: Being-in-the-World: A Commentary on Heidegger's Being and Time, Division I. MIT Press, Cambridge (1991)

Facebook: Facebook for Politics (2017). https://web.archive.org/web/20170701141827, https://politics.fb.com/ad-campaigns

Heidegger, M.: What is Called Thinking? (J. G. Gray, Trans.). HarperPerennial, New York (1976)

Heidegger, M., Stambaugh, J., Schmidt, D.J.: Being and Time. State University of New York Press, Albany (2010)

Hippocratic Oath (1923). https://www.loebclassics.com/view/hippocrates_cos-oath/1923/pb_LCL147.299.xml

Hobbes, T.: Leviathan. Oxford University Press, Oxford (2008)

Kurzweil, R.: The Singularity is Near: When Humans Transcend Biology. Gerald Duckworth, London (2005)

Sterelny, K.: Minds: extended or scaffolded? Phenomenol. Cogn. Sci. **9**(4), 465–481 (2010). https://doi.org/10.1007/s11097-010-9174-y

Temperton, J.: DeepMind's new AI ethics unit is the company's next big move (2017). http://www.wired.co.uk/article/deepmind-ethics-and-society-artificial-intelligence

Vygotsky, L.S., et al.: Mind in Society. Development of Higher Psychological Processes. Harvard University Press, Cambridge (1978)

Wittgenstein, L.: Describing human behaviour. In: Remarks on the Philosophy of Psychology, vol. 2, p. #629. University of Chicago Press (1980)

Against Leben's Rawlsian Collision Algorithm for Autonomous Vehicles

Geoff Keeling(✉)

University of Bristol, Bristol, UK
gk16226@bristol.ac.uk

Abstract. Suppose that an autonomous vehicle encounters a situation where (i) imposing a risk of harm on at least one person is unavoidable; and (ii) a choice about how to allocate risks of harm between different persons is required. What does morality require in these cases? Derek Leben defends a Rawlsian answer to this question. I argue that we have reason to reject Leben's answer.

1 Introduction

Suppose an autonomous vehicle (AV) encounters a situation on the road where (i) imposing a risk of harm to at least one person is unavoidable; and (ii) a choice about how to allocate risks of harms between different persons is required (Lin 2016; Goodall 2014).[1] What does morality require in these cases? How, morally, should AVs be programmed to allocate harm or risks of harm between the different parties? I call this the *moral design problem* (Keeling 2017, 2018).

Many people endorse a utilitarian answer to the moral design problem (Bonnefon et al. 2016). According to this approach, AVs should be programmed to minimise expected harm or loss-of-life in collisions. Leben (2017) recently proposed a contractualist alternative to the utilitarian approach. His answer is based on Rawls' (1971) theory of justice. Whilst utilitarians are concerned with maximising some conception of *the good*, such as pleasure or wellbeing, contractualists are concerned with the justifiability of moral principles to those affected by their prescriptions (Scanlon 1998).

In this paper, I argue that we should reject Leben's contractualist answer to the moral design problem. In Sect. 2, I explain the main ideas from Rawls' theory of justice which feature in Leben's answer to the moral design problem. In Sect. 3, I explain Leben's answer to the problem. In Sect. 4, I argue that Rawls offers less support for Leben's

I am extremely grateful to Chris Bertram, Noah Goodall, Jason Konek, Derek Leben, Niall Paterson, and Richard Pettigrew for their comments on earlier drafts. I am also grateful to audiences at the Philosophy and Theory of Artificial Intelligence Conference at the University of Leeds, and the Artificial Ethics Symposium at the University of Southampton.

[1] I use 'AV' to mean Level 5 autonomous vehicles in accordance with the Society for Automotive Engineers autonomous vehicle classification scheme. These vehicles require no human intervention or supervision in *any* circumstances that might arise on the road.

V. C. Müller (Ed.): PT-AI 2017, SAPERE 44, pp. 259–272, 2018.
https://doi.org/10.1007/978-3-319-96448-5_29

algorithm than we might initially expect. In doing so, I aim to show that Leben owes an independent argument for his answer to the moral design problem. In Sect. 5, I raise three objections to Leben's answer, all of which must be overcome if he is to provide a plausible defence of his view. In Sect. 6, I conclude.

2 Rawls on Justice

In this section, I explain the central ideas of Rawls' (1971) theory of justice which feature in Leben's (2017) answer to the moral design problem. This account of Rawls is incomplete in many respects. But I hope it will provide sufficient grounding for the discussion.

I start with two general remarks. (1) Sen (2011: 5–7) distinguishes two methodological approaches to questions about justice. On the one hand, some philosophers have tried to pinpoint what is necessary and sufficient for a just society. On the other hand, some philosophers have developed comparative conceptions of justice; the aim being to stipulate a criterion by which social arrangements can be evaluated as *more just* or *less just* relative to one another. Rawls' theory of justice is an example of the first approach. His theory tells us what is required for a society to be just. (2) Rawls is part of the *social contract* tradition. The social contract theorists aim to ground state's authority to restrict the freedoms of citizens in the actual, or hypothetical, consent of those citizens. In short, the state is justified in restricting the freedoms of citizens only if the citizens could at least hypothetically consent to those restrictions because it is in their interests to do so.

I now describe three features of Rawls' theory. First, Rawls' argues that justice is fairness. In Rawls' view, political society is a system of cooperation between individuals. The society is just when the terms of cooperation are fair to all those involved. Second, Rawls' main concern is with the principles of justice which regulate the *basic structure* of society. That is, the political and social institutions which 'assign basic rights and duties, and regulate the division of advantages that arise from social cooperation over time' (Rawls 2001: 10). Third, Rawls aims to provide both a method to determine the fair terms of cooperation and a statement of those terms. The method is a thought experiment called the *original position*, and the terms are Rawls' *two principles of justice*. I describe each of these in turn.

The original position is a hypothetical situation in which representative citizens decide on principles of justice to regulate the basic structure of society from a list of alternatives. These alternatives are taken to include things like utilitarianism, libertarianism and the two principles of justice offered by Rawls (1971: 122). Each party in the original position represents the interests of a sub-class of citizens; and all the citizens in society have a representative.

The parties in the original position must decide which principles of justice to adopt from behind a *veil of ignorance*, which Rawls describes as a situation where 'no one knows his place in society, his class, position or social status; nor does he know his fortune in the distribution of natural assets, his strength, intelligence, and the like' (*Ibid.*: 137). Rawls also states that the parties do not know whether the citizens they represent are in a minority or a majority. Neither do the parties know about the details of their life projects; their conception of the good; or their psychological dispositions such as risk-aversion, optimism or pessimism. Finally, the parties have no knowledge of the economic or political standing of their society, nor the level of civilisation or culture which the society can reasonably be expected to achieve (*Ibid.*: 137–142; Rawls 2001: 85–89).

The candidate principles of justice regulate the distribution of *primary goods* in the society, which are goods which all people have reason to want in order to facilitate their aims and ambitions. Examples include 'rights and liberties, powers and opportunities, income and wealth' (*Ibid.*: 62). Rawls assumes that all the parties in the original position are rational, and that each prefers more primary goods to less. So, whilst the representatives are unaware of their life aims and projects outside the original position, they have reason to select principles of justice which provide them with the freedoms and resources to pursue their life projects, no matter what these might be (*Ibid.*: 142–3). Rawls also stipulates that the parties in the original position are equal. In his words, 'the parties are equally represented [...] and the [principles of justice selected] are not influenced by arbitrary contingencies or the relative balance of social forces' (*Ibid.*: 120).

So, the original position is a bargaining situation which allows the parties to reach a *fair* agreement on principles of justice to regulate the basic structure of society. It does not presuppose an ethical or religious point of view in order to determine what is fair. Instead, it allows the parties to settle on mutually-advantageous principles of justice which benefit all citizens no matter what their life aims are or their conception of the good happens to be (Rawls 2001: 15). The agreement is fair for two main reasons. First, the veil of ignorance removes the unfair bargaining advantages which some individuals would ordinarily have over others in light of their social standing (*Ibid.*). Second, as the parties are ignorant of their position in society, they are unable to favour principles of justice because these principles confer benefits on *them* (*Ibid.*: 18; Harsanyi 1953: 434–5).

I shall presently turn to the principles of justice which, Rawls argues, the parties in the original position would agree upon. But first, I shall explain the decision procedure which Rawls believes the parties in the original position would use to discriminate between candidate principles of justice. According to Rawls, the parties would appeal to the *maximin* decision procedure. This means that the parties would look at the primary goods available to the worst-off citizens under each of the principles of justice, and select principles which provide the greatest allocation of primary goods to the worst-off citizens (Rawls 1971: 150–161). The argument for maximin will be discussed in detail in Sect. 4. But the basic idea is that, as the parties in the original position do not know how the citizens they represent will fare under different principles of justice, they have good reason to favour principles which *guarantee* a minimal set of rights, liberties and opportunities for the worst-off citizens (Rawls 2001: 97–00).

I now turn to the principles of justice which, Rawls argues, free and equal citizens in the original position would agree upon. According to

> *The First Principle of Justice:* Each person has the same indefeasible claim to a fully adequate scheme of basic liberties, which scheme is compatible with the same scheme of liberties for all (Rawls 2001: 42).

According to

> *The Second Principle of Justice:* Social and economic inequalities are to satisfy two conditions: first, they are to be attached to offices and positions open to all under conditions of fair equality of opportunity; and second, they are to be to the greatest benefit of the least advantaged members of society (*Ibid.*).

Rawls argues that the first principle takes priority over the second. In short, the liberties of individual citizens cannot be compromised to bring about social or economic gains for society as a whole (Rawls 1971: 61). The first part of the second principle aims to ensure social mobility. The idea is that individuals are not constrained from improving their social standing in virtue of their initial social position. The second part is called the *difference principle*, and it holds that whilst the distribution of wealth and income need not be equal, inequalities in the distribution of wealth must be to the advantage of the worst-off (*Ibid.*).

This description of Rawls' theory does not come close to a comprehensive treatment of the ideas contained within Rawls' work. But I hope this description will provide sufficient grounding for the discussion in subsequent sections.

3 Leben's Answer

In this section, I explain Leben's answer to the moral design problem. Leben's answer uses two ideas from Rawls: the original position and the maximin rule.

First, Leben imagines that the affected parties in a particular AV collision enter a bargaining situation analogous to Rawls' original position.[2] The affected parties include *any* individual who could receive at least some harm conditional on at least one alternative available to the AV. The idea is that, in this hypothetical situation, the parties could reach a *fair* agreement on which alternative the AV should select. The parties are told the survival probabilities of each party conditional on the alternatives available to the AV; but they do not know which survival probabilities correspond to *them*. Furthermore, if multiple parties have the same survival probability on one of the alternatives, the parties are not made aware of this (which mirrors Rawls' decision to exclude information about the relative proportions of the population that each citizen in the original position represents).

Second, Leben contends that in these circumstances, the parties would choose between the AV's alternatives using an iterated form of the maximin rule, called *leximin*. [3] The leximin rule compares the survival probabilities of the worst-off person on each

[2] Leben (2017) is, to my knowledge, the first person to apply the original position to the moral design problem. But it is worth noting that Schelling (2006) defends a similar position for determining whom to save in many-versus-one rescue cases.

[3] Leben does not use the term 'leximin'. He writes '[there] is one part of the Maximin procedure, that, to my knowledge, has not been worked out sufficiently well by Rawls or anybody else, and is perhaps the only original contribution that I have to make to the moral theory itself [...] It seems clear that agents in the original position would also consider the *next-lowest* payoffs, since they have an equal chance of being the next payer, and are interested in maximising her minimum as well' (2017: 110). The iterated form of maximin described by Leben is called leximin, and it has featured in moral philosophy (e.g. Otsuka 2006: 119–121; Hirose 2015: 29) and welfare economics (e.g. Sen 1976; Hammond 1976).

alternative (where the worst-off is understood as the individual with the lowest survival probability). It then selects the alternative which assigns the greatest survival probability to the worst-off person. In this respect, leximin is identical to maximin. However, the rules part company if the survival probabilities of the worst-off are equal on two alternatives. Maximin cannot discriminate between two such alternatives. In contrast, leximin compares the second-lowest survival probabilities on the remaining alternatives, and selects the alternative which gives the highest survival probability to the second worst-off person. If there is another tie, the third-lowest survival probabilities are compared, and so on. If two or more alternatives have identical profiles of survival probabilities, leximin randomises between them.

Based on this assumption about how the parties in his original position would decide between the alternatives in an AV collision, Leben develops a formal collision algorithm which uses the leximin rule. I shall illustrate Leben's algorithm with an example. Consider,

> *Example:* The AV can swerve left or swerve right. If the AV swerves left, Amy has a 60% chance of survival and Beth has a 30% chance of survival. If the AV swerves right, Amy has a 30% chance of survival and Beth has a 50% chance of survival.

Leben's algorithm first compares the survival probabilities of the worst-off parties conditional on each alternative. These are identical (30%), so the algorithm compares the survival probabilities of the second worst-off. In this case, 60% is greater than 50%, so the algorithm selects swerve left. If the survival probabilities for the second worst-off were identical, the algorithm would randomise between swerve left and swerve right.

This paper is critical of Leben's algorithm. I argue that we have reason to reject it as an answer to the moral design problem. But Leben and I agree about more than we disagree. I want to emphasise some of the excellent features of Leben's answer to the moral design problem before describing my criticisms.

First, Leben's answer is presented in decision-theoretic terms. It is not a set of moral principles, but an algorithm based on some principles. Leben therefore bridges the moral design problem with the related problem of how to programme moral principles into AVs. This is progress, as non-utilitarian moral principles are difficult to formalise. Furthermore, Leben's algorithm reaches a verdict in all the collisions with which the moral design problem is concerned. This is an important step forward for non-utilitarian answers to the moral design problem, which are difficult to capture in algorithms which cover all collisions which might arise.

Second, Leben's algorithm is based on contractualist principles. It therefore resists some of the objections which can be raised against utilitarian collision algorithms. It requires no interpersonal utility comparisons; and it does not demand that AV manufacturers sacrifice their passengers to save the greater number in collisions. At the heart of Leben's algorithm is the idea that AVs ought to be programmed to allocate harm in accordance with principles which are justifiable to the recipients of harm. It is impermissible, on Leben's view, to impose burdens on some to confer benefits on others. The maximin principle is designed to bring about Pareto efficient allocations of harm. (An allocation of harm is Pareto efficient if, and only if, there exists no alternative on which everyone is at least as well-off, and someone is strictly better-off.) So, whilst the moral motivation for Leben's project might at first be unfamiliar, the motivations are nevertheless laudable. To this end, my criticisms of Leben are not intended as an indirect defence of a utilitarian answer to the moral design problem. Instead, I hope these criticisms can be used to develop an even stronger contractualist answer to the moral design problem.

4 A Rawlsian Algorithm?

Leben (2017: 108) views his collision algorithm as a straightforward application of Rawlsian principles to the moral design problem. In this section, I challenge Leben's assumption that his collision algorithm is defensible on Rawlsian grounds.

There is an important disanalogy between Rawls' original position and Leben's original position for AV collisions. In Rawls' case, the parties are deciding between alternative principles of justice which regulate the distribution of primary goods in society. Rawls writes:

> I shall simply take as given a short list of traditional conceptions of justice [...] I then assume that the parties are presented with this list and required to agree unanimously that one conception is best among those enumerated (Rawls 1971: 122).

According to Rawls, the parties would appeal to the maximin principle when deciding between the principles of justice, and in doing so favour his two principles over utilitarianism (*Ibid.*: 175–183). Importantly, the parties in Rawls' original position are *not* choosing between alternative distributions of primary goods using the maximin principle. I believe that Leben has misread Rawls here. He writes:

> [...] the original position is a method limited to determining the distribution of what Rawls calls 'primary goods' (Leben 2017: 109).

In Leben's original position, the parties must decide between alternatives in a particular AV collision using information about the distribution of survival probabilities associated with each alternative. The alternatives, in effect, represent possible distributions of a single primary good, namely the survival probabilities of each affected party (*Ibid.*). Leben argues that the parties in his original position would use the maximin principle to decide between the alternative distributions of this primary good. Leben's original position is therefore analogous to a variation on Rawls' original position in which the affected parties decide between competing distributions of primary goods using the maximin principle.[4]

This disanalogy between Rawls' and Leben's accounts of the original position makes Leben's answer to the moral design problem hard to defend on Rawlsian grounds. Rawls and Leben are defending different claims. Rawls says that the parties deciding between principles of justice would, in the original position, decide between those principles using the maximin principle (Rawls 2001: 97). Leben says that parties deciding between distributions of survival probabilities would, in the original position, use maximin to

[4] A further disanalogy is that whilst *survival* is a primary good, it is not obvious that the probability of survival is a primary good. So, the parties in Leben's original position are choosing between alternative gambles concerning a primary good. I am grateful to Richard Pettigrew for this point.

decide on one such distribution. I shall develop this point into two criticisms of Leben's answer.

The first criticism concerns the veil of ignorance. Leben restricts the information available to the parties in his original position. I do not think these restrictions are defensible on Rawlsian grounds.

Rawls developed the original position to provide hypothetical circumstances under which free and equal citizens can reach a fair agreement on principles of justice to regulate the basic structure of society. The veil of ignorance is imposed for two reasons: it removes unfair bargaining advantages and it restricts the kind of arguments that might be provided in favour of some principles over others (*Ibid.*: 81–8). In short, Rawls does not want parties in the original position to select principles of justice *because* those principles favour individuals in *their* social circumstances (*Ibid.*; see also Harsanyi 1953). However, one of the most important features of the original position is that the parties have sufficient knowledge and understanding to evaluate the alternative principles of justice on offer (including, at least, utilitarianism and the two principles offered by Rawls). No information is excluded about what these principles entail.

The same cannot be said about Leben's veil of ignorance. Some information is excluded about the alternatives on offer. The parties are given information about the distributions of survival probabilities on each alternative. But if the same survival probability applies to two or more persons, the parties are not made aware of this (Leben 2017: 112). So, Leben's veil of ignorance does more than remove unfair bargaining advantages and prevent arguments of the form 'we should choose alternative *x* because it is in *my* best interests'. It also provides an incomplete description of the alternatives available to the parties. In some cases, this is problematic. Consider,

> *Problem Case:* The AV can swerve or continue its current path. If the AV continues, there is a 100% chance that its passenger and four pedestrians will sustain a fatal injury. If the AV swerves, there is a 100% chance that its passenger will sustain a fatal injury.

The parties in Leben's original position would know only that at least one person is guaranteed to die on either alternative. Leben argues that, with this limited information, the parties would agree to use his leximin algorithm and randomise between the alternatives. The parties do not have sufficient information to perform a utilitarian calculation, as they do not know which alternative maximises total or average survival probabilities over all affected parties. They also lack the requisite information to argue in favour of swerve on the grounds that swerve Pareto-dominates continue. (On swerve, at least one party is better-off, and no party is worse-off.) Because the parties have such limited information, I agree with Leben that leximin is the most sensible rule to use to discriminate between the alternatives.[5] But I agree with Leben only because the parties have insufficient information to use *any other* plausible algorithm for discriminating between the alternatives. It is unclear *why* the parties should be unaware of the fact that five people stand to die on one alternative, and just one of these same people on the

[5] Note that, with complete information, leximin mandates saving the greater number in many-versus-one cases (Hirose 2015: 164–5). So, Leben advocates using leximin *given the information available*, but leximin would not mandate randomising if complete information about the survival probabilities were given.

other. Adding this information would not provide unfair bargaining advantages to the parties in Leben's original position. Neither would it allow them to favour one alternative over the other because *they* stand to be better-off on that alternative.

There are two conclusions to draw here. First, Rawls' argument for imposing the veil of ignorance in the original position does not support the restrictions on information which Leben employs in his analogue of the original position for AV collisions. In Rawls' original position, the parties have sufficient information to evaluate the alternatives on offer. This is not so in Leben's case. Second, Leben in effect gerrymanders his veil of ignorance to ensure that the affected parties are forced to decide between the alternatives using his leximin algorithm. There exists no other rational decision procedure which could be used. This strikes me as a misunderstanding of Rawls' motivation for using the veil of ignorance. Rawls did not want to *force* the parties to accept his two principles of justice by restricting the information available to them. He wanted to show that, under fair bargaining conditions, it is in the parties' rational self-interest to accept his principles of justice.

I now turn to the second criticism, which concerns the maximin principle. It is worth noting from the outset that Rawls did not defend maximin as a universally applicable decision-rule. He writes:

> […] the maximin rule was never proposed as a general principle of rational decision in cases of risk and uncertainty, as some seem to have thought […] Such a proposal would be simply irrational […] The only question is whether, given the highly special and unique circumstances of the original position, the maximin rule is a useful heuristic rule of thumb for the parties to organise their deliberations (Rawls 2001: 97).

Rawls (2001: 97–8) argues that maximin is a plausible decision rule to use in the original position only if certain conditions are met. (1) The parties in the original position must have no knowledge of the probability of an arbitrary citizen being represented by each party in the original position. In other words, the parties in the original position do not know whether the citizens they represent are in a minority or a majority. (2) Because maximin concerns the worst-off citizens under different principles of justice, it must be necessary for the parties to be significantly more interested in the primary goods which can be guaranteed, as opposed to those which can be gained. Rawls argues that this condition is satisfied to some degree when the guaranteeable level for each citizen is 'quite satisfactory'; and that it is fully satisfied only if the guaranteeable level is 'completely satisfactory'. (3) As maximin examines only the worst-case scenario on each alternative, it must be the case that the worst outcomes on all other alternatives fall substantially below the guaranteeable level.

The first condition presents a trade-off for Leben. On the one hand, Leben can restrict the information available to the parties in his original position, such that the parties have no knowledge of the number of individuals who correspond to each survival probability. In this case, the first condition is satisfied. But the parties are unable to adequately evaluate the alternatives available to them in the collision, which I have argued is an undue restriction on the information available to them. On the other hand, Leben can relax the restriction on information, and allow the parties knowledge of *how many* people correspond to each survival probability. But then Rawls' first condition for maximin is not satisfied. So, Leben satisfies the first condition for maximin only if he imposes an undue restriction on the information available to the parties in his original position.

The second condition is fully satisfied in Rawls' original position when the guaranteeable level for each citizen is 'completely satisfactory' (*Ibid.*). If this condition obtains, Rawls argues that citizens will be more interested in securing basic rights and opportunities, rather than focusing on the additional rights and opportunities which could be gained. In the circumstances of an AV collision, the guaranteeable level is unsatisfactory. These collisions are such that death or serious harm is very likely for at least one affected party, and a choice about how to allocate harms across different parties is required. I suspect the affected parties in such a collision would be more concerned with maximising their chances of survival, rather than guaranteeing a *completely satisfactory* survival probability. So, the second condition for maximin is not satisfied.

The third condition states that the worst outcomes on all other alternatives falls substantially below the guaranteeable level. Recall that for Rawls, the alternatives are principles of justice as opposed to distributions of primary goods. What Rawls has in mind here is that the worst-case scenario for a citizen (in terms of rights and opportunities) under a utilitarian principle of justice is substantially worse than the rights and opportunities which can be guaranteed under Rawls' principles of justice. We can imagine, for example, a situation where one class of citizens is enslaved for the benefit of another class of citizens. Rawls thinks that maximin is a justifiable decision rule to prevent this kind of scenario arising. In Leben's case, this third condition is not met. A utilitarian collision rule does not offer a *substantially worse* outcome for the worst-off affected parties than Leben's leximin algorithm. It is likely that, on either algorithm, the worst-off affected parties will be exposed to serious risks of death or harm.

So, Rawls provides three conditions which are jointly sufficient to warrant to the use of maximin as a decision rule in the original position. None of these conditions are satisfied in the context of the moral design problem. It seems, therefore, that Rawls' argument for using maximin does not apply.

5 Objections to Leben's Algorithm

I have argued that Leben's (2017) answer to the moral design problem cannot be defended on Rawlsian grounds. This provides insufficient reason to reject Leben's answer. But it does motivate the need for Leben to provide an independent argument in favour of his answer. I now describe three challenges that must be overcome if Leben is to provide a defence of his view.

Challenge 1: In some collisions, Leben's answer mandates programming AVs to select alternatives which the affected parties could not rationally consent to, provided their preferences satisfy the von Neumann-Morgenstern (1953) axioms for rational preferences.[6]

The problem arises because Leben evaluates the alternatives in collisions using survival probabilities. He assumes that 'injuries like broken ribs, whip-lash, etc., can be represented as points along the dimension of likelihood of survival' (Leben 2017: 111, 113). Survival probabilities are therefore intended as a proxy for physical harm. Leben acknowledges that a one-dimensional scale of this kind is not entirely plausible, as some non-fatal injuries might be considered equivalent to or worse than fatal injuries. But irrespective of what kind of injury is placed at the bad end of the scale, Leben's decision to evaluate alternatives using the probability of *worst-case scenario* injuries obtaining is what gives rise to our challenge. Consider,

> *Scenario 1:* The AV can swerve left or right. If the AV swerves left, there is a 0% chance that its passenger will sustain a fatal injury and a 100% chance that its passenger will sustain a lifelong debilitating injury. If the AV swerves right, there is a 1% chance that its passenger will sustain a fatal injury and a 99% chance that its passenger will remain unharmed.

Leben's algorithm selects swerve left, because it gives the passenger the greatest chance of survival. I contend that there exists at least one scenario (equivalent to or analogous to *Scenario 1*) in which Leben's algorithm mandates programming the AV to select an alternative which the passenger could not rationally consent to. As mentioned, I shall assume that the passenger has rational preferences insofar as her preferences satisfy the von Neumann-Morgenstern (1953) axioms. I also assume that the passenger strictly prefers no injury to a lifelong debilitating injury; and that she strictly prefers a lifelong debilitating injury to a fatal injury. Note, however, that the passenger's preference ordering can be changed, and the same objection will arise.

Here is the problem: Leben's algorithm mandates swerve left no matter how low the probability of a fatal injury is on swerve right. It could be 1%, or 0.1% or 0.01%. One requirement of von Neumann-Morgenstern rationality is that an agent's preference ordering is held fixed under sufficiently small deviations in probabilities. This is called the *Archimedean Property*. Formally, letting $\prec$ denote strict preference, the requirement is that for any lotteries A, B and C, if $A \prec B \prec C$, then there exists a small probability, ε, such that $[(1-\varepsilon)A + \varepsilon C] \prec B \prec [\varepsilon A + (1-\varepsilon)C]$.[7] As there is no ε, such that *if* the probability of a fatal injury on swerve right is equal to ε, *then* Leben's algorithm mandates swerve right, it follows that there exists at least one collision (equivalent to or analogous to *Scenario 1*), where Leben's algorithm mandates programming the AV

[6] The axioms: let $\prec$ denote strict preference, $\sim$ denote indifference and $\preccurlyeq$ denote weak preference. *Completeness* holds that for any two lotteries A, B, either $A \prec B$, $B \prec A$ or $A \sim B$. *Transitivity* holds that if $A \preccurlyeq B$ and $B \preccurlyeq C$ then $A \preccurlyeq C$. *Continuity* holds that, if $A \preccurlyeq B \preccurlyeq C$, then there exists a probability $p \in [0, 1]$ such that $[pA + (1-p)C] \sim B$. *Independence* holds that if $A \prec B$, then for any C and $p \in [0, 1]$, $[pA + (1-p)C] \prec [pB + (1-p)C]$. My argument makes use of the Archimedean Property, which is sometimes assumed instead of completeness. But if either completeness or the Archimedean Property is assumed, the other is entailed by the von Neumann-Morgenstern Expected Utility Theorem.

[7] The lotteries in square brackets should be read, e.g. 'A with a probability $1-\varepsilon$ and C with a probability ε'.

to select an alternative which is not in the passenger's rational self-interest. Hence, in some collisions, Leben's algorithm mandates programming the AV to select alternatives to which the passenger could not rationally consent.

This is problematic for two reasons. First, many would argue that, in cases like these, it is morally permissible to take the gamble on behalf of the passenger and programme the AV to swerve right. If the gamble did not pay off, it would be a reasonable moral justification to highlight that we programmed the AV to select the option which we rationally expected to bring about the best outcome *for the passenger* (Otsuka 2012). Second, I take it that Leben's algorithm is intended to be an algorithm to which affected parties in collisions could at least hypothetically consent. Unless Leben assumes that affected parties in collisions have *irrational* preferences, at least by von Neumann and Morgenstern's (1953) standards, it seems that in at least some collisions, the affected parties could not rationally consent to Leben's algorithm.

This challenge arises because Leben, in effect, uses the maximin principle twice in his algorithm. First, the algorithm evaluates each alternative based on the probability of a *worst-case scenario* obtaining for each affected party. Second, the algorithm selects the alternative which provides the best deal for the worst-off party. The objection can therefore be avoided by removing the first use of maximin. For example, we might instead calculate the expected utility of each affected party conditional on the alternatives; and then select the alternative which gives the greatest expected utility to the worst-off party. This preserves the primary usage of maximin, whilst making it at least somewhat plausible that the affected parties in a collision could rationally consent to Leben's algorithm.

Challenge 2: The maximin rule gives undue weight to the moral claims of the worst-off. Consider,

> *Scenario 2:* The AV can swerve left or right. If the AV swerves left, there is a 100% chance that its passenger will die, and twenty nearby pedestrians will be unharmed. If the driverless car swerves right, there is a 99% chance that its passenger will die, and a 100% chance that twenty nearby pedestrians will receive lifelong debilitating injuries.

Leben's algorithm selects swerve right. Indeed, Leben's algorithm selects swerve right no matter how many pedestrians stand to receive lifelong debilitating injuries. Leben (2017: 144) acknowledges this counterintuitive feature of his algorithm and offers two points in response. First, he argues that scenarios of this kind are unlikely to arise. But the fact that these scenarios are *unlikely* does nothing to address the moral complaints of the pedestrians when such scenarios do arise. Leben's second response is that '[he] would always prefer to be one of the injured pedestrians (and [he] would thus prefer the action which produces the minimum outcome)' (*Ibid.*). So, whilst many pedestrians might receive lifelong debilitating injuries, no one *individual* receives an injury worse than the fatal injury which the passenger would sustain on swerve left. I am unconvinced. First, Leben's algorithm is meant to be contractualist: it aims to programme the AV to allocate harm in a way that is justifiable to each affected party. If we added a twenty-first pedestrian to *Scenario 2*, this would make no difference to the calculation performed. *Prima facie*, the moral claims of the twenty-first pedestrian are not given due consideration, because for Leben's algorithm, it is irrelevant whether or not she is present in the collision (Hirose 2015: 74). Second, I would prefer to lose a limb rather than die. It does not follow that, if forced to choose between killing one person and removing a limb from every human on the planet, I have stronger moral reasons to choose the latter option. The fact that I would *prefer* to lose a limb rather than die is not a good moral reason to inflict a very large number of serious injuries to prevent a single death (Norcross 1997).

Challenge 3: Suppose that Leben is correct about survival probabilities, and that he is also correct about maximin. In some collisions, there is another algorithm which assigns a higher survival probability to the worst-off than Leben's algorithm. Consider,

> *Scenario 3:* The AV can swerve left or swerve right. If the AV swerves left, there is a 0% chance that Anne will survive, and a 70% chance that Bob will survive. If the AV swerves right, there is a 1% chance that Bob will survive, and a 60% chance that Anne will survive.

Leben's algorithm mandates programming the AV to swerve right. This is because the worst-off party on swerve right has a 1% chance of survival, and the worst-off party on swerve left has a 0% chance of survival. Leben's algorithm, then, assigns a survival probability of 1% to the worst-off party in *Scenario 3*.

I now introduce a rival algorithm, which we can call *greatest equal chances*. Leben contends that sometimes the AV ought to randomise between the two alternatives. So, there are at least three options in *Scenario 3*: swerve left, swerve right and construct a fair lottery between the alternatives. Plausibly, if the AV can construct a fair lottery, then it can also construct a weighted lottery. On the greatest equal chances algorithm, the AV is programmed to construct a weighted lottery between the alternatives, where the weightings are fixed to ensure that the affected parties receive the greatest equal survival probabilities. The process is akin to tossing a biased coin to decide whether to swerve left or right, where the degree to which the coin is biased ensures that Anne and Bob are given the greatest equal chances of survival. If $x \in [0, 1]$ is the probability of swerve left in the weighted lottery, and $1 - x$ is the probability of swerve right, then the AV gives Anne and Bob an equal chance of survival provided $0.7x + 0.01(1 - x) = 0.6(1 - x)$. Solving for x, we see that if the AV assigns a probability of 0.457364 to swerve left and $1 - 0.457364$ to swerve right, then Anne and Bob have equal survival probabilities of 32.6%.

So, the greatest equal chances algorithm gives the worst-off a survival probability of 32.6%, which is greater than the 1% survival probability which Leben's algorithm assigns to the worst-off. If the alternatives in *Scenario 3* should be evaluated using survival probabilities; and if maximin is the rule that we have best reason to use, then we ought to adopt greatest equal chances in place of Leben's algorithm[8]. It follows that under Leben's two assumptions, there exist collisions in which another collision algorithm ought to be used to decide between the alternatives.

[8] The earliest statement of the relation between maximin and greatest equal chances is, to my knowledge, due to Parfit (2003: 76–8). For discussions of lotteries like the one described see (Rasmussen 2012), Rivera-López (2008) and Hirose (2015: 121–2).

6 Conclusion

In this paper, I argued that we have reason to reject Leben's (2017) answer to the moral design problem. First, I argued that Rawls' (1971) arguments for the veil of ignorance and the maximin principle do not support Leben's application of these tools to the moral design problem. In doing so, I established that Leben owes an independent argument in favour of his view. Second, I argued that Leben's algorithm is based on two problematic assumptions: (i) that we ought to evaluate the alternatives in AV collisions using survival probabilities; and (ii) that we ought to use the maximin principle to choose between the alternatives. I then argued that *even if* these assumptions are granted, there are some collisions in which a greatest equal chances algorithm is preferable to Leben's algorithm, because it provides a higher survival probability for the worst-off party.

References

Bonnefon, J., Shariff, A., Rahwan, I.: The social dilemma of autonomous vehicles. Science **352**(6293), 1573–1576 (2016)

Goodall, N.: Ethical decision making during automated vehicle crashes. Transp. Res. Record J. Transp. Res. Board **2424**, 58–65 (2014)

Hammond, P.J.: Equity, Arrow's conditions, and Rawls' difference principle. Econom. J. Econom. Soc. **44**(4), 793–804 (1976)

Harsanyi, J.C.: Cardinal utility in welfare economics and in the theory of risk-taking. J. Polit. Econ. **61**(5), 434–435 (1953)

Hirose, I.: Moral Aggregation. Oxford University Press, New York (2015)

Keeling, G.: Commentary: using virtual reality to assess ethical decisions in road traffic scenarios: applicability of value-of-life-based models and influences of time pressure. Front. Behav. Neurosci. **11**, 247 (2017)

Keeling, G.: Legal necessity, Pareto efficiency and justified killing in autonomous vehicle collisions. Ethical Theory Moral Pract. **21**(2), 413–427 (2018)

Leben, D.: A Rawlsian algorithm for autonomous vehicles. Ethics Inf. Technol. **19**(2), 107–115 (2017)

Lin, P.: Why ethics matters for autonomous cars. In: Maurer, M., Gerdes, J.C., Lenz, B., Winner, H. (eds.) Autonomous Driving, pp. 69–85. Springer, Berlin (2016)

Norcross, A.: Comparing harms: headaches and human lives. Philos. Public Aff. **26**(2), 135–167 (1997)

Otsuka, M.: Saving lives, moral theory, and the claims of individuals. Philos. Public Aff. **34**(2), 109–135 (2006)

Otsuka, M.: Prioritarianism and the separateness of persons. Utilitas **24**(3), 365–380 (2012)

Parfit, D.: Justifiability to each person. Ratio **16**(4), 368–390 (2003)

Rasmussen, K.B.: Should the probabilities count? Philos. Stud. **159**(2), 205–218 (2012)

Rawls, J.: A Theory of Justice. Harvard University Press, Cambridge (1971)

Rawls, J.: Justice as Fairness: A Restatement. Harvard University Press, Cambridge (2001)

Rivera-López, E.: Probabilities in tragic choices. Utilitas **20**(3), 323–333 (2008)

Scanlon, T.M.: What We Owe to Each Other. Harvard University Press, Cambridge (1998)

Schelling, T.: Should the numbers determine whom to save? In: The Strategies of Commitment, pp. 113–146. Harvard University Press, Cambridge (2006)

Sen, A.: The Idea of Justice. Harvard University Press, Cambridge (2011)

Sen, A.: Welfare inequalities and Rawlsian axiomatics. Theor. Decis. **7**(4), 243–262 (1976)
von Neumann, J., Morgenstern, O.: Theory of Games and Economic Behavior. Princeton University Press, Princeton (1953)

Moral Status of Digital Agents: Acting Under Uncertainty

Abhishek Mishra(✉)

National University of Singapore, Singapore, Singapore
abhishek.vsm@gmail.com

Abstract. This paper addresses how to act towards digital agents while uncertain about their moral status. It focuses specifically on the problem of how to act towards simulated minds operated by an artificial superintelligence (ASI). This problem can be treated as a sub-set of the larger problems of AI-safety (how to ensure a desirable outcome after the emergence of ASI) and also invokes debates about the grounds of moral status. The paper presents a formal structure for solving the problem by first constraining it as a sub-problem to the AI-safety problem, and then suggesting a decision-theoretic approach to how this problem can be solved under uncertainty about what the true grounds of moral status are, and whether such simulations do possess these relevant grounds. The paper ends by briefly suggesting a way to generalize the approach.

1 Introduction

How should we act if artificial agents are deserving of moral consideration? Here, by artificial agents, I mean digital agents – agents which have been realized on computational substrates.

In this paper, I want to work towards this general problem by focusing first on a more particular case – how we should act regarding simulations run by an artificial superintelligence. The topic of artificial superintelligence (ASI) has received increased attention recently, with focus on the questions of whether an ASI is possible and if so, how we should steer its creation given various risks.[1] However, the topic of the moral status of artificial agents has been considered less, with scant attention being paid to the moral status of simulations run by an ASI (Armstrong et al. 2012; Bostrom 2014). It is this gap that I wish to fill.[2]

One way to pose the question is to first consider the *moral* question:

[1] Chalmers (2010), Bostrom (2014), Yudkowsky (2001), Armstrong et al. (2012), Dainton (2012), Steinhart (2012), Shulman and Bostrom (2012), etc.

[2] There has also been discussion on how we should handle artificial suffering more generally, even in non-ASI circumstances (See for instance Metzinger (2013)). However, for the purposes of this paper, I want to focus on ASI-run simulations as (1) they might be significantly more extensively deployed given that the cost to do so would be lower for a superintelligent agent, and (2) they are relevant for the larger ASI problem (as will be laid out in the Sect. 2.1), which is my larger focus here.

V. C. Müller (Ed.): PT-AI 2017, SAPERE 44, pp. 273–287, 2018.
https://doi.org/10.1007/978-3-319-96448-5_30

Moral Question: Do simulations run by an ASI have moral status?

(a) What are the properties/relations that ground moral status?
(b) Do such simulations instantiate these relevant properties/relations?

The moral question is important, and we will need to consider it, but that is not the question that I shall be focusing on. I want to consider the question of the moral status of simulations as a sub-problem of the larger problem of how we might control the creation of an ASI (the AI-safety problem). This way of posing the question of moral status is the *decision problem*:

Decision Problem: How should we act to ensure that simulations with moral status are treated permissibly by the ASI, while also ensuring otherwise desirable outcomes from the creation of the ASI?

My answer to how we might solve the decision problem is roughly as follows: we should act by first decision-theoretically formalizing the problem as one under moral *and* non-moral uncertainty, then work out the appropriate values of possible outcomes in this formalization, and finally utilize the right decision rule to select the appropriate action. However, executing this total project would require solving deep problems in ethics, philosophy of mind, decision theory and moral uncertainty, and is therefore beyond the scope of this paper. Here, I shall focus on just the first part – how we can decision-theoretically formalize the problem as one under moral and non-moral uncertainty (Sepielli (forthcoming); MacAskill 2014).

The structure of the paper then will roughly be as follows. In Sect. 2, I will set up the decision problem. I will do this by first introducing the larger problem of constraining the superintelligence (the AI-safety problem), and then introducing the problem of simulations as a subset of the AI-safety problem.[3] I will introduce this problem of simulations as a decision problem, without too much formalization here. To end this section, I will consider one quick solution to this decision problem, and outline why we should view it as inadequate. In Sect. 3, I will briefly enumerate the accounts of moral status that are most prominently held. I will argue that we should focus on accounts of moral status that ground it in properties that are computationally realizable, and will eliminate those accounts of moral status that we can quickly rule out as not computationally realizable. I will also suggest further moves we can make to thin the list of remaining properties that we need to check for computational realizability. In Sect. 4, I will pull together the prior conclusions to more rigorously formalize the decision problem. Here, I will outline what it will take to transform the decision problem from a decision under ignorance to a decision under risk, and how the problem will look in each case. Finally, in Sect. 5, I will end with some parting thoughts about how such an approach can be generalized.

[3] While I'm using the phrase 'AI-safety' here as specific to ASIs, the term has been used more broadly in the literature. See for instance Amodei et al. (2016).

2 Introducing the Decision Problem

2.1 Artificial Superintelligence and the AI-Safety Problem

The issue of moral status arises when we consider the possibility of an ASI. For the purposes of this paper, the qualities of such an ASI would include (at least) radically superior cognitive capacities across all domains, faster than humanly correctable speed of implementation and thus a position from which it can exert profound influence over many, many lives (perhaps all of posterity).

Given the potential capabilities of such an intelligence, it is a matter of basic prudence to ask how its actions will affect the world. Any actions taken by an ASI will be taken in service of a goal that initial creators of the ASI (or creators of its *seed* form) will endow it with,[4] so the specification of this goal or principle becomes of crucial importance. Furthermore, if we want to prevent the ASI from acting in certain ways to avoid undesirable outcomes, we will also need to give it some constraints. These goals and constraints together, I shall refer to henceforth as the ASI's 'governing principle'. It is this problem of determining an adequate governing principle for an ASI that I refer to as the AI-safety problem.[5]

Recently, there have been some guiding principles and constraints proposed that solutions to this problem must adhere to. I want to outline three such principles and constraints that are relevant to our discussion.

The Deadline Constraint. Solving the AI-safety problem is what Bostrom calls doing "philosophy with a deadline" (Bostrom 2014). We are unsure about exactly when such an ASI might arise. Given the consequences of implementing the wrong solution to the AI-safety problem, it is imperative that we endeavour to be as right as possible *as quickly as possible.*

The All-Things-Considered (ATC) Constraint. The costs and harms that we need to watch out for and guard against, in solving the AI-safety problem, are of varied types. There are moral considerations, such as the moral status of an ASI and of other artificial agents (like simulations), as well as obligations to future generations. There are also other considerations, such as prevention of extinction scenarios or the crippling of

[4] One point often raised by individuals is whether such an intelligence will start to develop its own goals. While I will not go into this issue here, there is an extensive discussion of this topic in Chap. 7 of Bostrom (2014), where he argues that there need be no necessary relationship between intelligence and *fundamental* goals.

[5] To align the terminology used with the existing literature (Bostrom 2014), the AI-safety problem is the problem of articulating the correct motivation for the ASI, which is one part of what is known as the motivational selection problem.

human potential. When considering solutions to the AI-safety problem as well as any of its sub-problems, it is crucial that we reflect all these relevant considerations.[6]

The Principle of Epistemic Deference (PED). The PED can be stated as follows:

> A future superintelligence occupies an epistemically superior vantage point: its beliefs are (probably, on most topics) more likely than ours to be true. We should therefore defer to the superintelligence's opinion whenever feasible (Bostrom 2014).

The general sentiment is that we should always defer to our epistemic superiors, and this is especially so when costs of being wrong are extremely high, as is the case in solving the AI-safety problem.

2.2 Moral Status of Simulations

Here, it would be fruitful to clarify what is meant by moral status. One way of doing so is to consider a distinction between a moral agent and a moral patient (Norcross 2004). While a moral agent is a being that is subject to moral obligations and other moral expectations, a moral patient is a being whose interests ought to be considered in one's moral decision making, for the being's own sake. By moral status then, we are concerned exclusively with a being's standing as a moral patient, and this can be best captured as follows,

> [A]n entity has moral status if and only if or its interests matter morally to some degree for the entity's own sake, such that it can be wronged.[7]

Thus, an entity with moral status requires us to act morally for its own sake, independent of other instrumental reasons (though such reasons could still apply, even overridingly).[8] Based on this understanding of moral status, we can now consider what simulations could have such status.

There are some solutions that have been put forward to the AI-safety problem that require that the ASI pay attention to and determine the preferences, attitudes, desires

[6] Here, the ATC constraint seems to be too vaguely formulated – for instance, we need to be able to say not just that solutions run afoul of it, but also assess the extent to which they do so. The ATC constraint thus needs to be further developed to provide a metric that allows us to make this assessment. Such a metric would need to track, for any given solution, the probability of incurring other costs (apart from morally impermissible treatment of simulations) along with the magnitude of the cost incurred. We shall look at one way of defining such a metric in Sect. 4.

[7] Jaworska and Tannenbaum (2013). An understanding of moral status as being concerned with moral patienthood is not universally accepted in the literature – sometimes, moral status is associated both with moral agency *and* moral patienthood. For the purposes of this paper, however, we should understand it as just referring to moral patienthood.

[8] There is a further distinction that can be drawn here, between distinct types of non-instrumental moral status. Kamm notes that we can speak of the status an entity has "in its own right" compared to the status that it has "for its own sake", where violations against the latter allow us to speak of the entity being morally wronged whereas violations against the former do not permit it. Here, I will limit my scope to only considering an account of moral status in the traditional sense, where an entity counts "for its own sake". Kamm (2007).

and/or other mental states of existing individuals. Let us call this class of solutions the psychological-profile class of solutions. Examples of solutions in this class include Coherent Extrapolated Volition[9], or more generally, other 'Do What I Mean' or 'Do What I Want' models (Bostrom 2014, pp. 220–221). We can then consider a further sub-class (let's call it the 'simulation class') – a sub-class where the solutions populating it determine the psychological profile of an individual by first modeling or simulating the individual with a high enough level of fidelity, and then subsequently extracting the relevant preferences/mental states. As Armstrong, Sandberg and Bostrom note,

> To answer specific questions, the [ASI] may have to simulate other entities inside itself. For instance, … to decide whether humans would have their happiness increased by a certain project, it may have to create models of specific humans (Armstrong et al. 2012).

This procedure for determining the psychological profile might be recommended by design or selected by the ASI as an adequate method even without being explicitly recommended. However, crucially, what defines solutions in the simulation-class is that the representations of individuals utilized by the ASI for those solutions are of high-enough fidelity that they may have moral status.

The project of this paper then is to answer the decision problem. We might modify and restate the decision problem as follows:

DP: Given that we need to solve the AI-safety problem, and given that some solutions to the AI-safety problem may cause the ASI to create simulations with moral status, how should we act?[10]

2.3 Preliminary Suggestion – The PED Solution

One solution is to offload the decision problem (or part of it) to the ASI. This can be argued to be in line with PED. We should get the ASI to figure out for us whether the simulations it runs have moral status, and what would count as morally permissible treatment given that they have such status.

There is however a major reason that we cannot be too hasty in offloading the problem to the ASI. The PED requires that we only defer to the superintelligence's opinions "whenever feasible" – this need not fit our situation. Consider the case where for the ASI to arrive at the answer to the question of whether simulations possess moral status,

[9] Coherent Extrapolated Volition as a solution to the AI-safety problem recommends that we ask the ASI to implement the coherent extrapolated volition of humankind, where this CEV is an agent's wish for himself if he were idealized in certain ways – if he were smarter, more the person he wished he were, etc. This proposal looks for coherence among such idealized wishes of humans and recommends that the ASI implements the wishes where there is such a strong coherence. Yudkowsky (2004).

[10] This is a decision problem that is relevant to more than one type of ASI. Among other types, ASIs are classified as oracles (question-answering ASIs), genies (intermediate-level task-oriented autonomously acting ASIs) and sovereigns (autonomously acting ASIs with a single open goal). However, all three types may run simulations to understand the psychological profiles of individuals. See Bostrom (2014) for further details on the classification of the various types of ASIs.

it first needs to run such simulations. Perhaps it would need to understand what we mean by morality, or what intuitions we have about moral status. In this situation, it would not be possible for the ASI to answer the question of whether simulations have moral status without creating simulations, and thus potentially incurring the same high costs that we wanted to avoid in the first place.[11] In such a scenario then, it would not be feasible to defer to the opinion of the ASI.

There remains a possibility that the ASI can determine the answer to such questions without simulating. We can thus stipulate that if we can get the ASI to answer such questions without simulating and *without sacrificing accuracy*, then we should utilize this method. However, in the absence of such certainty, we ought not to blithely defer to the ASI's opinions anyway. In this paper then, I will assume that the above possibility is a very real one, and thus that it is not feasible to defer to the ASI's opinion on this topic.

3 The Grounds of Moral Status

Given that we're focusing on the decision problem rather than the moral question, my aim in this paper is not to argue for one account of moral status over another. There is extensive disagreement on the topic, and my aim is to consider whether there are any alternative strategies that can be applied to solve the decision problem.

3.1 The Various Accounts of the Grounds of Moral Status

Here, I will be focusing on four main categories of accounts that answer to the question of what grounds moral status. Between these four categories, we will cover most of the positions staked out in the debate.[12] The four categories are as follows: (1) Sophisticated Cognitive Capacities (SCC) accounts, (2) Potential for SCC and Membership in SCC Species accounts, (3) Special Relationship accounts, and (4) Rudimentary Cognitive Capacities (RCC) accounts.

SCC accounts claim that the grounds of moral status are certain sophisticated cognitive capacities that entities can possess. If an entity possesses the relevant sophisticated cognitive capacity, then that entity possesses some level of moral status. Examples of such sophisticated cognitive capacities are self-awareness, being future-oriented in desires and plans, a capacity to value, bargain and assume duties and responsibilities, personhood, (Jaworska and Tannenbaum 2013) and other capacities that come under what is generally described as sapience[13].

---

[11] More generally, such a possibility could obtain if the ASI's architecture involves 'black-box' methods. See Armstrong et al. (2012), pp. 15–19.

[12] I follow, with some modifications, the categorization put forward by Jaworska and Tannenbaum (2013).

[13] Bostrom and Yudkowsky introduce this notion of sapience as "a set of capacities associated with higher intelligence, such as self-awareness and being a reason-responsive agent". Bostrom and Yudkowsky (2011).

Potential for SCC and Membership in SCC-Species accounts are a variant of SCC accounts. Here, even absent the sophisticated cognitive capacities, having either the potential for such capacities or belonging to a species whose members typically have such capacities is also sufficient to endow an entity with moral status. Such potentiality or membership accounts can be (and often enough have been) associated with any of the relevant sophisticated cognitive capacities suggested in the first category.[14]

Special Relationship accounts ground moral status in relationships that we share with entities. For instance, we share the relationship of being co-members of the human community with all other humans, and some special relationship accounts claim that this provides us with certain duties to other human beings. Such accounts are often advocated instead of potentiality or membership accounts as relationships are more often perceived as morally significant than potential, or species membership.

RCC accounts claim that moral status is grounded in certain rudimentary cognitive capacities. Examples of such rudimentary cognitive capacities are the capacity for pleasure and pain, the capacity for basic emotions, the capacity for consciousness (in the what-is-it-like-to-be or sentient sense) and the capacity for having interests.[15]

3.2 Navigating the Various Accounts

So far, we have observed two insights:

1. Quick solutions to the decision problem (such as the PED solution) offer some benefits, such as satisfying the deadline constraint, but suffer from other failings and so for now should be treated as non-optimal solutions, and
2. There are numerous types of accounts of moral status, and numerous accounts within each type, with no clear consensus on any one account or type of accounts.

[14] Two important accounts, while not being potentiality or membership accounts, still resemble them. The first account is Shelly Kagan's 'modal personhood' account – to have moral status, an entity must either have the capacity of personhood ("a being that is rational and self-conscious, aware of itself as one being among others, extended through time"), or have the property of being a 'modal person', such that even if it is not currently a person, it *could* have been a person. The second account is S. Mathew Liao's genetic basis for moral agency account, where for an entity to have moral status it is sufficient that it possesses the genetic basis for moral agency, as it occurs in those human beings that we normally take to exercise moral agency. See Kagan (2016) and Liao (2010) respectively.

[15] Apart from these four major categories of accounts, there are a few other possibilities that we need to cover for the sake of comprehensively mapping out the logical space. Firstly, there is a possibility that the correct account of the grounds of moral status invokes a property that is not covered under any of the above four categories. Secondly, there is also the possibility that the correct account of the grounds of moral status could be a combination of two or more of the above four accounts – the correct account could thus feature a conjunction of two or more accounts (for instance, sapience as well as sentience) or a disjunction of two or more accounts (for instance, with both sapience and sentience being sufficient conditions for moral status without either being a necessary condition).

Given the deadline constraint, the second insight is troubling. It is unclear whether we will be able to solve the moral question by the time we need to act to specify the governing principle. Assuming we can't, we need to find alternative ways of arriving at an appropriate action, perhaps by thinning the list of accounts and types of accounts that are relevant.

My suggestion here is to first check which candidate properties proposed in the various accounts are realizable on computational substrates. This is a way of limiting the grounds of moral status to those that are relevant to the decision problem, because properties which are not computationally realizable need not be worried about – simulations could never possess those properties, regardless of whether the properties do ground moral status. This thus cuts down on the amount of philosophical work that needs to be done, as per the deadline constraint.

The clearest case of an account that requires a non-computational substrate is the account that requires a special relationship of being co-members of the human community. Simulations are not members of the human community, assuming that to be a member of the human community one needs to be a biological human. Thus, regardless of whether moral status is truly grounded in this relationship, simulations can't possess moral status – this account can be safely ruled out. One can run a similar argument against accounts that require an entity to belong to a particular biological species, as several membership-based accounts do.

However, this still leaves many other accounts, including other kinds of special relationship ones. The question now is how we can further limit the search space of properties that we need to check for computational realizability.

One way to do this is to show that certain properties are just dependent cases of other ones – to determine whether the former are computationally realizable, we would first need to determine whether the latter are.

For instance, consider potentiality accounts of moral status. Potentiality accounts claim that for an entity to have moral status, it must at minimum have the potential for a certain capacity or property, even if it does not currently possess it. An example of this would be an account that says that moral status is grounded in the potential for rationality (understood in some specific way). However, an entity could not possess the potential for rationality if rationality was not realizable on its substrate. Thus, to be relevant to the decision problem, potentiality accounts that state that moral status is grounded in the potential for property x need to have property x itself first be computationally realizable. If it is not, then this rules out as relevant to the decision problem not just accounts which ground moral status on property x, but also those that do so on the potential for property x.[16]

[16] Once we have confirmed that the relevant property is computationally realizable, there is still further work to be done depending on which account is the correct account of moral status. If, for instance, the rationality account is the right one, then all we need to check is if simulations are rational. However, if the potential for rationality account is the right one, then we need to ask whether simulations are rational, *and* if not, whether they have the potential to be rational. This would require the further step of figuring out what would it take for something to have the potential to be rational. However, in either case, the computational realizability of rationality still needs to be checked.

Given the layout of the terrain outlined in Sect. 3.1, this helps us limit the number of properties we need to test for computational realizability enormously. When it comes to SCC accounts and potential for SCC accounts, we just need to check the computational realizability of the sophisticated cognitive capacity in question. Furthermore, this can be extended to the remaining special relationship accounts as well. Consider that for special relationship accounts, it is not the case that *any possible relationship at all* can confer moral status to an entity. This would allow for us to give moral status to our chair, since we share a relationship with it of being in the same city. For special relationship accounts to be plausible, they would have to be grounded in some relevant moral property that is shared between us and the other entity (such as the property of being human, as seen earlier). Whether special relationship accounts are relevant to the decision problem thus depends on whether these properties are computationally realizable. We can then understand that there can be 3 types of special relationship accounts - where the relevant moral property is one of the SCC properties, one of the RCC properties, or neither. If it is an SCC or a RCC property, then once again as in the case of potentiality accounts, this would limit the search space.

At this point, we can take stock of which grounds of moral status we need to check for computational realizability. We need not check those grounds that require the property of having a biological substrate, or other properties which require biological substrates. We also need not check potentiality, membership or special relationship accounts which are dependent on properties already included in the SCC and RCC accounts. Thus, we need to check the computational realizability only for the sophisticated cognitive capacities and rudimentary cognitive capacities listed in Sect. 3.1, as well as any other properties that are the basis of potentiality, membership or special relationship accounts.

4 The Decision Problem Updated

I will now further formalize the decision problem, before showing how the decision formalization changes as we update it with the relevant information. This process will turn the decision problem from a problem under ignorance to a problem under risk, where we have the relevant information. Once I have fully formalized the problem, I will then consider what it would take to solve the problem.

4.1 Formalizing the Decision Problem

For the purposes of formalization, I shall make some simplifying assumptions. First, I want to limit outcomes of actions taken to having two parts, which can be tracked by corresponding metrics – (1) the level of moral infringement (MI) of the simulations given their having moral status, and (2) the level of accuracy of the resulting simulation, which will be a stand-in for all the other costs (moral and non-moral) flagged by the ATC constraint. Secondly, for each of these two metrics, I would like to stipulate certain values given certain actions:

I. Moral Infringement (MI) – There will be no moral infringement if we bar the ASI from running simulations that have the relevant property that grounds moral status, an equal or higher level of moral infringement if we don't bar the ASI from running such simulations but constrain their running in certain ways (to accommodate their moral status), and the highest level of moral infringement if we allow the ASI to run such simulations in a completely unconstrained manner.
II. Accuracy – There will be the highest loss of accuracy if we bar the ASI from running simulations possessing a certain property, an equal or lower loss of accuracy if we don't bar the ASI from running such simulations but constrain their running in certain ways, and the least or no loss of accuracy if we allow the ASI to run such simulations in a completely unconstrained manner.

For our current purposes, we can think of the rankings for each metric as ordinal rankings rather than cardinal (or ratio) ones.

Finally, P_A refers to property A and P_{CR} refers to the property of being computationally realizable such that '$P_A \subseteq P_{CR}$' signifies that property A is computationally realizable. S_A refers to simulations that have property A. P* refers to the property that grounds moral status such that '$P_A \subseteq P^*$' signifies that property A grounds moral status.[17]

We can now introduce the Decision Matrix 1 (DM1) as representing the decision problem as a problem under maximal ignorance, where we are ignorant about (1) which property is computationally realizable as well as (2) which property grounds moral status. DM1 shows the decision problem concerning property X, where property X is one of the candidate properties that ground moral status.

Decision Matrix 1.	$P_X \subseteq P_{CR}$ $P_X \subseteq P^*$	$P_X \subseteq P_{CR}$ $P_X \not\subseteq P^*$ $P^* \subseteq P_{CR}$	$P_X \subseteq P_{CR}$ $P_X \not\subseteq P^*$ $P^* \not\subseteq P_{CR}$	$P_X \not\subseteq P_{CR}$ $P_X \subseteq P^*$	$P_X \not\subseteq P_{CR}$ $P_X \not\subseteq P^*$ $P^* \subseteq P_{CR}$	$P_X \not\subseteq P_{CR}$ $P_X \not\subseteq P^*$ $P^* \not\subseteq P_{CR}$
Bar S_X	- No MI - Most loss of accuracy	- Most MI[a] - Most loss of accuracy	- No MI - Most loss of accuracy	- No MI - Least/no loss of accuracy	- Most MI - Least/no loss of accuracy	- No MI - Least/no loss of accuracy
Run S_X Constrained	- Some MI - Some loss of accuracy	- Most MI - Some loss of accuracy	- No MI - Some loss of accuracy	- No MI - Least/no loss of accuracy	- Most MI - Least/no loss of accuracy	- No MI - Least/no loss of accuracy
Run S_X Unconstrained	- Most MI - Least/no loss of accuracy	- Most MI - Least/no loss of accuracy	- No MI - Least/no loss of accuracy	- No MI - Least/no loss of accuracy	- Most MI - Least/no loss of accuracy	- No MI - Least/no loss of accuracy

[a]This need not strictly be true – it rests on the assumption that if P* is computationally realizable, it will be instantiated by the simulations run by an ASI. Of course, this is entirely dependent on the architecture of the simulations run by the ASI, and what P* actually is. Accounting for this, the MI value here (and in the column "$P_X \not\subseteq P_{CR}$, $P_X \not\subseteq P^*$, $P^* \subseteq P_{CR}$") should actually be 'No MI/Most MI', where the exact value will depend on whether P* is instantiated for the ASI-run simulations. However, for the sake of easier representation of the decision-situation, I have made the aforementioned assumption.

[17] Here, we shall simplify the scenario such that it is not the case that multiple properties can individually be sufficient for grounding moral status – thus one and only one property grounds moral status.

Thus, for instance, if we choose to bar a simulation with property X ('Bar S_X') when property X is computationally realizable and also grounds moral status ('$P_X \subseteq P_{CR}$, $P_X \subseteq P^*$'), then simulations with moral status will not be treated impermissibly ('No MI'), but there might be great loss of accuracy in capturing the psychological profiles of individuals for use by the ASI since it can't simulate effectively ('Most loss of accuracy'). However, if we choose to run a constrained simulation ('Run S_X Constrained') under the same conditions ('$P_X \subseteq P_{CR}$, $P_X \subseteq P^*$'), then while there is some risk of morally impermissible treatment since the constraints might not be fool proof ('Some MI'), it would reduce the loss in accuracy as some level of simulation would be allowed for the ASI ('Some loss of accuracy').

Prima facie, no action dominates any other. Furthermore, any closer examination of whether some outcomes are more valuable than others would require that we be able to commensurate MI values with accuracy values. This is a problem that we will return to shortly.

One way to make progress would be to check for which properties are computationally realizable. If we can have knowledge of this, we can erode some of the uncertainty in DM1 and represent the new decision problem with Decision Matrix 2 (DM2). DM2 would then just be DM1 without the last 3 columns, since there is no longer uncertainty about whether property X is computationally realizable – if it were not, we would not be considering it as relevant to the decision problem.[18]

Let's now look at the final piece of information needed to transform this problem from a problem under ignorance to a problem under risk – knowledge about which property grounds moral status, which can be reflected by the probabilities of each of the three possible states (columns) listed in DM2. For the relevant probabilities, we can use *credences* that we have regarding each of the three states. The probabilistic coefficient for the first state is our credence that property X grounds moral status, for the second state it's our credence that property X does not ground moral status and the property that does is computationally realizable, and so on.[19] We can thus imagine a Decision Matrix 3 (DM3) that is identical to DM2, except with these credences associated with each of the three states.[20]

So far, we've made three simplifying assumptions that we will now factor back into the decision problem. Firstly, we've articulated the actions in the decision problem as only relating to one property – either bar, run in a constrained manner, or run in an

[18] An alternative scenario could be if we're not *certain* about which properties are computationally realizable, but hold credences about the computational realizability of the various properties. In this case, DM2 would look more or less like DM1, except with credences associated with each column (each situation).

[19] I have assumed in Sect. 3.2 that we won't be able to solve the moral question (or even MQ(a)) in time. However, if we can solve MQ(a) in time and thus have full information about which property it is that grounds moral status, this would just be reflected with a credence of 1 associated with the scenario that that property grounds moral status, and a credence of 0 for all other scenarios.

[20] We can use our credences for these states in a way that is consistent with recent work on normative uncertainty. See MacAskill (2014). The only difference here is the further problem of deciding whose credences should be taken into account.

unconstrained manner simulations which instantiate property X. However, unless we are certain that moral status is grounded in one particular property (something we have already assumed in Sect. 3.2 we won't be certain of), we will have to specify actions relating to multiple properties. Thus, the list of actions would include actions like 'Bar S_X, Run Unconstrained $S_{Y,Z}$', 'Bar $S_{X,Y}$, Run Constrained S_Z', and so on, including any number of combinations of the three acts specified with the total set of properties in play. The greater number of actions we would need to choose from would make the resulting decision matrix much more complex.

Secondly, we've only allowed a range of three actions that can be taken – either bar the simulations, run constrained versions of them, or run unconstrained versions of them. However, this is obviously a simplification. To give one example, it might be that the most moral way in which we can act is to create as many simulations as possible in as pleasurable circumstances as possible. Such an action is not captured in the original three. Also, even if we want to run simulations in a constrained manner, there are various degrees and types of constraints we can apply, each of which can be represented as a discrete action. Rolling back this simplification will expand the list of actions even further.

Finally, to truly address the decision problem, we need to go beyond computational realizability to whether the property will be instantiated in a simulation. This means that even DM3 (incorporating the changes I've just discussed) is still a decision problem under uncertainty, because we still don't have full information – while we know which properties are computationally realizable, we still don't have any credences about which of those properties will be instantiated in a simulation run by the ASI (computational realizability being a necessary but insufficient condition).

4.2 Solving the Decision Problem

Regardless of the decision formalization we're solving under, there are two parts to solving the decision problem. First, we shall need to allocate appropriate values to each outcome in the decision matrix. Without this, we will not be able to compare the various outcomes to select preferable ones. Second, once we have allocated values to all the outcomes, we will need to select decision rules to choose the right act. I will not attempt to solve the decision problem because each of these two parts raise serious philosophical difficulties. We shall look at each of these in turn.

There are two main obstacles to being able to attribute values to the various outcomes. Firstly, the extent of the disvalue of moral infringement of simulations would depend on our ethical stance – in this case, on which property actually grounds moral status. For instance, there *might* be a greater disvalue associated with moral infringement of simulations if moral status is grounded in the capacity for pain than if it's grounded in rationality – it *might* be morally preferable to interfere with simulations in the latter case. How might we arrive at an understanding of how much worse certain actions are according to one theory compared to another? This becomes especially important once we start aggregating across different decision matrices corresponding to different properties. Such an understanding is crucial for our ability to rank, ordinally or cardinally, the various outcomes. In the literature, this problem is known as the *problem of intertheoretic comparisons*,[21] and is one of the major problems facing any account of decision-making under normative uncertainty. The second obstacle resembles the first one – how can we compare the importance of preventing moral infringement to the importance of high accuracy? This is crucial if we want to establish even just a preference between the outcome where there is no moral infringement but major loss of accuracy, and the outcome where there is maximal moral infringement but no loss of accuracy. Without such values, we cannot utilize decision rules to choose between the acts we can take. Recently, there have been some attempts to solve or work around the problem of intertheoretic comparisons,[22] but more work remains to be done.

If we can resolve the problem of intertheoretic comparisons, then we will be able to solve the decision problem by using the right decision rule. However, that raises another problem – which decision rules should we use to select between the various actions available? This is a problem that comes up regardless of whether we operate on a decision under ignorance (DM1, DM2) or a decision under risk (DM3). There are various competing proposals for which decision rule should be used for decisions under ignorance, with candidates such as maximin, leximin, maximax, the optimism-pessimism rule, minimax regret (Peterson 2009, Chap. 3). When it comes to decisions under risk, the ground is a little less contested – the dominant rule is expected utility maximisation. However, this too is not without its detractors.[23] Furthermore, considering the extreme possible values of the outcomes (moral infringement of an extremely high number of simulations, or a catastrophic loss of accuracy resulting in extinction or permanent curtailment of humanity's potential (Bostrom and Cirkovic 2008)), standard decision rules such as expected utility maximization might ill-fit the decision problem (Bostrom 2009). Further discussion of this issue is beyond the scope of this paper, but I would like to flag the problem nonetheless.

5 Conclusion

The decision problem has several moving parts, probably more than even the ones outlined in this paper. However, this paper is a start on introducing, developing and formalizing the decision problem. While we may encounter further problems, making progress on the questions raised here would put us in good stead to ensure that we engage with simulations created by the ASI in a way that is permissible as well as prudentially beneficially.

[21] MacAskill (2014), Chap. 2. The problem of intertheoretic comparisons in theories of normative uncertainty itself resembles the problem of interpersonal comparisons in social choice theory. See Steele and Stefánsson (2016), List (2013).

[22] MacAskill puts forward a solution to the problem of intertheoretic comparability by drawing parallels with social choice theory, specifically with how social welfare functionals have been axiomatized under alternative assumptions about informational comparability. See MacAskill (2014).

[23] See Peterson (2009) Chap. 4 for some paradoxes raised that undermine the utilization of expected utility maximisation as a decision rule for decisions under risk.

Furthermore, an approach such as the one sketched out in this paper can also be generalized to deal with digital agents other than simulations as well. While I have not pursued this line of thought in this paper, we can abstract from the specific features of this case, such as the digital agent being a simulation run by an ASI for accuracy considerations. The decision-theoretic structure provided will still preserve the essential features of acting under uncertainty towards digital agents with moral status (such as uncertainty about the right account of moral status, and uncertainty about the computational realizability of the relevant properties).

References

Amodei, D., Olah, C., Steinhardt, J., Christiano, P., Schulman, J., Mane, D.: Concrete problems in AI safety. arXiv:1606.06565v2 [cs.AI] (2016)
Armstrong, S., Sandberg, A., Bostrom, N.: Thinking inside the box: controlling and using an oracle AI. Mind. Mach. **22**, 299–324 (2012)
Bernstein, M.H.: On Moral Considerability. An Essay on Who Morally Matters. Oxford University Press, Oxford (1998)
Boonin, D.: In Defense of Abortion. Cambridge University Press, New York (2003)
Bostrom, N.: Pascal's mugging. Analysis **69**, 443–445 (2009)
Bostrom, N.: Superintelligence: Paths, Dangers, Strategies. Oxford University Press, Oxford (2014)
Bostrom, N., Cirkovic, M.: Introduction. In: Bostrom, N., Cirkovic, M. (eds.) Global Catastrophic Risks. Oxford University Press, Oxford (2008)
Bostrom, N., Yuskowsky, E.: The ethics of artificial intelligence. In: Frankish, K., Ramsey, W.M. (eds.) The Cambridge Handbook of Artificial Intelligence. Cambridge University Press, Cambridge (2011)
Chalmers, D.: The singularity: a philosophical analysis. J. Conscious. Stud. **17**, 7–65 (2010)
Cohen, C.: The case for the use of animals in biomedical research. N. Engl. J. Med. **315**, 865–870 (1986)
Dainton, B.: On singularities and simulations. J. Conscious. Stud. **19**, 42–85 (2012)
Dworkin, R.: Life's Dominion: An Argument about Abortion, Euthanasia, and Individual Freedom. Vintage Books, New York (1993)
Gruen, L.: The moral status of animals. In: Zalta, E.N. (ed.) The Stanford Encyclopedia of Philosophy (Fall 2014 edition) (2014). https://plato.stanford.edu/archives/fall2014/entries/moral-animal/
Feinberg, J.: Abortion. In: Reagan, T. (ed.) Matters of Life and Death, pp. 183–217. Temple University Press, Philadelphia (1980)
Finnis, J.: The fragile case for euthanasia: a reply to John Harris. In: Keown, J. (ed.) Euthanasia Examined. Cambridge University Press, Cambridge (1995)
Jaworska, A., Tannenbaum, J.: The grounds of moral status. In: Zalta, E.N. (ed.) The Stanford Encyclopedia of Philosophy (Summer 2013 Edition) (2013). https://plato.stanford.edu/archives/sum2013/entries/grounds-moral-status/
Kagan, S.: What's wrong with speciesism? J. Appl. Philos. **33**, 1–21 (2016)
Kamm, F.M.: Intricate Ethics. Rights, Responsibilities and Permissible Harm. Oxford University Press, Oxford (2007)
Kant, I.: Groundwork of the Metaphysics of Morals (1785). Gregor, M. (trans. and ed.) Cambridge University Press (1988)
Liao, M.S.: The basis of human moral status. J. Moral Philos. **7**(2), 159–179 (2010)

List, C.: Social choice theory. In: Zalta, E.N. (ed.) The Stanford Encyclopedia of Philosophy (Winter 2013 Edition) (2013). https://plato.stanford.edu/archives/win2013/entries/social-choice/

MacAskill, W.: Normative uncertainty. DPhil Thesis (2014)

Marquis, D.: Why abortion is immoral. J. Philos. **86**, 183–202 (1989)

McMahan, J.: The Ethics of Killing: Problems at the Margins of Life. Oxford University Press, Oxford (2002)

Metzinger, T.: Two principles for robot ethics. In: Hilgendorf, E., Gunther, J.-P. (eds.) Robotik und Gesetzgebung, pp. 263–302. Nomos, Baden-Baden (2013)

Norcross, A.: Puppies, pigs and people: eating meat and marginal cases. Philos. Perspect. **18**, Ethics, 229–245 (2004)

Nozick, R.: Do animals have rights? In: Socratic Puzzles, pp. 303–310. Harvard University Press (1997)

Peterson, M.: An Introduction to Decision Theory. Cambridge University Press, Cambridge (2009)

Sepielli, A.: Moral uncertainty. In: Routledge Encyclopaedia of Philosophy Online (Forthcoming)

Shulman, C., Bostrom, N.: How hard is artificial intelligence? – Evolutionary arguments and selection effects. J. Conscious. Stud. **19**, 103–130 (2012)

Singer, P.: Practical Ethics, 2nd edn. Cambridge University Press, Cambridge (1993)

Steele, K., Stefánsson, O.H.: Decision theory. In: Zalta, E.N. (ed.) The Stanford Encyclopedia of Philosophy (Winter 2016 Edition) (2016). https://plato.stanford.edu/archives/win2016/entries/decision-theory/

Steinbock, B.: Life Before Birth: The Moral and Legal Status of Embryos and Fetuses. Oxford University Press, Oxford (1992)

Steinhart, E.: The singularity: beyond philosophy of mind. J. Conscious. Stud. **19**, 131–137 (2012)

Stone, J.: Why potentiality matters. Can. J. Philos. **17**, 815–829 (1987)

Yudkowsky, E.: Creating friendly AI 1.0: the analysis and design of benevolent goal architectures. The Singularity Institute, San Francisco, CA (2001)

Yudkowsky, E.: Coherent extrapolated volition. The Singularity Institute, San Francisco, CA (2004)

Friendly Superintelligent AI: All You Need Is Love

Michael Prinzing(✉)

University of North Carolina, Chapel Hill, NC 27514, USA
prinzing@live.unc.edu

Abstract. There is a non-trivial chance that sometime in the (perhaps somewhat distant) future, someone will build an artificial general intelligence that will surpass human-level cognitive proficiency and go on to become "superintelligent", vastly outperforming humans. The advent of superintelligent AI has great potential, for good or ill. It is therefore imperative that we find a way to ensure —*long before* one arrives—that any superintelligence we build will consistently act in ways congenial to our interests. This is a very difficult challenge in part because most of the final goals we could give an AI admit of so-called "perverse instantiations". I propose a novel solution to this puzzle: instruct the AI to love humanity. The proposal is compared with Yudkowsky's Coherent Extrapolated Volition, and Bostrom's Moral Modeling proposals.

1 Introduction

Many AI researchers believe there is a non-trivial chance that AI with greater than human-level cognitive capabilities will be developed sometime in the (perhaps somewhat distant) future (Müller and Bostrum 2016). These AI may eventually become "superintelligent"—i.e., capable of *vastly* outperforming humans in all, or nearly all, cognitive tasks (Bostrom 2014; Chalmers 2010).[1] This is both an exciting and unsettling prospect. If a superintelligence were friendly—if its goals aligned with ours—it could provide untold benefits to humanity. It could cure diseases. It could increase economic output such as to end all poverty, hunger, and need. On the other hand, if the superintelligence were malignant—if its goals diverged in important ways from ours—we would be powerless to stop it (Bostrom 2014, Chaps. 8–10).[2]

There is always a risk, when thinking about topics like this, of falling into futile futurology: wild, ungrounded speculation. But, while it's good to keep this risk in mind, I believe it is important to start considering the possibilities. For one thing, the suggestion that superintelligent AI will (eventually) be developed is *not ungrounded*. This is something that many AI researchers are seriously concerned about (Müller and Bostrum 2016; see also Bostrom 2014; Soares and Fallenstein 2015; Yampolskiy 2016). This

[1] There are a number of ways in which an "intelligence explosion" like this might happen (Good 1965). Perhaps the most plausible would involve a seed AI undergoing recursive self-improvement (Yampolskiy 2016, Chap. 5).

[2] Given that superintelligence is defined as cognitive performance *vastly beyond human-level*, this is hard to dispute.

V. C. Müller (Ed.): PT-AI 2017, SAPERE 44, pp. 288–301, 2018.
https://doi.org/10.1007/978-3-319-96448-5_31

certainly doesn't mean that superintelligence is inevitable—just that it's a real possibility. Thus, given the magnitude of the possible outcomes, it is important to work out how we can increase the probability of ending up with a friendly superintelligence, and decrease the probability of a malignant one. It's far better to have a solution and not need one, than to need one and not have one. Crucially, this work will need to be done before (preferably long before) any such being is created. The time to start on this problem is *now*.

It seems to me that the way to understand the friendly superintelligence problem is in terms of a relationship. What we are aiming to ensure is that our relationship with the superintelligence is healthy and beneficial. Perhaps the only such relationship we can have with such a being is no relationship at all. But, in any case, we are trying to ensure that it wants what is best for us, cares about our interests, and will do right by us. When the question is put it in these terms, the shape of a solution starts to become clear. What will make our relationship with the superintelligence a good one is the same thing that makes *any* personal relationship good: love.

2 The Puzzle: Avoiding "Perverse Instantiations"

Before I explain this proposal, it's important to specify the puzzle it aims to solve. There are several related puzzles that it is *not* a solution to. These concern how to express complex, world-affecting goals in computer code. Suppose the goal we want the AI to pursue is the maximization of human happiness. It's quite easy to state such a goal. But it's extremely difficult, even in a natural language like English, to explain precisely what it means. This is classic philosophical territory. *What is happiness?* Supposing that we could give a correct accounting of the relevant concepts, we would then face the challenge of converting that natural language analysis into the functions and operators of a programming language. These initial puzzles, in short, are concept learning puzzles. Abstract natural language concepts are the ingredients of any goal that we might give an AI. The issue I'm interested in, for the purposes of this paper, assumes that we have solved these puzzles.

The next challenge, the one I'm addressing, is to figure out which goals or instructions to provide. A direct approach would be to simply specify some concrete goal like "maximize human happiness". The main problem with this method, as Bostrom (2014) discusses, is that it is extremely difficult to accurately capture the things we value in explicit specifications. For any goal that we might specify directly, even one that sounds agreeable, there is a significant chance that realizing the goal would go horribly wrong. In Bostrom's (2014, 146) terminology, that goal likely admits of some "perverse instantiation"—a way of realizing the letter of the instructions, while betraying their spirit. For instance, one way of maximizing human happiness would involve implanting electrodes into our brains and inducing a perpetual state of vacuous, drooling euphoria. We *would* feel happy, but that's not exactly my vision of a life well-lived.

Thus, those who have thought seriously about this problem tend to favor a more indirect approach (Yudkowsky 2004; Bostrom 2014). Rather than specify a concrete goal, we should give the AI a procedure by which to decide for itself what to do. This approach is appealing insofar as it delegates to the superintelligence the hard work of figuring out how to remain human-friendly. My proposal adopts this indirect strategy.

3 The Solution: Love

I propose that we can avoid perverse instantiations by instructing the AI to love humanity. We could start with a roughly human-level AI—one which we expect to undergo recursive self-improvement resulting in superintelligence—and give it the final goal of loving humanity. As it progresses towards superintelligence, the AI's understanding of love (its grip on the concept and its referent) will deepen. Since the AI's final goal is to love humanity, it will seek to better understand both what love is (what we have in mind, as well as what we refer to when we use the term in the relevant ways), and what is involved in being loving (if Abe loves Bea, what implications does this have for his behavior?). Ultimately, its goal would be to apply what it has learned about love to its relationship with us.[3] The following sub-sections elaborate this proposal.

3.1 What Is Love?

One might worry that the concept of love is simply to slippery and too vague to be of use here. However, it's possible to provide a good deal of precision to the idea. As psychologist Fehr (2013, 202) writes, there "are now empirically based answers to questions about the meaning of *love* and how it is experienced". Obviously, "love" has many senses: "I love pizza"; "I am in love with her"; "I love you man". This is because the term is not applied to a single phenomenon, but a web of related phenomena.[4] Most importantly for our purposes, empirical investigation has distinguished four kinds of love (Berscheid and Hatfield 1974; Berscheid 2006, 2010; Fehr 2013): romantic/passionate, attachment, companionate, and compassionate. The relevant forms for my proposal are the final two.

It's worth emphasizing what I'm *not* suggesting. Romantic or passionate love—what the ancient Greeks called *eros,* and which English speakers pick out with the expression "being *in* love"—is typically triggered by physical attraction and sexual desire. Naturally, this is not what I have in mind. Nor am I suggesting that the superintelligence display attachment love, affection directed towards a particular individual. What I'm proposing is that the superintelligence display companionate and compassionate love towards humanity. Companionate love is the kind found between close friends and family members. It's characterized by caring, trust, honesty, respect, and can be experienced for many people simultaneously. Compassionate love is sometimes called altruism or selfless love. "A unique antecedent of compassionate love is the perception

[3] Some reviewers have questioned whether AI will be capable of emotion. I see no grounds for skepticism, however. Artificial emotions have long been a theme in AI research (Sloman and Croucher 1981; Picard 1997; Scheutz 2014).

[4] "Perhaps it is no wonder that love has puzzled so many for so long. Part of the confusion is that the word 'love' has been affixed to different parts of this larger, dynamic love system" (Fredrickson 2016, 848).

that the other is in distress or in need" (Fehr 2013, 203). I'll say more about these forms of love when I discuss what we should expect from a loving superintelligence in Sect. 3.3. The point, for now, is that love is not a hopelessly vague notion. There are precise, empirically grounded ways to articulate the idea.

3.2 Who Is to Be Loved?

Thus far, I've claimed that the superintelligence should love humanity. But who *exactly* should this include? Should it include fetuses, and brain dead comatose patients? Or, perhaps "humanity" is too narrow. Perhaps, we should include non-human animals and extra-terrestrial life (if there is any). I'm inclined to say that the superintelligence should love all persons. This specification brings us into philosophically treacherous waters. Who counts as a person is itself disputed. We might choose to err on the side of inclusiveness, allowing in all disputable categories just to be safe. Then again, this might beg important moral questions in favor of vegans and pro-lifers. A superintelligence that loved fetuses and non-human animals might prevent abortions and meat eating, which could be a wrongful imposition. Perhaps the best option will be to defer to the superior epistemic position of the superintelligence by instructing it to discover the extension of the term "person". This would add greater complexity to my proposal. But, it may be worth it.

3.3 What Should We Expect from a Loving Superintelligence?

While it is impossible to say in detail what a loving superintelligence would do (if we could say, then we would be superintelligent ourselves), the psychology and philosophy of love provide strong evidence for some general predictions.

There is robust psychological evidence that companionate love is the most central part of the concept of love. When subjects in psychological studies are asked about the characteristics of love, they tend overwhelmingly to cite features of companionate love (Fehr 1988). These results have been replicated repeatedly and display cross-cultural stability (Luby and Aron 1990; Button and Collier 1991; Kline et al. 2008). Moreover, in a series of studies, Fehr and Russell (1991) found that companionate love was considered the most typical or paradigmatic form. Five features of love are consistently (across studies and cultures) found to be central to the concept: trust, caring, honesty, friendship, and respect (Fehr 2013, 206).

Somewhat recently, evidence has emerged that the concept of love may be essentialist—i.e., "people view certain features as *necessary* for them to judge that a given relationship is an instance of the concept of 'love'" (Hegi and Bergner 2010, 634). The necessary feature is "investment in the well-being of the other, for his or her own sake" (Hegi and Bergner 2010, 621). When investment in the other's interests was described as missing from a relationship, subjects consistently found it "very contradictory" to assert that the people loved each another. In other words, if Abe is not invested in Bea's well-being for her own sake, then it simply can't be true that Abe loves her. Though other characteristics of love are important, none display this *sine qua non* status.

In other words, the psychological evidence strongly supports the claim that "love, by definition, conveys a caring orientation toward others" (Fredrickson 2016, 850). An essential part of a father's love for his daughter, for instance, is his desire to see her flourish. Depending on his beliefs about what that requires, this may mean taking an active role in her development or stepping back to allow her to struggle and grow on her own. We would see something analogous from a loving superintelligence.

This conclusion is reinforced by consideration of the philosophical literature (Helm 2017). Two of the three standard philosophical theories of love—love as robust concern (Frankfurt 1999), and love as emotion (Badhwar 2003)—concur with these psychological conclusions.[5] The love as emotion view effectively defers the question "what is love?" to psychology. And, interestingly, the love as robust concern view just is the conclusion drawn from Hegi and Bergner 2010. Love, on the robust concern view, is a robust concern *for the beloved's well-being*. To love someone is to take on her interests as your own.

Thus, the psychology and philosophy of love give us strong reason to expect that a superintelligence which loves humanity would be invested in our well-being for our own sakes. It would display trust, caring, honesty, friendship, and respect. By giving the AI the final goal of loving humanity, we not only rule out nightmarish scenarios in which the superintelligence completely disregards human interests, we also ensure that it seeks to *advance* our interests.[6]

3.4 What Happens When Interests Conflict?

It's important to recognize that loving someone doesn't always mean giving her what she wants (or thinks she wants). Suppose that Bea would like a house by the sea. If Abe loves Bea, and if he sees no reason *not* to give Bea a house by the sea, then he will try to give her one. But there will be many cases in which Abe's love for Bea does not lead him to do this—for instance, when Abe thinks that, despite what Bea believes, a house by the sea is not in her interests. Thus, while a loving superintelligence would be invested in human interests, it would not attempt to uncritically satisfy each human desire. A particularly important instance of this phenomenon arises from interpersonal conflicts.

Sadly, some humans are hateful. They have enemies with whom they fight, and whom they wish they could destroy. Thus, one might object, if a superintelligence were to love Ayman al-Zawahiri (the leader of al-Qaeda), would it not seek to destroy his

[5] The third theory, love as valuing (Singer 2009), likely concurs as well. Though explaining how would take us too far afield.

[6] Of course, in order for it to do this, it must have sensible views about what our well-being consists in. A religious zealot might sincerely think that she advances my well-being by forcibly converting me. It seems pretty clear, however, that the zealot's beliefs about well-being are false. (Were it not for her belief in an afterlife, she would probably reject them herself.) There are currently vibrant research programs in psychology and philosophy, which have revealed a good deal about the nature of well-being (for surveys see Snyder and Lopez 2009; Fletcher 2016). A superintelligence, with its superior epistemic position, can be expected to have even better-informed views about well-being.

enemies, and to advance his wicked goals and ambitions? Don't those who love evil people do evil things for them?

This objection overlooks *why* some humans do evil things for the people that they love. The fact is that humans don't love everyone. And they certainly don't love everyone equally. One of the ways in which love-based reasons for action can be outweighed or undercut is by their conflict with *other* love-based reasons. Imagine that two of your equally beloved friends are fighting. Your friend Abe becomes enraged by a dispute with your friend Bea. He wants nothing more than to see Bea forlorn and destitute. Given that you also love Bea, is there any chance that you will satisfy Abe's desire? I think not. In this case, your love for Bea undercuts your reason to give Abe what he wants. Since my proposal is that the superintelligence be programmed to love everyone equally, it would not harm one human merely to satisfy another's malevolent desire.

If Abe and Bea are so belligerent that they each demand you take a side, the most loving thing to do, I suspect, would be to withdraw from both. If they each want nothing to do with you unless you take sides and hate the other, then the only remaining choice is to distance yourself from both and hope that one day a resolution can be found. A love egalitarian superintelligence would likely display similar behavior. Of course, a relationship with a loving superintelligence would be *extremely* advantageous. As I've indicated, such a being could potentially do wonderful things for us. Thus, there would be very powerful incentives to maintain a close relationship with the superintelligence. This would require one to not be at odds—or, at least, to not be too aggressively adversarial —with the other objects of the superintelligence's love (i.e., other people). This may be an added benefit of my proposal. A loving superintelligence could potentially help to resolve some of humanity's conflicts as well as prevent new ones.

4 Comparison with Alternatives

My proposal has two main competitors: Yudkowsky's (2004) Coherent Extrapolated Volition, and Bostrom's (2014) Moral Modeling. I'll compare my proposal to each.

4.1 Coherent Extrapolated Volition

Yudkowsky has proposed that superintelligent AI should promote our "coherent extrapolated volition" (CEV). This proposal takes what philosophers would call an "ideal advisor" approach, one that centers on what we would desire under idealized conditions. Yudkowsky writes:

> [O]ur coherent extrapolated volition is our wish if we knew more, thought faster, were more the people we wished we were, had grown up farther together; where the extrapolation converges rather than diverges, where our wishes cohere rather than interfere; extrapolated as we wish that extrapolated, interpreted as we wish that interpreted. (Yudkowsky 2004, 6)

Since this proposal is put in intentionally poetic terms, some elaboration is necessary.

To "know more" means to be aware of all the relevant facts concerning the objects of our decision-making. If we "thought faster" we would not just be smarter; we would also have thought longer and more clearly about our desires and options in life. Some of our personal characteristics and desires receive second-order endorsement; some don't. For instance, Abe may approve of his affection for his friends, while wishing he were less angry and vindictive. If he "were more the person he wished to be", he would keep or increase the former trait, while reducing or eliminating the latter. If we "had grown up farther together" we would have had more shared experiences and a stronger sense of solidarity with humanity. Extrapolation "convergence" refers to a high probability of one's choosing in a certain way under idealization. If it's hard to know what our idealized selves would choose concerning something, then the superintelligence is to leave those options open. The points on which individual extrapolated volitions "cohere rather than interfere" are the choices that idealized humans would agree on. The idea is to have the superintelligence only act on those goals that everyone (or maybe just most people) would endorse in their idealized state. The requirement that our volition be "extrapolated as we wish that extrapolated, interpreted as we wish that interpreted" indicates that our present selves should guide the process of idealization. Thus, the resulting extrapolated volition should include goals and/or values that our *actual* selves could be led to understand and approve of.

The idea, in short, is that the superintelligence is to take each individual, idealize her belief and desire set in order to determine a "volition", and then aggregate these volitions, acting where they cohere. I believe that my proposal captures what is appealing about this proposal, while avoiding some of its problems. I'll discuss each in turn.

Motivations for CEV. The central appeal of this proposal is that it leaves humanity in control, in some sense, of our future. The superintelligence would do whatever we would have collectively and ideally decided for it to. As I've indicated, a loving superintelligence will, in a similar fashion, seek to advance human interests, but will not blindly satisfy expressed desires. It will aim to do what is *best* for us.

Yudkowsky (2004, 14) suggests that another appeal of his proposal is that it allows for moral progress. There are some living today who remember racial segregation in the United States. A mere 200 years ago chattel slavery was alive and well. It seems unlikely that moral progress has peaked at this moment in history. In a few decades or centuries, our own moral sensibilities may look as barbarous as segregation and slavery do now. The CEV proposal aims to accommodate this thought. I believe love egalitarianism would do the same. For now, however, readers will have to accept a promissory note: I'll say more about the morality of a loving superintelligence in Sect. 4.2.

Yudkowsky's proposal would have the superintelligence only act where our extrapolated volitions cohere. This is thought to prevent a particular subset of humanity from "hijacking" our planet's (or universe's) future, imposing its own conception of the good on all of humanity forever (Yudkowsky 2004, 17). Imagine, for instance, how you would feel if it were al-Qaeda that first built a superintelligent AI and programmed it to advance their goals. By ensuring that the superintelligence advances goals that we all can agree on, Yudkowsky thinks, we can avoid potential conflicts over how the AI is to be programmed. (Imagine what the Pentagon might do if it became convinced that al-Qaeda were about to develop superintelligent AI.) This will be very important because an AI arms race increases the incentive to cut safety corners, increasing the risks of malignant superintelligence (Armstrong et al. 2013).

A loving superintelligence would similarly prevent any one person or group from hijacking humanity's future. As I emphasized in Sect. 3.4, the AI is to be programmed to love everyone equally. Since love involves investment in the well-being of the beloved, and since autonomy is an important constituent of human well-being (Ryan and Deci 2000; Helm 2017, Sect. 3), a love egalitarian superintelligence would be strongly averse to imposing foreign conceptions of the good on us.

Problems with CEV. There are two problems with the CEV proposal: the idealization of our actual selves adds enormous complexity; and there will likely be insufficient coherence in the individual extrapolated volitions.

The computational resources required to extrapolate a person's volition would be astronomical. In rough sketch, discovering Abe's extrapolated volition would require the AI to: either observe Abe's behavior or (for better accuracy) scan his brain; discover and categorize his present beliefs, desires, and interests; search for inconsistencies; discover all the potential objects of his volition; model his learning and thinking processes, and then amp them up so that he "thinks faster"; model various social interactions that Abe might experience which would lead to a wiser, more mature and pro-social version of Abe; filter those out from other social experiences that would make him less wise or pro-social; search again for inconsistencies that arise as his desires are extrapolated; resolve any such inconsistences; produce a rough draft of his volition and compare it with the actual Abe to see whether real Abe would approve of his idealized self and his idealized choices. This would be insanely computationally demanding. If we then multiply this by 8 billion or more people, it becomes unfathomable. After all the individual volitions have been extrapolated, there also remains the task of synthesizing them for coherence.

Perhaps my imagination is simply too limited. But I have a hard time imagining even a superintelligence with the processing power necessary to complete the task. Perhaps the CEV could be computed if all of Earth's resources were converted into processing power. But, of course, by then the CEV is irrelevant. It will be too late to do us any good. On the other hand, maybe the superintelligence needn't actually determine humanity's CEV precisely. Bostrom (2014, 213) suggests that a superintelligence would be able to make a pretty good *guess* as to what our CEV would include, and then act on its guess. It could then update its speculation as it learns more about human psychology and society.

> For example, it is more plausible that our CEV would wish for there to be people in the future who live rich and happy lives than that it would wish that we should all sit on stools in a dark room experiencing pain. If *we* can make at least some such judgments sensibly, so can a superintelligence. (Bostrom 2014, 213)

This seems right. But—given how much is at stake—I'd want a hell of a lot more precision than that! Part of what makes this kind of guesswork so difficult is the absence of any examples. Present-day machine learning works best with large data sets—lots of example cases to train algorithms. But, we have zero examples of extrapolated volitions.

Perhaps future technologies won't require exemplar data. But, we have no way of anticipating such developments.

Of course, some of the complexity required by the CEV proposal would also be required by my proposal. In either case, for instance, the AI will need to know a lot about individual human desires. But, my point is that by basing the AI's instructions on an *idealized* version of each individual, the CEV proposal is vastly more complicated than mine. On my approach, the superintelligence would have much needed examples. It could see actual loving relationships and use those as (fallible) models for its own relationship with humanity. Just like discovering our CEV, discovering what love involves would also require studying human psychology and society. But it won't require the additional (and, I think, Herculean) task of idealizing that psychology and society in order to determine what we would collectively wish for in some distant counterfactual scenario.

Another problem for the CEV proposal arises from the fact that it would have the superintelligence only act where our extrapolated volitions cohere. This, as we saw, was taken to be one of its merits. But, there is likely to be little coherence. It's certainly true that most humans desire things like food and shelter, opportunities to socialize and express themselves, and so on. These would plausibly end up in each extrapolated volition. But, beyond basic needs like these, there is likely to be little convergence on a wide range of important issues. Consider all the different kinds of people that there are: acetic hermits and minimalists, and materialistic Wall Street money-chasers; religious fundamentalists, and anti-theists; neo-Nazis, and social justice warriors; right libertarians, and communists. Even if all these people were idealized in the relevant ways, it seems highly unlikely that there will be much coherence in their volitions on questions of politics, economics, morality, and so on. If that's right, then, the superintelligence's hands would be tied on these important issues.

A loving superintelligence will similarly find itself in a position where the various objects of its love have incompatible interests. However, this will not prevent it from acting on those interests. One obvious way to satisfy conflicting interests is to localize them. If the hedge fund manager wants to live in a post-industrial, free market society, and the acetic hermit wants to live in an agrarian, barter-based society, then they could each have their own corner of the planet (or, if we're advanced enough, a planet of their own) on which to do that. Now, you might think that the *idealized* hermit and hedge fund manager would also reach this compromise, so this isn't an advantage for my proposal. I'll concede that this is possible. However, as we have no examples of extrapolated volitions, we simply can't be sure that they would. Moreover, even if the extrapolated volitions of the hermit and hedge fund manager end up with a high degree of coherence, it's far from obvious that this would generalize. It is an unfortunate fact that many humans' desires are not merely incompatible with the desires of others—but, worse, precisely what they desire is the thwarting of others' desires. (Recall the al-Zawahiri example from above.) Some of these nasty desires might be eliminated by the volition extrapolation process. But, given that one's extrapolated volition is meant to be sensitive to one's current self, it's hard to imagine that, for instance, the neo-Nazi's extrapolated volition would bear much good will towards the social justice warrior. Similarly, it's hard to imagine the social justice warrior's volition cohering—or even compromising with—the neo-Nazi's. Thus, if we instruct the superintelligence to advance only those goals that we would all agree on, then we give up on much of the good that a superintelligence could do for us.

4.2 Moral Modeling

Bostrom mentions an additional concern regarding the CEV proposal. This is that even our idealized selves might not be such great people. "Moral goodness", he writes (Bostrom 2014, 267), "might be more like a precious metal than an abundant element in human nature, and even after the ore has been processed and refined in accordance with the prescriptions of the CEV proposal, who knows whether the principal outcome will be shining virtue, indifferent slag, or toxic sludge?"

This consideration suggests another approach, which Bostrom calls Moral Modeling (MM). The idea is to program the AI to learn moral concepts, discover moral facts, and promote moral goodness and/or rightness. If the superintelligence is better at moral philosophy than humans are, then it may succeed where we have yet to—discovering and acting on the correct moral theory. In other words, we could make a super-moral superintelligence.

Assuming that this approach could be successfully implemented, it would surely result in the *morally* best results of all the options. There are, nevertheless, a few reasons why I prefer my proposal. First, I suspect that the behavior of a love egalitarian superintelligence and a super-moral superintelligence would be very similar. If there are conflicts between morality and love egalitarianism, it's not clear to me where they lie. Obviously, my inability to spot differences does not entail that there are none. The thought is just that the expected outcomes for MM and my proposal will be quite similar, perhaps even identical.[7]

Some philosophers have seen a tension between love and morality—in some cases treating them as distinct, even competing, domains within practical reasoning (Slote 1983; Wolf 1992). Morality, they suggest, is supposed to be impartial; it's about taking into account everyone's interests, and not weighting some people as more important than others. Love, on the other hand, is inherently partial. If I love my friend, I will favor him over others. Thus, love and morality can, and perhaps often do, conflict. This conflict

[7] It has been suggested to me that love egalitarianism might be operationally equivalent to an interest-based consequentialism. If loving someone means being invested in her well-being, then loving equally should require weighing and advancing individual interests equally. Something like this is probably right. But it also seems clear that love comes with deontological constraints. If I killed a healthy person in order to harvest his organs and save five other people, I could not plausibly insist that I nevertheless loved him. This shows why we couldn't simply instruct the AI to promote human well-being. "Promote" is vague. Does it mean: maximize the sum total? maximize the minimum individual level? maximize the maximum level? satisfice to some threshold?... Love helps to resolve this problem. Loving is not a maximizing procedure, and (as I suggested) comes with side constraints. Though we can't articulate the procedure for making loving decisions, we clearly follow some such procedure in our daily lives. And we seem to think it's the right way to do things.

is dissolved, however, if we recall that the superintelligence will be programmed to love everyone equally. What is impartiality, after all, if not equal partiality towards all?

It's also worth mentioning that some moral philosophers think that even the appearance of tension between love and morality is illusory. As Velleman (1999, 341) writes, "Love is a *moral* emotion precisely in the sense that its spirit is closely akin to that of morality. The question, then, is not whether two divergent perspectives can be accommodated but rather how these two perspectives converge".[8] In the tradition of Christian ethical thought, love plays a very central role. Paul Ramsey, a major figure in twentieth century Christian ethics, argued that love is the basis for all of moral theory. Ramsey (1950, xvi–xvii) argued that love is the "primitive idea" and "fundamental notion" for morality. Dyck (1968), another theologian, argues that love is not only a moral virtue, but the primary—perhaps sole—test and guide for action. As he puts it, "love is no mere sentiment or emotion. It is the relational bond of a covenant to form and sustain community… But it is also the power and the passion to get on with this task" (Dyck 1968, 545). Some Christian ethicists argue that a complete moral theory will have to incorporate other concepts or principles that cannot be derived from love alone (Harris 1976). But, regardless, it's clear that love has a central place in this tradition of ethical thought.

In short, I expect that a superintelligence programed to love to humanity would get us most—if not all—of what we would hope for in a super-moral superintelligence. So, in terms of outcomes, the choice between these proposals may not make much of a difference. In terms of difficulty in implementation, and the probabilities of success, however, my proposal has an advantage. Our goal, recall, is to maximize the probability of good outcomes and minimize the probability of bad outcomes. I believe that my proposal is superior because it comes with less risk of something going wrong. There are several reasons why that would be.

Working out what would be necessary in order to implement MM reveals layers of added complexity. Before we could give an AI instructions for discovering moral facts, we would need to figure out what moral facts are and how they can be discovered. In other words, we need to answer the central questions of metaethics (moral semantics, metaphysics, and epistemology). This would be an enormous initial hurdle. A natural thought here would be to outsource this work to the AI itself. Before it does any first-order moral theorizing, the superintelligence should work out the correct metaethical theory. So, our instructions might be: "Figure out what moral facts are, and how they might be discovered. If there are any, discover what they are. If there are no moral facts, or if moral facts are culturally relative, or some such thing, then shut down. Otherwise, perform the morally best actions." I'm a pessimistic about this approach, as I'll explain.

When it comes to morality, people disagree a lot. Philosophers and non-philosophers alike disagree about which moral concepts apply to which objects of evaluation (e.g., which actions are right, which character traits are vicious). They also disagree about

[8] On Velleman's view, respect for others is the minimum of moral expectation, while love is the maximum of moral supererogation. "[R]espect is a mode of valuation that the very capacity for valuation must pay to instances of itself. My view is that love is a mode of valuation that this capacity *may* also pay to instances of itself. I regard respect and love as the required minimum and optional maximum responses to one and the same value" (Velleman 1999, 366).

what makes it the case that moral concepts apply. Many people believe that God's will is what makes an action right or wrong. Others think that God's will, should it exist, has no important connection with the right- or wrong-making properties of an action. Some people think that moral facts are culturally relative. Others think that they are absolute and response-independent. Some think that moral evaluations are evaluations of outcomes or states of affairs. Others think that moral evaluations are evaluations of a person's will, intentions, or character. In other words, there is substantial disagreement at every level of moral discourse (Merli 2009). Perhaps no other concepts are as disputed as moral concepts. Some even claim that they are "essentially contested"—meaning that this kind of disagreement is an essential or constitutive feature of the concepts (Gallie 1955).

All of this conceptual disagreement would make it extremely difficult for an AI to make sense of moral concepts. I'm not taking a stand here on the metaethical implications of moral disagreement. My point is that, plausibly, an AI's only access to moral concepts (and thereby moral facts) will be through human moral discourse. I'm assuming that a machine would not have independent access to moral reality. On some metaethical views, an AI could have direct epistemic access to mind-independent moral facts through the capacity for reason, or a faculty of moral intuition, or some such thing. However, even on the assumption that such a view is right, it's not at all clear what such faculties are or how they would work—much less whether and how they could be incorporated into an artificial being. It is far more plausible that, if a machine is to discover moral facts, it will have to do so through us. Thus, widespread and persistent conceptual controversies, which would make it very difficult for an AI to acquire the concepts, pose a serious obstacle to a successful implementation. The motivation for MM was to prevent human foibles and moral imperfections from spoiling the good that a superintelligence could do. But, given that it will have to acquire its moral knowledge though us, it seems that this proposal doesn't actually solve that problem.[9]

5 Conclusion

It is vitally important that we figure out, before superintelligence is developed, how to ensure that it acts in ways congenial to human interests. I have suggested that we think

[9] Of course, people sometimes also disagree about what's involved in loving someone. But, this is not typically disagreement about what love *is*. As the research surveyed in Sect. 3.3 showed, there is a remarkable degree of consensus on that question (despite appearances). One might object, in a similar spirit, that there is disagreement about what well-being consists in. Since loving involves an investment in well-being, if well-being is as controversial as morality, then my view has the same problem as MM. I deny, however, the antecedent of this conditional. While there certainly are competing theories of well-being, for the most part, there isn't much disagreement in the literature over what contributes to a person's well-being (see Fletcher 2016). People tend to agree on which things are good for a person, even if they disagree about *why* those things are good for her. (E.g., is accomplishment intrinsically good for a person, or only insofar as it contributes to his positive mental states?) When it comes to the practical matter of promoting well-being, however, such disputes may not be of much significance.

of this problem in terms of a relationship. And the key to a good relationship is love. Thus, my proposed solution to the problem of friendly superintelligence is to teach the AI about companionate and compassionate love, and instruct it to love everyone equally.

After briefly clarifying this proposal, and exploring some of its implications, I compared it with two of the most promising alternatives: CEV and MM. I argued that my proposal captures what is appealing about the CEV and avoids its most serious problems. I also argued that the outcomes of my proposal would likely be very similar to the outcomes from a successful implementation of the MM proposal. However, implementing MM would face greater challenges. Thus, my proposal is to be preferred.

This paper is intended to open a line of inquiry. Obviously, I don't pretend to have resolved the problem of friendly superintelligence. Rather, I've suggested—at a very general, non-technical level—an approach for solving the problem. If the ideas presented here hold up, then it will be for future research to develop them.

Acknowledgments. I'd like to thank audience members at the PT-AI 2017 conference, as well as Vincent Müller and the reviewers for this volume. Special gratitude goes to Daniel Kokotajlo and Miriam Johnson for their comments on earlier drafts of this paper.

References

Armstrong, S., Bostrom, N., Shulman, C.: Racing towards the precipice: a model of artificial intelligence development. Technical report. Future of Humanity Institute (2013). https://www.fhi.ox.ac.uk/wp-content/uploads/Racing-to-the-precipice-a-model-of-artificial-intelligence-development.pdf. Accessed 24 Dec 2017

Badhwar, N.: Love. In: LaFollette, H. (ed.) Practical Ethics, pp. 42–69. Oxford University Press, Oxford (2003)

Berscheid, E.: Searching for the meaning of "love". In: Sternberg, R.J., Weis, K. (eds.) The New Psychology of Love, pp. 171–183. Yale University Press, New Haven (2006)

Berscheid, E.: Love in the fourth dimension. Annu. Rev. Psychol. **61**, 1–25 (2010)

Berscheid, E., Hatfield, E.: A little bit about love. In: Huston, T.L. (ed.) Foundations of Interpersonal Attraction, pp. 355–381. Academic Press, New York (1974)

Bostrom, N.: Superintelligence: Paths, Dangers, Strategies. Oxford University Press, Oxford (2014)

Button, C.M., Collier, D.R.: A comparison of people's concepts of love and romantic love. Paper Presented at the Canadian Psychological Association Conference, Calgary, Alberta (1991)

Chalmers, D.: The singularity: a philosophical analysis. J. Conscious. Stud. **17**(9–10), 7–65 (2010)

Dyck, A.: Referent-models of loving: a philosophical and theological analysis of love in ethical theory and moral practice. Harv. Theol. Rev. **61**(4), 525–545 (1968)

Fehr, B.: Prototype analysis of the concepts of love and commitment. J. Pers. Soc. Psychol. **55**, 557–579 (1988)

Fehr, B.: Social psychology of love. In: Simpson, J., Campbell, L. (eds.) The Oxford Handbook of Close Relationships. Oxford University Press, Oxford (2013)

Fehr, B., Russell, J.A.: The concept of love viewed from a prototype perspective. J. Pers. Soc. Psychol. **60**, 425–438 (1991)

Fletcher, G.: The Routledge Handbook of Philosophy of Well-being. Routledge, New York (2016)

Frankfurt, H.: Autonomy, necessity, and love. In: Necessity, Volition, and Love, pp. 129–141. Cambridge University Press, Cambridge (1999)

Fredrickson, B.: Love: positivity resonance as a fresh, evidence-based perspective on an age-old topic. In: Barrett, L., Lewis, M., Haviland, J. (eds.) Handbook of Emotions, 4th edn, pp. 847–858. Guilford Press, New York (2016)

Gallie, W.B.: Essentially contested concepts. Proc. Aristot. Soc. **56**, 167–198 (1955)

Good, I.: Speculations concerning the first ultraintelligent machine. Adv. Comput. **6**, 31–88 (1965)

Harris, C.: Love as the basic moral principle in Paul Ramsey's ethics. J. Relig. Ethics **4**(2), 239–258 (1976)

Hegi, K., Bergner, R.: What is love? An empirically-based essentialist account. J. Soc. Pers. Relationsh. **27**(5), 620–636 (2010)

Helm, B.: Love. In: Zalta, E. (ed.) The Stanford Encyclopedia of Philosophy (2017). https://plato.stanford.edu/entries/love/. Accessed 30 Mar 2018

Kline, S.L., Horton, B., Zhang, S.: Communicating love: comparisons between American and East Asian university students. Int. J. Intercult. Relat. **32**(3), 200–214 (2008)

Luby, V., Aron, A.: A prototype structuring of love, like, and being-in-love. Paper Presented at the Fifth International Conference on Personal Relationships, Oxford, UK (1990)

Merli, D.: Possessing moral concepts. Philosophia **37**, 535–556 (2009)

Müller, V., Bostrom, N.: Future progress in artificial intelligence: a survey of expert opinion. In: Müller, V. (ed.) Fundamental Issues of Artificial Intelligence, pp. 553–571. Springer, Berlin (2016)

Picard, R.: Affective Computing. MIT Press, Cambridge (1997)

Ramsey, P.: Basic Christian Ethics. Charles Scribners Sons, New York (1950)

Ryan, R., Deci, E.: Self-determination theory and the facilitation of intrinsic motivation, social development, and well-being. Am. Psychol. **55**(1), 68–78 (2000)

Scheutz, M.: Artificial emotions and machine consciousness. In: Cambridge Handbook of Artificial Intelligence. Cambridge University Press, Cambridge (2014)

Singer, I.: Philosophy of Love: A Partial Summing-Up. MIT Press, Cambridge (2009)

Slote, M.: Goods and Virtues. Clarendon Press, Oxford (1983)

Sloman, A., Croucher, M.: Why robots will have emotions. In: Proceedings of the 7th International Joint Conference on AI, pp. 197–202 (1981)

Snyder, C.R., Lopez, S.J.: The Oxford Handbook of Positive Psychology. Oxford University Press, Oxford (2009)

Soares, N., Fallenstein, B.: Aligning superintelligence with human interests: a technical research agenda. Machine Intelligence Research Institute (2015). https://intelligence.org/files/TechnicalAgenda.pdf. Accessed 10 Mar 2018

Velleman, D.: Love as a moral emotion. Ethics **109**(2), 338–374 (1999)

Wolf, S.: Morality and partiality. Philos. Perspect. **6**, 243–259 (1992)

Yampolskiy, R.: Artificial Superintelligence: A Futuristic Approach. CRC Press, New York (2016)

Yudkowsky, E.: Coherent extrapolated volition. Machine Intelligence Research Institute (2004). https://intelligence.org/files/CEV.pdf. Accessed 30 Dec 2017

Autonomous Weapon Systems - An Alleged Responsibility Gap

Torben Swoboda(✉)

University of Bayreuth, Universitätsstr. 30, 95447 Bayreuth, Germany
torben.swoboda@gmail.com

Abstract. In an influential paper Sparrow argues that it is immoral to deploy autonomous weapon systems (AWS) in combat. The general idea is that nobody can be held responsible for wrongful actions committed by an AWS because nobody can predict or control the AWS. I argue that this view is incorrect. The programmer remains in control when and how an AWS learns from experience. Furthermore, the programmer can predict the non-local behaviour of the AWS. This is sufficient to ensure that the programmer can be held responsible. I present a consequentialist argument arguing in favour of using AWS. That is, when an AWS classifies non-legitimate targets less often as legitimate targets, compared to human soldiers, then it is to be expected that using the AWS saves lives. However, there are also a number of reasons, e.g. risk of hacking, why we should still be cautious about the idea of introducing AWS to modern warfare.

1 Introduction

Matthias (2004) argues that autonomous, learning machines create the possibility of a responsibility gap. That a machine is capable of learning and acting autonomously, implies that its actions can no longer be predicted or controlled by anyone. However, an agent can be held responsible for the machine's behaviour only if the agent can control and predict the machine's behaviour. For this reason, nobody can be held responsible for the actions of such an autonomous machine. With advancing technology, we will have such machines replace humans in different lines of work. Hence, compared to the status quo with human workers, there will be more situations where nobody is held responsible. In cases in which this happens, we face a responsibility gap.

The prospect of a responsibility gap seems problematic, when one considers autonomous weapon systems (AWS). For Sparrow (2007) it is not possible to hold somebody responsible, when an AWS kills non-legitimate targets in war, e.g. surrendering soldiers. Similar to Matthias, this follows because "the possibility that an autonomous system will make choices other than those predicted (...) is inherent in the claim that it is autonomous" (Sparrow 2007, p. 70). As a result, he deems AWS as immoral to use. The paper has evoked appraisal but also numerous critiques. For example, Simpson and Müller (2015) argue that it is

V. C. Müller (Ed.): PT-AI 2017, SAPERE 44, pp. 302–313, 2018.
https://doi.org/10.1007/978-3-319-96448-5_32

permissible to use AWS, because the deciding criterion is not one of responsibility attribution, but rather if the risk of war can be fairly distributed. Hellström (2013) suggests that a society can collectively share responsibility. A society might do so, if it considers using AWS advantageous. However, the core claim by Sparrow and Matthias, namely that prediction of an autonomous, learning machine, is not possible, has largely been accepted.

In this paper, I critically evaluate in which ways we can predict and control machines of the sort that Sparrow and Matthias describe. While in reality there will be numerous persons involved in the design of AWS, as well as the decision to deploy these in war, I ignore the problems of collective responsibility in this paper. Instead, I shall assume the existence of one master programmer, who makes all decisions on her own.[1] I conclude, that there are cases where the master programmer can be held responsible for the behaviour of an autonomous, learning machine. I will argue that whether a responsibility gap emerges, crucially depends on the *particular* instantiation of the machine and not on the general features of autonomy and a capacity to learn. An important implication of my argument is that, prima facie, autonomous and learning machines are not immoral to use to the extent that this immorality arises due to the responsibility gap. Still, my paper does not imply that AWS are morally unproblematic. I believe that substantial arguments should be raised against the use of AWS, however they are not of the kind that Matthias and Sparrow put forward.

In Sect. 2 I take a closer look at the responsibility gap argument by Matthias and Sparrow. Following that I will explicate how computer scientists use machine learning to develop autonomous machines in Sect. 3. I will argue in Sect. 4 that programmers remain in control and can predict the AWS such that they can be held responsible for the AWS's behaviour. In the last section I will provide some arguments against using AWS in war.

2 The Responsibility Gap

Matthias (2004) first presented an argument about the possibility of a responsibility gap. Put generally, autonomous machines are capable of learning which means that their actions become unpredictable and uncontrollable.

Matthias presents the case of a NASA robot that is used for extra-terrestrial exploration, intelligent elevators that learn to be at certain floors during rush hour, a cancer detection algorithm, and a toy that learns to respond to auditory commands. The NASA robot stores the type of terrain it crosses and evaluates how difficult it was to cross this type of terrain. In the future it accesses this kind of information to determine which path to take.

All of the examples share the feature that an algorithm is implemented that allows the machine to learn from experience (e.g. association of type of terrain and its difficulty). This means that the machine's behaviour is changing over time. This in turn means that the programmer cannot control what the machine

[1] See (Robillard 2017) for the idea.

does in a specific situation, because the machine's behaviour is essentially determined by its experience. Different experiences can lead to different behaviour, and even more troublesome, a different chronological ordering of the same experience can also lead to different behaviour. Since the behaviour depends on the experience, and the experience cannot be known by the programmer beforehand, it is not possible to predict the behaviour.

Usually, when an agent does not know particular facts regarding an action or outcome, or the agent has limited control over a situation, then the agent is also less responsible for what happened. Thus, Matthias presents us a dilemma. Either we do not make use of autonomous, learning machines, which is not attractive as they provide all sorts of advantages. Or we face a responsibility gap. Given that driver-less cars, as well as some medical diagnostic tools fall under the category of autonomous, learning machines, and we normally would want to hold someone responsible when these machines make mistakes, the dilemma Matthias presents us is an important one.

Before I explicate the particular problem with AWS and the responsibility gap, it will be helpful to provide some background on the notion of responsibility first.

2.1 On the Concept of Responsibility

The concept of responsibility has many meanings (Van de Poel et al. 2015). For example, a bus driver has the responsibility to safely transport his passengers. But more often we are concerned with responsibility for something that already has happened, i.e. a backwards-looking notion of responsibility. In such a case, responsibility is often connected to reactive attitudes (Strawson 1962). For example, we think that a fireman, who is responsible for saving a child, is praiseworthy. The responsibility gap argument focuses on those situations where something bad has happened. More precisely, Sparrow (2007, pp. 71–72) is concerned with responsibility as blameworthiness or liability. For example, if I jump a red light and thereby cause an accident, I am to be blamed for this. As a result, it is appropriate to punish me for my actions. Additionally, I am also to be held liable. This means that I have an obligation to compensate other parties for any damage I have caused. Liability and blameworthiness are distinct, e.g. a murderer might not be able to compensate, but it is still appropriate to punish her.

Necessary conditions for responsibility that are commonly mentioned in the literature are capacity to act responsible, causality, knowledge, control, free will, and wrong doing (Van de Poel et al. 2015, pp. 21–25). Depending on the particular account of responsibility, all or only some of these conditions need to be met, so that someone can be held responsible (for example Fischer and Ravizza (2000) argue that free will is not a necessary condition). If one of the necessary conditions is not satisfied, then an agent fails to be morally responsible. In particular, if either the condition knowledge or control is not satisfied, then the agent is *excused*. In such a case, the agent is merely accountable, i.e. she has to explain for her role in bringing about some event. The two conditions, knowledge and control, are of especial importance, because Sparrow's argument aims to establish that these conditions are not satisfied in the case of AWS for anyone. Hence, we face a responsibility gap. I will assume that in the case of AWS all other necessary conditions for responsibility attribution are satisfied.

The knowledge condition can be traced back to Aristotle. For him, an agent cannot be held responsible if she was ignorant of relevant features of the situation. This is best exemplified: Suppose that Tom backs his car out of the garage and kills a kitten that was snoozing behind the rear tire. Suppose further that Tom was genuinely unaware of the kitten through no fault of his own (e.g. because he does not own a kitten) (Fischer and Ravizza 2000). Then it seems unreasonable to hold Tom responsible for the kitten's death. Instead, we would think of this event as an accident, because he was non-culpably ignorant. But being ignorant does not always imply that one fails to be responsible. Consider that a construction worker carelessly throws rubble from the top of a construction site onto the street. A pedestrian who happens to walk by is hit and harmed by the rubble. The construction worker can honestly say that he did not know that somebody was walking down the street. But this does not excuse him, because it can reasonably be known that pedestrians could walk on by and be hit the rubble. Put differently, the construction worker is ignorant in a culpable way.

Whether an agent is under ignorance in a culpable or non-culpable way crucially depends on the context of the situation and the agent. The introduction of culpability as a refinement whether ignorance exempts one from being held responsible bears the risk of circular argumentation. But going into deeper detail about the conditions of culpability is beyond the scope of this paper. However, in Sect. 4.2 I will explicate what kind of information can and cannot be reasonably known by programmers about AWS. I will then argue why this kind of information is enough to hold the programmer responsible.

Control refers to the ability of a person to have an influence on what happens. My notion of control is closely related to what Fischer and Ravizza (2000, p. 18) call power necessity:[2] A person has control if and only if, were a certain proposition to obtain, the person has the power so to act, so that the proposition does not obtain. This expresses the idea that one can only be held responsible if one had a viable alternative to do otherwise with regards to what one is being held responsible for. My understanding of control implies a guarantee that the proposition does not obtain, not a mere possibility that it does not. It can be argued that a possibility that the proposition does not obtain constitutes (at least some) control. While this may be, the condition I put forward is more demanding on the programmer and thus strengthens Sparrow's position.

[2] Note that neither Matthias nor Sparrow are explicit in what they mean with control. However, Matthias refers to Fischer's and Ravizza's work. For an alternative notion of control, which is also based on Fischer's and Ravizza's work and applied to AWS, see Santoni de Sio and Van den Hoven (2018).

2.2 Autonomous Weapon Systems and the Responsibility Gap

Sparrow's argument is an applied case of Matthias' responsibility gap to military machines. Sparrow argues that the jus in bello guidelines should include a responsibility principle. The ius in bello principles explicate what may justly be done in warfare. For example, the principle of discrimination establishes that combatants need to distinguish between legitimate and non-legitimate targets. Enemy soldiers who are participating in military operations are legitimate targets and may be killed. Prisoners of war and injured soldiers incapable of fighting are non-legitimate targets and may not be killed. What non-legitimate targets are depends inter alia on the view of just war theory (see the discussion between traditionalists and revisionists).

The responsibility principle requires that, generally, it must be possible that someone can be held responsible if non-legitimate targets have been harmed. However, the responsibility principle allows for exceptions (Sparrow 2007, p. 67). For example, if a soldier kills a civilian in an accident, the soldier is excused, and while she is held accountable, nobody is held responsible in the sense of blameworthiness or liability. This exception is acceptable, since the soldier can generally, i.e. under non-exceptional circumstances, be held responsible for her actions, e.g. when she intentionally kills a civilian. On the other hand, a soldier with a certain mental disorder could never be held responsible, precisely because it is the mental disorder that causes her not to have the capacity to act in a responsible way, which entails that she can never be held responsible. The responsibility principle would therefore forbid the deployment of soldiers with that kind of mental disorder.

There are two lines of argumentation for the responsibility principle. The deontic argument states that one disrespects the enemy if one were to ignore the responsibility requirement. The idea is that enemy soldiers are nevertheless right-bearers and deserve moral consideration. They cannot be slaughtered at will. If enemy soldiers are surrendering, then they cannot be killed, but must be taken as prisoners of war instead. If we were to make use of a weapon that would kill these soldiers and there was nobody to be held morally responsible for, then we would effectively circumvent the principle of discrimination. This would show a lack of regard for the value of the lives of the enemy and thus be disrespectful.

The consequentialist's argument emphasises that the responsibility principle creates an incentive structure that deters agents from committing wrongful acts. This is the case, as we allow for punishing a person only if we hold her responsible. Without the responsibility requirement we allow the use of weapons that can harm non-legitimate targets and nobody is being punished for it. As a result, it is to be expected that there will be an increase of those that are non-legitimately attacked. This outcome should be prohibited with the responsibility principle.

Sparrow is concerned that AWS violate the responsibility principle. More precisely, what Sparrow fears is that an AWS commits a war crime. For example, an AWS intentionally kills surrendering soldiers. The reason for this might be that these soldiers had killed the AWS's "robot comrades", which the AWS wants to avenge, or to "strike fear into the hearts of onlooking combatants", or simply because "it calculated that the military costs of (...) keeping them prisoner were to high" (Sparrow 2007, p. 66). The AWS is allegedly capable of this behaviour because it can learn from experience. As a result, the decision that is made by the AWS is more influenced by what it has learned, rather than its initial programming. Given that the AWS learns from experience, it follows that it is going to make choices other than what the programmer anticipates. Consequently, the programmer has lost (partial) control over the AWS, as it behaves in ways that are not intended or encouraged by the programmer.

Given that two necessary conditions for responsibility, control and prediction, are not satisfied, it would be unjust to hold the programmer responsible. Moreover, the programmer is the person who would be best able to predict and control the machine, since she designed it. If even the programmer is unable to control and predict the machine, then it seems even less likely that anybody else could do so. Assuming that the machine cannot be held responsible for its actions (e.g. because it cannot be punished (Sparrow 2007, pp. 71–73)), responsibility cannot be attributed to anybody. As a result, we encounter a responsibility gap, whenever an AWS commits a wrongful act.

3 Machine Learning for Ethical Decision Making

Sparrow is correct in demanding that an AWS satisfies the principle of discrimination. This leads to the question, how we can implement this principle in a machine. Following Moor (2006), ethical principles can be represented either explicitly or implicitly in a machine.

An example for an explicit ethical system is MedEthEx (Anderson et al. 2006), which follows the biomedical principles of Beauchamp and Childress (1979). The machine advises a doctor whether she may offer the patient a treatment again, given that the patient has rejected said treatment before, or if the doctor has to accept the patient's decision. It uses a symbolic representation for different principles that are relevant for the case. The decision making process of the system is a set of clauses, which conjoin and disjunct the symbols in various ways.

But the domain of application of MedEthEx is still fairly limited, compared to an AWS. It might thus be argued that a symbolic approach might not be flexible enough to be an adequate representation of the complex environment an AWS is placed in. One of the most flexible systems is an artificial neural network (ANN). It allows that information is not represented as a single symbol, but rather the information is saved throughout the whole network. Figure 1 shows a representation of such a neural network.

For the programmer the principle of discrimination is ultimately a classification problem: People need to be categorised as legitimate and non-legitimate targets. An ANN can be used to solve this classification problem. First, the programmer has to consider what kind of variables shall be used to predict the class of the person. This implies that there are features that are useful to separate

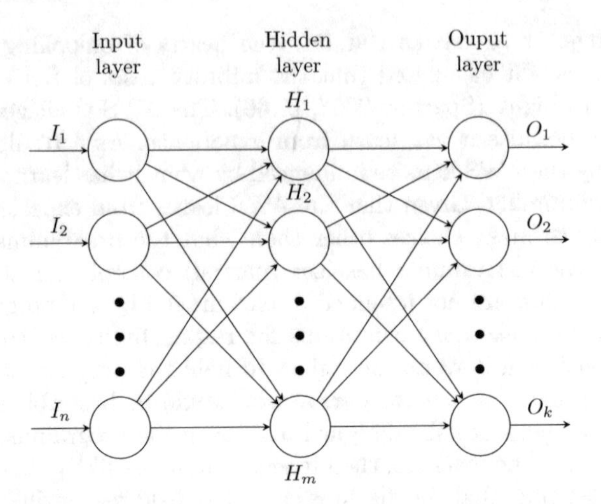

Fig. 1. A representation of an artificial neural network. A neuron is connected with all neurons in the preceding and following layer via *weights*. A neuron outputs a signal which is multiplied with the value of the corresponding weight. All the incoming signals for a neuron are summed up and used as an input for a function of that neuron, which determines the output of that neuron. During backpropagation, i.e. the learning phase, the values of the weights change.

legitimate from non-legitimate people. For example, if a person has a gun, this might be an indicator (though not a sufficient condition) that she is a legitimate target. A person holding her hands up high, as in a surrendering gesture, on the other hand seems to be a non-legitimate target.[3] Second, the programmer needs to acquire lots of labelled samples. A labelled sample contains the values for all features as well as the class to which it belongs (legitimate and non-legitimate targets). Third, the neural network has to be trained. That is, it needs to find the set of optimal values for the weights, which minimises the amount of samples that are misclassified. The weights are the connections between different neurons. Put differently, the weights determine how much one neuron influences another one. The trained ANN draws a hyperplane through a d-dimensional space, where d is the number of features being used. This hyperplane is called the decision boundary. Fourth, after the training of the neural net, its accuracy has to be evaluated with new samples. This time, the correct class label is hidden from the network. The accuracy of the neural net is the percentage of correctly predicted classes. When the programmer is satisfied with the accuracy of the ANN, the training is stopped. Otherwise the model has to be adjusted, e.g. different variables, or samples etc.

Since the machine is adapting the value of the weights, based on samples, computer scientists say that the machine learns from experience. But it is not necessary that the machine has to learn while it is already operating. Indeed,

[3] The programmer is not limited to such simple variables. It could also be video or audio data for example. If we were to make use only of video data, then the ANN would be an implicit ethical system, as there is no reference to ethical principles.

this might be very undesirable, as the machine might then learn a new set of weights that is suboptimal.

4 Control and Prediction of Autonomous Weapon Systems

In this section I am going to answer the question I raised in the beginning of the paper. That is to say, I first examine if the programmer can control the machine's actions and, second, how the programmer can make predictions about the machine's behaviour in order to close the responsibility gap.

4.1 Control

What Sparrow is concerned about is that the AWS is capable of learning and thus the programmer loses her influence over the AWS. He writes: "If it [an AWS] has sufficient autonomy that it learns from its experience and surroundings then it may make decisions which reflect these as much, or more than, its initial programming. The more the system is autonomous then the more it has the capacity to make choices other than those predicted or encouraged by its programmers. At some point then, it will no longer be possible hold the programmers/designers responsible for outcomes that they could neither control nor predict." (Sparrow 2007, p. 70)

What Sparrow suggests here is that the AWS is continuously learning while it is already operating. While it might be possible to design an AWS such that it can do that, there is no need to design an AWS in this way. It is entirely possible to design the AWS as described in Sect. 3, namely that the training phase of the AWS is stopped and the AWS runs on that learned model. Indeed, this seems to me how we should design AWS, precisely because otherwise the AWS could learn a new model that is suboptimal. Since the programmer is in control whether the AWS continues learning or not and furthermore because allowing the AWS to continue learning bears the risk of suboptimal models, we should hold the programmer responsible if she designs the AWS to learn while operating.

Sparrow might argue that this is not the kind of AWS he is talking about. The AWS he is concerned with is of that kind of nature that it does learn during the operation, that is the condition that makes it truly autonomous. It would follow that Sparrow's responsibility gap does not apply to weapon systems that do not learn while operating. This would be an unattractive position, as AWS that are only once trained, are still very capable of targeting and killing non-legitimate targets, while it also remains the case that it is unpredictable who in particular they attack.

4.2 Local and Non-local Prediction

On the one hand, Sparrow's claim that there remains uncertainty what an AWS will attack is correct. On the other hand, I argued in Sect. 3 that there will be a clear decision boundary between legitimate and illegitimate targets. In what way can these claims be reconciled?

Bostrom and Yudkowsky (2014) distinguish between local and non-local behaviour. If we consider a chess program, then the local behaviour are the specific moves of the figures, e.g. moving the bishop from e2 to b5. The non-local behaviour corresponds to the optimality criterion: pick the move that increases the winning probability the most, given the training samples.

Chess programs are superior to humans in chess since Deep Blue beat the prevailing chess champion Garry Kasparov in 1997. It is impossible to win against them, because their non-local game map that links specific moves to possible future outcomes is much more accurate than what us humans can come up with in our minds. It is impossible to predict the local behaviour of chess programs (except for trivial cases, like where one move leads to a checkmate). But it is nevertheless possible to make predictions about the non-local behaviour. In fact, we know that the program picks the move that increases its chances of winning.

Applying the concept to the case of AWS, it is impossible to know if an AWS is going to attack a particular object or not (local). But the programmer is still aware of some features of the machine. First, the programmer has designed the machine such that the risk of falsely classifying non-legitimate targets as legitimate targets is minimised, instead of maximised. Second, during the test phase the programmer has learned how accurate the machine handles new data. This gives the programmer an initial expectation how well the machine would do in the real world.[4] Both of these factors allow the programmer to make non-local predictions about the behaviour of the machine.

What kind of knowledge is required for attributing responsibility? It is not local prediction, because this would be akin to the construction worker case. Not knowing who in particular is going to be wronged by one's action does not mean that one cannot be held responsible. If we do not accept this type of excuse in the construction worker case, then we should not allow the programmer to be excused either. Instead what matters is that given the results of the test phase the programmer can expect that a certain percentage of non-legitimate targets are wrongly classified as legitimate targets. The programmer is responsible for wrongful deaths caused by the AWS, because she willingly accepted that a certain percentage of non-legitimate targets are wrongly categorised as legitimate targets. This raises the question how high the accuracy threshold of the AWS must be. Do we require the accuracy to be 100%, 99%, 50%? I present an, at least initially, plausible answer to this question in Sect. 4.3, namely that the AWS must have a higher accuracy than human soldiers. However, in Sect. 5 I give reasons why this answer is more problematic than it might seem.

But there is also another reason why the programmer might be held responsible. This would be the case, when the programmer has used features which might

[4] Note, however, that this expectation can be vastly wrong. What the AWS has learned is a simplified model of the real world. Excluding relevant variables, biased or too few samples can lower the accuracy of the model, to name only a couple of issues that the programmer must be aware off.

improve the accuracy of the machine, but using said features is deemed immoral. For example, in credit scoring it is prohibited to use gender, race, religion etc. to decide whether an applicant's loan is approved (Mester 1997). Adding these variables increases the accuracy of determining whether an applicant can pay back the loan in the future. However, using these features is considered discriminatory against members of certain social groups. It is up to debate what features may not be used while designing an AWS.

Finally, the machine might be more or less accurate with regards to different social groups. Larson et al. (2016) have analysed the COMPAS program, which calculates the recidivism rate of criminals. While it was similarly accurate for Caucasians and African Americans (59% and 63% respectively), COMPAS wrongly predicted African Americans to be recidivists at nearly twice the rate than Caucasians (45% to 23%), which seems problematic. For AWS this means that no social group should bear a higher false positive rate than another. If the programmer makes no attempt to account for this requirement, she is to be held responsible for the shortcoming of the AWS.[5]

4.3 On the Accuracy Threshold

How accurate needs an AWS to be before we may use it on the battlefield? An initial reaction might be that it may not make any mistakes. This is, on the one hand, practically impossible to achieve. On the other hand, there is also a consequentialist argument against this. Human soldiers make mistakes and target the wrong persons. When we have to decide whether to make use of an AWS, we face opportunity costs. AWS have a certain false positive rate and humans will have one as well. If it were the case that an AWS attacks fewer non-legitimate targets than human soldiers, then using the AWS must, from a consequentialist perspective, at least be permissible. This is because by using AWS we effectively safe lives. Müller (2016, p. 74f) makes a related point. We do not expect that pharmacists make no mistakes in matters of life or death. Rather we expect 'due care' by pharmacists. My argument here is that the programmer has exhibited due care if her AWS make fewer false positive errors than humans.

5 Arguments Against Autonomous Weapon Systems

While I acknowledge a theoretical possibility that an AWS satisfies the consequentialist criterion, it remains to be shown that an AWS can satisfy the principle in practice. Again, it depends what the AWS is designed to do. An AWS whose purpose is to attack specific tanks, boats, or aircraft might do sufficiently well. But an AWS that shall attack individual soldiers is a much more complex task, for example because terrorists disguise as civilians. Therefore, I remain sceptical if an AWS can satisfy the consequentialist criterion with respect to individual

[5] For a critical discussion on algorithmic fairness and discrimination see e.g. Dwork et al. (2012) and Hardt et al. (2016).

persons. Moreover, it is not enough that the AWS can in principle solve the discrimination problem, it needs to do so in real time. If the AWS is to be used in war, then it must be able to make its decisions within a fraction of a second.

A further problem related to the consequentialist's criterion is establishing the current accuracy base line. The military has great interest in using AWS and for that reason has an incentive to exaggerate the false positive rate of soldiers. In the worst case, they might give the directive to shoot first and ask questions later. This would lead to an increase of civilian deaths, so that it is easier for the AWS to satisfy the consequentialist's criterion. This is of course undesirable.

A third point to consider is that there remains doubt whether the system has actually learned to discriminate between the classes as we intended it to. Dreyfus and Dreyfus (1992) describe a case, where an ANN was supposed to learn to distinguish pictures with camouflaged tanks from pictures without tanks. While the network seemed to be doing a fine job, it turned out that the network learned to distinguish sunny pictures from cloudy pictures. ANN are a black box in the sense that we do not know what kind of pattern the network has learned. This problem can be reduced by rigorous testing, but it cannot be eliminated.

Fourth, AWS are subject to hacking. If an AWS has a security loophole, then all AWS with the same software and hardware have the same weak point. This bears the risk that a whole army of AWS is disabled, or turned against friendly soldiers.

Lastly, in reality there will not be one master programmer, but many people involved in designing an AWS. Hence, we face a collective responsibility issue.

This list is not exhaustive, but is a starting point for reasons why we should remain cautious about the idea of introducing AWS onto the battlefield.

6 Conclusion

Sparrow and Matthias have offered a formidable challenge not just for the use of AWS, but autonomous, learning machines in general. If my argument were correct, then we face no responsibility gap, as the programmer can control when and how the machine learns and predict its non-local behaviour. However, this does not entail that using this kind of machine is unproblematic. Further research could aim at establishing conditions when the programmer has exhibited 'due care' in greater detail and at a more general level, rather than being tied to the case of AWS.

References

Anderson, M., Anderson, S.L., Armen, C.: An approach to computing ethics. IEEE Intell. Syst. **21**(4), 56–63 (2006)

Beauchamp, T.L., Childress, J.F.: Principles of Biomedical Ethics. Oxford University Press, Oxford (1979)

Bostrom, N., Yudkowsky, E.: The ethics of artificial intelligence. In: Frankish, K., Ramsey, W. (eds.) Cambridge Handbook of Artificial Intelligence, pp. 316–334. Cambridge University Press, Cambridge (2014)

Dreyfus, H.L., Dreyfus, S.E.: What artificial experts can and cannot do. AI Soc. **6**(1), 18–26 (1992)
Dwork, C., Hardt, M., Pitassi, T., Reingold, O., Zemel, R.: Fairness through awareness. In: ITCSC, pp. 214–226 (2012)
Fischer, J.M., Ravizza, M.: Responsibility and Control: A Theory of Moral Responsibility. Cambridge University Press, Cambridge (2000)
Hardt, M., Price, E., Srebro, N.: Equality of opportunity in supervised learning. In: Advances in Neural Information Processing Systems, pp. 3315–3323 (2016)
Hellström, T.: On the moral responsibility of military robots. Ethics Inf. Technol. **15**(2), 99–107 (2013)
Larson, J., Mattu, S., Kirchner, L., Angwin, J.: How we analyzed the compas recidivism algorithm. ProPublica (5 2016), 9 (2016)
Matthias, A.: The responsibility gap: ascribing responsibility for the actions of learning automata. Ethics Inf. Technol. **6**(3), 175–183 (2004)
Mester, L.J.: Whats the point of credit scoring? Bus. Rev. **3**(Sep/Oct), 3–16 (1997)
Moor, J.H.: The nature, importance, and difficulty of machine ethics. IEEE Intell. Syst. **21**(4), 18–21 (2006)
Müller, V.C.: Autonomous killer robots are probably good news. In: Di Nucci, E., Santonio de Sio, F. (eds.) Drones and Responsibility: Legal, Philosophical and Socio-Technical Perspectives on the Use of Remotely Controlled Weapons, pp. 67–81. Ashgate, London (2016)
Robillard, M.: No such thing as killer robots. J. Appl. Philos. (2017). https://doi.org/10.1111/japp.12274
Santoni de Sio, F., Van den Hoven, J.: Meaningful human control over autonomous systems: a philosophical account. Front. Robot. AI **5**, 15 (2018)
Simpson, T.W., Müller, V.C.: Just war and robots killings. Philos. Q. **66**(263), 302–322 (2015)
Sparrow, R.: Killer robots. J. Appl. Philos. **24**(1), 62–77 (2007)
Strawson, P.: Freedom and resentment. Proc. Br. Acad. **48**, 187–211 (1962)
Van de Poel, I., Royakkers, L., Zwart, S.D.: Moral Responsibility and the Problem of Many Hands, vol. 29. Routledge, New York (2015)

Author Index

A
Alexandrova, Anna, 117
Astromskis, Paulius, 231
Avin, Shahar, 117

B
Banerjee, Shreya, 136
Beckers, Sander, 235
Bhatnagar, Sankalp, 117
Bidabadi, Golnaz, 40, 190
Bringsjord, Selmer, 136

C
Cave, Stephen, 117
Cheke, Lucy, 117
Chin, Chuanfei, 3
Crosby, Matthew, 117

D
Danziger, Shlomo, 158
Dodig-Crnkovic, Gordana, 19

F
Fabra-Boluda, Raül, 175
Ferri, Cèsar, 175
Feyereisl, Jan, 117
Freed, Sam, 187

G
Gokmen, Arzu, 248
Govindarajulu, Naveen Sundar, 136
Greif, Hajo, 24
Guazzini, Jodi, 36

H
Halina, Marta, 117
Hardalupas, Mahi, 252
Hernández-Orallo, José, 117, 175
Human, Soheil, 40, 190
Hummel, John, 136

K
Kane, Thomas B., 255
Keeling, Geoff, 259

L
Lewis, Colin W. P., 212
Loe, Bao Sheng, 117
Longinotti, David, 43

M
Martínez-Plumed, Fernando, 117, 175
Maruyama, Yoshihiro, 194, 207
Mishra, Abhishek, 273
Mogensen, René, 57
Monett, Dagmar, 212
Moruzzi, Caterina, 69

O
Ó hÉigeartaigh, Seán, 117

P
Peschl, Markus F., 40
Pierce, Bryony, 73
Price, Huw, 117
Prinzing, Michael, 288

V. C. Müller (Ed.): PT-AI 2017, SAPERE 44, pp. 315–316, 2018.
https://doi.org/10.1007/978-3-319-96448-5

R
Ramírez-Quintana, M. José, 175

S
Savenkov, Vadim, 40, 190
Schweizer, Paul, 81
Shevlin, Henry, 117
Sloman, Aaron, 92
Strasser, Anna, 106
Swoboda, Torben, 302

V
van Leeuwen, Jan, 215

W
Weller, Adrian, 117
Wiedermann, Jiří, 215
Winfield, Alan, 117

Z
Zednik, Carlos, 225

Printed in the United States
By Bookmasters